LIZARDS

ERIC R. PIANKA AND LAURIE J. VITT

LIZARDS

WINDOWS TO THE EVOLUTION OF DIVERSITY

WITH A FOREWORD BY
HARRY W. GREENE

UNIVERSITY OF CALIFORNIA PRESS
BERKELEY   LOS ANGELES   LONDON

THE PUBLISHER GRATEFULLY ACKNOWLEDGES THE
GENEROUS CONTRIBUTION TO THIS BOOK PROVIDED BY
THE GENERAL ENDOWMENT FUND OF THE UNIVERSITY
OF CALIFORNIA PRESS ASSOCIATES.

University of California Press
Berkeley and Los Angeles, California

University of California Press, Ltd.
London, England

Facing the title page: a grand female *Chamaeleo minor*
(photo by R.D. Bartlett); table of contents: *Enyalioides
palpebralis* (photo by Laurie Vitt)

Library of Congress Cataloging-in-Publication Data

Pianka, Eric R.
    Lizards : windows to the evolution of diversity /
Eric R. Pianka and Laurie J. Vitt.
        p.   cm. — (Organisms and environments ; 5)
    Includes bibliographical references (p.) and index.
    ISBN 0-520-23401-4 (cloth : alk. paper)
    1. Lizards.   I. Vitt, Laurie J.   II. Title.   III. Series.
QL666.L2 P54   2003
597.95—dc21   2001004910

Printed and bound in Italy
12   11   10   09   08   07   06   05   04   03
       10   9   8   7   6   5   4   3   2   1

The paper used in this publication meets the minimum
requirements of ANSI/NISO Z39.48-1992 (R 1997)
(Permanence of Paper).♾

ORGANISMS AND ENVIRONMENTS

Harry W. Greene, Consulting Editor

CONTENTS

TABLES AND FIGURES

FOREWORD

HARRY W. GREENE

Lizards: Windows to the Evolution of Diversity is the fifth volume in the University of California Press series on organisms and environments. Our unifying themes are the diversity of plants and animals, the ways in which they interact with each other and with their surroundings, and the broader implications of those relationships for science and society. We seek books that promote unusual, even unexpected connections among seemingly disparate topics, and we want to encourage projects that are special by virtue of the unique perspectives and talents of their authors. Arizona grasslands, Bornean treeshrews, Seri ethnoherpetology, the amphibians and reptiles of Baja California, and the American bison have been the subjects of previous volumes in the series.

Lizards occur in all but the highest and coldest places on earth, and some tropical rain forests and deserts have several dozen species at a single locality. They come in many sizes, from a Caribbean island gecko that when curled up would barely cover a small coin to Komodo monitors so large they readily evoke images of prehistoric saurians. Various lizards use winglike flaps to glide through tropical forest canopies, strong claws to dig burrows in prairie sod, and fringed toes to run bipedally over windblown sand dunes. Some species subsist on a diet of nothing but leaves and flowers, others eat only ants, and a few take animals as large as deer for prey. Even the social systems of lizards vary widely, with some species that associate with others of their kind only to mate, for a few minutes each year, and others that remain in family groups for at least several months. Some lizards have lifelong mates, whereas members of certain all-female species have no mates and reproduce asexually, in the complete absence of males. Not surprisingly, all this diversity means that lizards fascinate biologists of virtually every stripe, and these

reptiles have long served as highly useful study subjects in fields ranging from biochemistry and physiology to anatomy and psychology. Among vertebrates, lizards rival insects and birds in the frequency with which they are studied by ecologists and evolutionary biologists, and many people around the world simply enjoy seeing them.

In some ways this is a book about two curious boys who grew up chasing lizards, learned to carefully ask questions of nature, traveled all over the world in search of answers, and have now synthesized their collective life's work to date. Both Eric Pianka and Laurie Vitt began graduate school in the state of Washington and did doctoral research on North American desert lizards. Both have gone on to become internationally respected scholars. Apart from these similarities, the geographic and taxonomic backgrounds they developed as scientists are distinct and largely complementary. Soon after graduate school Pianka took his fascination with arid lands to the great central deserts of Australia and then to southern Africa's Kalahari; he has since spent several decades concentrating on the skinks and monitors of those regions. Vitt, however, shifted his focus from the desert to a variety of neotropical faunas, particularly the extremely diverse reptile assemblages at several Brazilian localities. Together Pianka and Vitt have without doubt seen and studied, in more different places and in more different ways, more kinds of lizards and more things about lizards than any other two humans in history. In the course of research careers dedicated to one particular group of organisms, each has made fundamental contributions to science, especially in the areas of behavioral, community, and evolutionary ecology.

Lizards provides a survey of unprecedented depth and breadth. In the first set of chapters Pianka and Vitt summarize the classification and evolutionary history, anatomy, behavior, and ecology in prose that is accessible to the general reader; they also synthesize a wealth of details on those topics in ways that inspire new and provocative generalizations. Next the authors explore lizard biology on a group by group basis, with emphasis on structural peculiarities and natural history. In the penultimate chapter, Pianka and Vitt integrate lizard biology and taxonomic diversity on a global basis, with the goal of understanding historical differences between the two major evolutionary lineages—iguanas and their kin on the one hand; geckos, skinks, monitors, and their relatives on the other—and their component groups. The final chapter assesses interactions between lizards and people, including our use of lizards as cultural icons, food, leather, and pets. Pianka and Vitt conclude that although human activities are drastically reducing the diversity of life on earth and threaten many species of lizards with extinction, some of them will likely survive long after we have run our course.

Books like this are at the heart of modern biology and conservation. Although hypothesis testing is often heralded as the essence of science, in fact discoveries of new organisms, of new things about organisms, and of previously unseen patterns in nature are what chronically revise the questions that intrigue biologists. By accounting for what we now know about a particularly widespread, diverse, and ecologically important group of vertebrates, *Lizards* will undoubtedly generate many new research programs. Furthermore, while providing readers with an accurate, up-to-date view of the lives of these wonderful organisms, Pianka and Vitt illuminate the human side of science, their personal transformations of childhood curiosity into the passions and scholarship of accomplished professionals. With their stories, the myriad details of natural history, and a spectacular array of lively photographs, *Lizards* will make it possible for anyone with an interest in these animals to better understand and appreciate them.

ACKNOWLEDGMENTS

No book like this could be written without help—and we had a lot of it! We first thank the Big 12 Faculty Exchange Program for providing funds that allowed us to visit each other's campus to work. At various stages, many friends and colleagues reviewed portions of the book or answered our sometimes naive questions. We can't thank all of you enough for catching many errors and in some instances helping to reorganize our thinking. These people are Roger A. Anderson, Robin Andrews, Daniel Beck, Jonathan Campbell, Robert Espinoza, Richard Etheridge, Lee Fitzgerald, Darrel Frost, Carl Gans, Paul Gier, Harry Greene, David Hillis, Wendy Hodges, Ray Huey, Vic Hutchison, Bryan Jennings, Dennis King, David Kizerian, Randy Nydam, William Presch, Kevin de Queiroz, Tod Reeder, Douglas Ruby, Shawn Sartorius, Kurt Schwenk, and Martin Whiting. Aaron Bauer and William E. Cooper provided detailed comments on the entire book and discussed many issues with us at length.

We asked many of our friends for photographs and received many more than we could possibly use. We thank the following people for providing us the opportunity to use their photos, whether we did or not: José Pedro Sousa do Amaral, Chris Austin, Teresa Cristina S. Avila-Pires, D. G. Barker, Richard Bartlett, Daniel Beck, Chris Bell, Andreas Brahm, Rafe Brown, Janalee Caldwell, Jonathan Campbell, Ellen Censke, Y. N. Cheng, Harold Cogger, Luis Coloma, Alain Compost, Laura Cunningham, C. H. Diong, Ken Dodd, Travis La Duc, Richard Durtsche, Walter Erdelen, Dante Fenolio, Paul Gier, Larry Gilbert, Frank Glaw, Lee Grismer, Wolfgang Grossman, David Hillis, Wendy Hodges, Walter Hödl, Marinus Hoogmoed, Jeff Howland, Ray Huey, Das Indraneil, Bryan Jennings, Daryl Karns, Michael Kearney, Boris Klusmeyer, Tim Laman, Bill Lamar, Adam Leache, Bill Leonard, Rolf Leptien, Jonathan Losos, Bill Love, Bill Magnusson, Otavio Marques, Brad Maryan, Chris Mattison, Ray Mendez, P. le F. N. Mouton, Bob Murphy, John Murphy, Petr Necas, Mark Oshea, David Pearson, Gretchen Pianka, Louis Porras, Chris Raxworthy, Stephen Richards, Gordon Rodda, Jim Rorabaugh, Herb Rosenberg, Paddy Ryan, Ivan Sazima, Wolfgang Schmidt, Cecil Schwalbe, Kurt Schwenk, Stephen Secor, Glen Shea, Monica Swartz, K. A Tung, Martin Whiting, Mark Wilkinson, Steve Wilson, and Frank Yuwono.

We thank University of California Press editors Doris Kretschmer and Rose Vekony for guiding our book through the narrow and treacherous straits of preparation, reviews, acceptance, launch, and production. Nola Burger came up with a great design.

We are especially grateful to our copy editor, Anne Canright, who taught us that, as scientists, we tend to write things backwards, ending with our conclusions and major points, rather than beginning with them. By putting such important concepts up front, Anne streamlined and greatly improved our presentation. All we can say is, Anne, you sure can write!

Finally, we thank the lizards of the world for making our lives so interesting.

INTRODUCTION: THE LOGIC OF BIOLOGY

This green crested lizard, *Bronchocela cristatella*, lives in tropical rain forest of Southeast Asia along with other agamids, gekkonids, and skinks. (L. Lee Grismer)

In 1993, while assembling the third symposium volume on lizard ecology (Vitt and Pianka 1994), we resolved one day to write a semipopular book on lizards in a coffee table format. A few years later, publication of the elegant book *Snakes: The Evolution of Mystery in Nature* (Greene 1997) prompted us to propose *Lizards: Windows to the Evolution of Diversity* as a companion volume to the University of California Press. Snakes are merely one group of very specialized lizards, and other lizards certainly deserve equal consideration. Lizards, moreover, are much more diverse than snakes and can be exploited to explain and understand a great deal of basic ecology.

We both began studying lizards in the U.S. desert southwest more than thirty years ago. Pianka extended his studies to include Africa and Australia, and Vitt has worked in the southeastern United States, as well as Central and South America. Between us, we have studied and photographed members of almost all lizard families and many subfamilies. Moreover, our perspective on the biology of lizards is based on extensive firsthand personal experience with the animals in their natural habitats. Most popular books currently on the market were written by authors with little field experience and are oriented toward young children or the pet trade. Many contain numerous mistakes and misinformation. Our intent here is to provide an up-to-date, more scientific reference for lay people, somewhat comparable to the now out-of-date lizard section in Karl P. Schmidt and Robert F. Inger's *Living Reptiles of the World* (1957).

People often ask, "What good are lizards?" to which we respond with "What good are people?" Such anthropocentrism is abhorrent. Lizards have as much of a place on the planet as any living creature, including humans. Indeed, they have successfully inhabited Earth for much longer than humans have—lizards will undoubtedly persist long after humans and most other mammals have gone extinct. Lizards are spectacular products of natural selection and have diversified to fill an amazing variety of ecological niches. They are extremely good "model" organisms for study, and understanding their ecology and diversity can be exceedingly informative. What we have learned about lizards is applicable to nearly every conceptual area in modern biology; indeed, in many cases development of entire fields of biology had their origins in the study of lizards. Because many lizards are quite beautiful, they are very popular as pets among herpetoculturists around the world, and some people make their living by breeding many species of lizards in captivity for resale. For us to answer the question "Why lizards?" requires a brief journey through our independent but convergent pasts.

LIZARDS IN NATURE

To begin our journey through the world of lizards we discuss evolutionary history of lizards, placing them within what we call tetrapod vertebrates, animals with backbones

that have—or at least whose ancestors had—four limbs. We also introduce the current families of lizards placed in a phylogenetic context: a tree of families, not to be confused with a family tree but conveying somewhat similar information. Our tree of families shows historical ancestor-descendant genealogical relationships among groups of lizards discussed, whereas a family tree shows genealogical relationships among close relatives.

From the perspective of every individual animal, Earth is a complex place, and the biotic environment increases that complexity manyfold. We examine how lizards get around in this complex world, meeting the challenges of structurally diverse habitats and experiencing changes in temperature and moisture on daily and seasonal bases. Lizards eat primarily arthropods, but some have shifted to other invertebrates, vertebrates, and even plants. While some are strict herbivores, others kill and feed on vertebrates much larger than themselves by slicing through leg muscle, arteries, and veins with knifelike teeth and then tracking down their dying prey by following its scent trail. Lizards in turn are preyed upon by a vast number of other animals, ranging from large invertebrates such as scorpions to organisms that are many times larger than they are, such as hawks and eagles. Lizards are attacked from underground, across the ground, and from the air. Lizards interact with each other as well (some lizards are predators of others). Often, spectacular visual displays serve as signals between individuals. Males may engage in contests comparable to the imagined head butting of such extinct monsters as *Pentaceratops,* just on a more miniature scale. In many cases, a plethora of chemical signals allow lizards to determine species, sex, and sexual receptivity. All such behaviors are oriented toward a single simple goal: sending genes into future generations.

Lizards interact with a wide diversity of other organisms as key components of ecological communities. Not only do their positions in food webs vary depending on the organization of those communities, but their individual natural histories determine to a large extent what they are capable of doing in their own corners of the world. Species interactions have clearly resulted in some of what we call structure in present-day lizard assemblages, but evolutionary history of different lizard groups has also played a role in determining what possibilities are open. All these topics are addressed in chapters 2–7.

In chapters 8–13 we sail through various lizard families, detailing what we see as some of the most interesting traits and peculiarities of each family, always in an evolutionary perspective. We first consider iguanians, a group represented over much of the planet by species that typically sit on perches from which they ambush prey. Although these lizards have chemosensory organs, their use of chemicals is, for the most part, restricted to social behavior (herbivores are exceptions). Included are some of the most bizarre lizards, the true chameleons of Africa and Madagascar. Moving slowly through their arboreal habitats, these spectacular lizards have eyes that move independently and protrusible sticky tongues longer than their bodies that can literally be hurled at an unsuspecting insect, picking it off of a branch. Color patterns and chameleons' ability to change colors are rivaled by few other living vertebrates.

From iguanians we move into geckos and their relatives, Australian pygopodids (sometimes called flap- or scaly-footed lizards), then on to so-called worm lizards

ERIC R. PIANKA

When I was about six years old, in the mid-1940s, my family drove east from our hometown, Yreka, near Oregon in far northern California. We went across the U.S. to visit our paternal grandparents, German immigrants who lived in Illinois. Somewhere along the way, in the South, we stopped at a roadside park for a picnic lunch. There I saw my first lizard, a gorgeous, green, sleek, long-tailed arboreal creature (later I determined that this must have been an *Anolis carolinensis*) climbing around in some vines. My uncle and I did our utmost to capture that lizard, but all we were able to get was its tail. I still remember standing there holding its twitching tail, wishing intensely that it was the lizard instead. About a year later, back in California, I caught my first garter snake, which I tried to keep as a "pet," but it soon escaped (snakes aren't pets, but they are escape artists!). In the third grade, I discovered a captive baby alligator in the classroom next door. I was transfixed by that alligator and stood by its aquarium for hours on end reveling in its every move. As a little boy, without knowing anything about science, I was fascinated with biology. Years later, in graduate school, I discovered the layers in the biological cake, and eventually I went on to earn a Ph.D. and, later, my D.Sc. as an ecologist.

Yreka, the seat of Siskiyou County, lies in the shadow of Mount Shasta. In those days it was a sleepy little town of about 3,500 people, surrounded by relatively pristine wilderness. My boyhood was rich, replete with the great outdoors and lots of adventure. It was an easy walk out our back door to a variety of relatively undisturbed natural habitats. After school, my brothers and I roamed the juniper-covered hills around town. We lived outside and knew virtually every hectare within several kilometers of town. I collected everything from rocks to insects (especially butterflies) to bird eggs to lizards and snakes, including rattlesnakes. My collections were semiprofessional, too, with every specimen labeled with its scientific name. I went to great trouble to build a glass display case to house my collection of bird eggs, and we went to even greater lengths to find the nests of all the local species, ranging from hummingbirds to great horned owls. I kept snakes and lizards alive, and when they died, pickled them as proper museum specimens, complete with records and field notes. My mother was exceedingly supportive. She didn't particularly like snakes, let alone venomous ones, yet she always encouraged me to follow my interests wherever they took me. She allowed me to keep Esmeralda, a 1.5-m boa constrictor, in our house. How many mothers would allow a son to chill a live rattlesnake in the family refrigerator to cool it down for photography? Her only condition was that she wasn't going to enter the kitchen, let alone open the fridge, until I declared it "all clear."

When I was thirteen I found an unexploded bazooka shell that nearly killed me when it blew up in our front yard. I had gangrene and lost 10 cm of shin bone in my left leg, spent many months in the hospital having restorative surgery (bone and skin grafts), and was on crutches and in casts or a leg brace all through high school. Summers were usually devoted to reconstructive surgery, grafting skin and bone. At one point, to transfer a pellicle of flesh and skin from my right calf "donor area" to cover the extensive area of fragile thin scar tissue on my left leg, I actually had my legs sewn together for nearly two months. I developed a love-hate relationship with my own leg. More than once I would have willingly chosen death over the pain I had to suffer. Such liabilities can be transformed into assets, however. From all that hardship and suffering emerged no small measure of endurance and strength: I learned to tolerate physical discomfort, became self-reliant, and developed fortitude and independence, as well as an astonishingly strong will, all of which have helped me to become a successful field biologist. My disability has not prevented me from putting many kilometers on my game leg, and I have led

continued on next page ▶

a very rewarding life in spite of being a gimp. I would not wish such hardship on anyone, but it did build character and make me much more introspective and intellectually inclined than I might otherwise have been.

I joined the American Society of Ichthyologists and Herpetologists as a Life Member while I was still a sophomore in high school. I had nurtured a dream all through high school, my plan for the future: when I graduated, I was going to take my brother (he was my "legs") and go to Mexico to catch lizards and snakes. My parents humored me along in this fantasy, probably thinking that it would never come to be. But I persisted, planning it all out. Besides typing, Spanish, and English, the most important class I took in high school was auto shop: my class project involved overhauling the engine in my old brown 1948 DeSoto. (Auto mechanical skills should be a prerequisite for anyone who plans to survive fieldwork in remote areas.) I worked and saved every cent I could for the trip. Once I got there, I was stunned by the wide open wild spaces and beauty of the desert southwest. Everywhere we stopped or camped, I found myself enthralled with and awed by the lizards—zebra-tailed lizards, whiptail lizards or racerunners, horned lizards, chuckwallas, desert iguanas, leopard lizards and gorgeous bright-green collared lizards. We drove south beyond

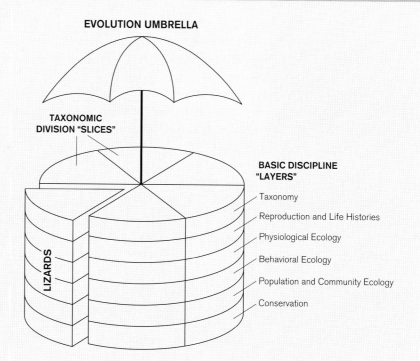

The "biological cake" shows the intersection of taxon-based science (slices) and concept-based science (layers). Neither is complete without the other. Evolution is the overarching organizing principle.

Mexico City, looking for boas and collecting D.O.R. ("dead on road") lizards and snakes. I returned to Mexico again during my sophomore year in college and published my first scientific paper on those Mexican specimens when I was only a junior.

Once I was invited to give a seminar at the Naples Marine Institute. Struggling to think of a way to justify talking about desert lizards to an audience inclined toward marine biology, I came up with an analogy: "lizards are just spectacularly beautiful terrestrial fish." Surprisingly, my marine biologist audience seemed to accept this. I enjoy catching lizards, or "lizarding," as much as or more than many people enjoy fishing, and the catch is, to my mind, much more interesting.

and blind lizards. Geckos are perhaps best known for their abilities to produce sounds and walk up walls, but not all geckos do either of these. A large number of geckos have shifted their primary activity to the night and have been observed on walls of buildings by nearly every human who has ever traveled to the tropics. Rafting on natural floating bits of habitat or on man-made structures has carried some throughout the Pacific islands. Geckos too can be spectacular in coloration and morphology. Pygopodids, worm lizards, and blind lizards are either legless or have only remnants of legs. Some are snakelike (some pygopodids), whereas others (such as worm lizards) are so strange in their locomotion that only seeing is believing. One Mexican worm lizard even has small, gopherlike front limbs that it uses to widen its burrows and pull itself through, but it has no hind limbs at all!

From geckos and their relatives, all rather sedate in terms of their locomotory abilities, we turn to racerunners and their allies. This group includes some of the most conspicuous lizards in most habitats of the world. They are active during the day, alert, fast moving, and always appear to be warily watching the observer. Chemical communication is important in their social behavior, and the same chemosensory system is used both to find and keenly discriminate among prey. Some of these lizards are very tiny, living in multilayered leaf litter of Amazon rain forest. Others are a meter or more in total length, basking on treetops overhanging waterways in the Amazon system and dropping to the water to forage. Once in water, some fold back their legs and swim like snakes only to emerge and climb back up into the treetops. Some are emerald green, appearing to glow as the sun is reflected from their velvety skin. Possibly related, night lizards are quite secretive, generally small-bodied animals living in crevices or under skeletons of long-dead plants. Individuals of some of these very tiny species can live more than ten years!

The next radiation of lizards we examine, collectively known as skinks, is the largest family of lizards. This worldwide group appears to have had an extremely complex history. With few exceptions, skinks in the New World are quite similar morphologically when color is taken away, whereas morphological diversity in Old World skinks rivals that of all other lizards combined. Striking color patterns with high-contrast stripes, blue and red tails, and brilliant head coloration of males associated with social behavior exist. Like racerunners and their allies, skinks have highly developed chemosensory systems used in both social communication and prey detection and discrimination. Like geckos, skinks have made their way to isolated Pacific islands and in some instances occur at remarkably high densities.

The next four lizard families we consider are more sedate in their behaviors than most skinks and racerunners. Girdled (cordylid), African plated (gerrhosaurid), alligator and glass (anguid), and knob-scaled (xenosaurid) lizards appear to move at relatively slow rates while foraging, almost a cruising gait. Girdled and knob-scaled lizards are often associated with rock crevices, remaining hidden much of the time. Their flattened bodies and in some instances highly modified dorsal scales allow them to enter deep crevices and hold themselves in should a predator try to remove them. In many respects, African plated lizards are reminiscent of some large skinks. Alligator lizards, perhaps best known by herpetologists and herpetoculturists for their vice grip–like bite, move about rather slowly on their small legs, with the long tail following in a snakelike manner. Their close relatives, glass lizards, are renowned for their ability to break their long tails into many highly motile pieces, distracting the attention of a would-be predator from their body.

The final chapter on lizard diversity focuses on the earless monitor and Gila monsters and their allies, the dramatic monitor lizards. Possibly a relative of ancestors to snakes, the earless monitor of Borneo remains poorly known. Gila monsters, in addition to having skin resembling intricate patterns of beadwork, are best known because they are the only venomous group of lizards (excluding snakes). Most of the time, Gila monsters are active within burrows of desert rodents, but early spring finds them active on the surface in the Sonoran Desert. Monitor lizards no doubt are the basis for dragon stories. Some, such as the Komodo dragon, are large and actually prey on animals larger than themselves, making these lizard predators of particular concern to people living nearby. Their long, yellow forked tongue, extruded and waved about like a snake's tongue, certainly brings to mind the dragons of mythology and their breath of fire.

So, these are the lizards, from tiny to gigantic, from drab to remarkably beautiful, from totally harmless to venomous, one of the most diversified of all groups of terrestrial vertebrates. Once our journey through extant lizards is completed, we consider in more detail some things that render lizards unique, paying additional attention to that radiation within lizards known as snakes. In closing, we consider interactions between humans and lizards. Lizards have held a prominent place in indigenous art and myth, have been hunted and eaten, have been raised for food, have become the basis for a multimillion dollar pet trade, and are important components of most natural ecosystems. If you haven't yet had an intimate relationship with a lizard, you should!

LOGIC IN THE NATURAL WORLD

All biologists "discovered" many important biological phenomena as kids, based on their own personal direct observations of animals either in the field or in cages. When he was just seven years old, Laurie Vitt pronounced Montana milksnakes to be coral snake mimics, having seen coral snakes only in pictures and without a clue as to what mimicry in the biological sense really was. We learned that each species has its own distinctive habits, places, and times of activity. Trying to catch lizards and having them escape leaving behind their twitching tails taught both of us about strategies for surviving predator attacks. We both noticed variation among individuals of a given species in their abilities to escape. It was a minor intellectual step to deduce that faster lizards, which were harder to catch, had a survival advantage. We already knew that some traits were heritable, that is, passed on from one generation to the next, simply because we had observed litters of fam-

LAURIE J. VITT

My early interest in "herps" began when, at about the age of four, I collected my first snake, a large adult *Thamnophis sirtalis* east of Billings, Montana. During my childhood I discovered huge overwintering aggregations of garter snakes early in spring, and I wondered why the large "balls" of snakes seemed to consist mostly of males. Elsewhere I discovered piles of prairie rattlesnakes with a few bullsnakes and western racers mixed in. Most exciting for me was the discovery of milksnakes under sandstone surface rocks. When I found that milksnakes would not eat frogs or mice, I was forced to collect nearly everything that walked or crawled to offer my first milksnake. Among those creatures were fence lizards, which milksnakes do eat—my first introduction to the keen chemosensory system of squamate reptiles.

When I was twelve years old my family packed up and moved to Tracy, California. There I explored the canyons of Corral Hollow southwest of Tracy, where I discovered alligator lizards, skinks, side-blotched lizards, horned lizards, and a host of snakes that I had never seen. Every new animal I encountered was a major find. One spring day in 1958 I collected a beautiful adult patch-nosed snake, which eventually escaped from one of my terraria only to be run over by a passing car. I found its flattened body two days after it escaped. (Just a few years ago the first documented record of a patch-nosed snake from Corral Hollow was published, based on a shed skin.) I also saw, but was unable to catch, an adult blunt-nosed leopard lizard in the east end of

Corral Hollow. I recall thinking it had to be a giant exemplar of the common side-blotched lizard species, because it had tiny scales. Years later I discovered the *Western Field Guide* by R. C. Stebbins and recognized the lizard that had burned such a clear image in my mind. No blunt-nosed leopard lizard has ever been collected in Corral Hollow, which was no doubt one of its last outposts following the agriculturization of the San Joaquin Valley.

From Tracy we moved to east Sacramento where I discovered even more reptiles and amphibians. By this time I was as interested in lizards as in snakes. I was exposed to the world diversity of reptiles and amphibians when I began working for a pet dealer named Frank Buck in Carmichael. There I saw my first *Dracaena,* a lizard I have been lucky enough to observe in the field only a few times since. After just three years in Sacramento we moved to Bellingham, Washington. I had to be dragged kicking and screaming because I knew that reptile diversity was much lower in the Pacific Northwest. During my senior year in high school I searched every hillside and canyon in the Bellingham area trying to satisfy my thirst for reptile discoveries. Under a sandstone rock on a sunny hillside I found my second rubber boa (I had collected one in California during a scout camp at Carson Pass), then my third, fourth, and ultimately twelfth: I found eleven within thirty minutes!

Later that year I met Ernie Wagner at the Woodland Park Zoo, who took me to remarkable rattlesnake dens in eastern Washington and southern Idaho. At

about the same time I became sidetracked when I discovered that twelve years of musical training put me in an excellent position to enter the rock music business, a vocation that would finance my entire college career. I was a mediocre undergraduate, partly because I found most general education courses to be based on false premises and partly because I was deeply involved in music, spending much of my time traveling with bands.

When I returned to Western Washington University (at the time, Western Washington State College) for a master's program, I had decided that I really wanted to be a herpetologist and, in particular, a university professor. I was unsure whether to pursue snakes or lizards as a focus of study, however, reaching a decision only when my master's advisor, Herbert A. Brown, convinced me that it would be easier to conduct field research on lizards. And the rest is history: although I am still fascinated by snakes, the remarkable diversity of life histories, behaviors, morphologies, and adaptations of lizards continues to keep my interest.

I have no problem justifying what I do. While most people on the planet revolve their activities around a single species, *Homo sapiens,* I am interested in understanding not only a portion of the natural world that led to the evolution of our species but also, more broadly, the complex ecological networks that sustain all life on Earth as we know it. The biology of lizards is a window through which we can peek at the evolutionary history of life. I can't imagine anything being more exciting.

ily pets. Also, we couldn't help but notice that children of most of our neighbors looked much more like their own parents than like ours.

Because we were just kids, we didn't formalize our understanding of nature. Some of our discoveries were already well known, but others had not yet been worked out. Thus, as children, we discovered evolution by natural selection only to later learn that Charles Darwin had beaten us to the punch—but we knew things must work that way. For most of us who entered the biological arena from a lifetime of observing animals in natural settings— the environments to which they have become adapted— evolution by natural selection was intuitively obvious: there simply could be no other logical explanation for the amazing diversity that surrounded us. What Darwin discovered on his voyage was exactly what we encountered growing up; our "voyages" were simply more localized. Things we discovered in our youth often caused us to ask the very questions that continue to fascinate us today.

The logic of the natural world, a commonsense logic that anyone can easily grasp, is extremely sound and powerful. Sadly, few people today enjoy the advantages of childhoods like we had. Nowadays, most people grow up in an artificial world and in social organizations based on a complete lack of this sensible natural logic. People are actually encouraged to make absurd assumptions, some of which now threaten our very existence. Our entire economy is based on the flawed premise of chain letters, Ponzi growth schemes doomed from the outset. Economic growth on a global scale requires population growth; after all, someone has to purchase items produced. Research and development, production, marketing, and purchasing all require people. Many fail to appreciate that such growth, albeit seemingly attractive in the short term, has dire long-term ecological consequences, which are only now beginning to be appreciated. Biology teaches us that everything has a carrying capacity dictated by space and availability of resources: growth and economies simply cannot continue to expand indefinitely. Potential consequences of unlimited population growth have been known since the late 1700s when Malthus demonstrated that, unchecked, populations grow exponentially. Yet *Homo sapiens* has proceeded as though the "rules" of the natural world apply to everything except humans. As children, we watched oranges first become infected with small spots of fungus, then become completely overtaken by the fungus encapsulated with fruiting bodies producing spores, and finally, if our parents didn't throw them out first, they dried up into blackened amorphous masses. Examining those dried-out remains of oranges carefully— like any good future biologist—we discovered that the fungus was gone. No doubt some spores found other habitable fruit on which to test the limits of resources again, but the fungus on the original orange had depleted all its available resources and then gone extinct. Our "orange," however, is not in a basket of other oranges. There is only one planet Earth. Unlike the fungus, we have nowhere else to go. Consequently, our survival depends on careful management of the limited resources we have. This is the logic of children and the logic of the natural world upon which we all depend.

Throughout this book, we continually return to questions we asked as children in an attempt to encourage our readers to open their minds and to ask questions. We hope our readers will be convinced to appreciate, embrace, and search for this powerful biological logic.

A chuckwalla, *Sauromalus varius*,
overlooks the Sea of Cortez,
Gulf of California. (Chris Mattison)

Part One

LIZARD LIFESTYLES

CHAPTER 1

EVOLUTIONARY HISTORY AND PHYLOGENY

The leaf chameleon, *Brookesia stumpfi*, from Madagascar is terrestrial and does not have a prehensile tail like chameleons in other genera. It's very hard to find in dead leaves! (Bill Love)

Knowing exactly which species one is referring to is critical to any discussion of the evolution and natural history of lizards. Common names—"blue-belly," for example—though often evocative, are approximate, even potentially confusing, since they could apply to many different species. Such inaccuracy is easily resolved by the system that scientists use to name species, a nomenclature (developed by the Swedish botanist Carolus Linnaeus [1707–78], and therefore known as the Linnean system) that has been standardized by an international commission.

According to this system, each species has a unique scientific name, derived from Greek or Latin and usually descriptive, which is used by scientists everywhere. Take, for example, *Homo sapiens: Homo* (Latin for "human") is the genus to which we belong, and our species is *sapiens,* Latin for "wise" or "knowing." Note that genus and species are italicized, while higher-order designations are capitalized. To be distinctive, moreover, animal family names always end in *-idae,* while subfamily names end in *-inae.* Armed with a few rules, anyone can instantly recognize families, subfamilies, genera, and species names.

During the last third of a century, biologists have adopted a whole new approach to systematics based on relatedness, or phylogenetic affinity, phylogenetic systematics. According to this system, groups must be monophyletic, which means that each recognized group must include an ancestor plus all of its descendants. A monophyletic group is called a "clade," and each member is a taxon. An entire clade can be a taxon within a more inclusive clade (for example, the genus *Sceloporus* within the subfamily Phrynosomatinae). To apply this criterion to classification of organisms, biologists require a family tree showing degree of relatedness among all taxa descended from a common ancestor. Such a genealogical tree is called a phylogeny.

Consider evolution of vertebrates (fig. 1.1). The ancestor of all vertebrates was a fish (technically, that makes us fish, too), which gave rise to lungfish and the ancestor of the first tetrapods (meaning "four legs"), amphibians. So-called "reptiles," birds, and mammals all descended from ancestors of amphibians. Hence, lungfish are the sister group to tetrapods (all other terrestrial vertebrates). Within tetrapods, amphibians are the sister group to all other tetrapods (known as Amniota in reference to a membrane in the amniotic egg). Because the traditional group "reptiles" (turtles, crocodilians, lizards, and snakes) does not include birds—which, however, arose from within reptiles—it does not constitute a monophyletic group, but is instead paraphyletic (meaning it does not contain all descendants of a common ancestor). "Reptiles" is therefore an invalid grouping under the rules of modern phylogenetic systematics. This problem is solved by calling birds reptiles! Exactly the same problem occurs within lizards because we have traditionally considered "snakes" a distinct group, but in fact snakes originated from within lizards.

Nearly all estimates of phylogenies are no more than

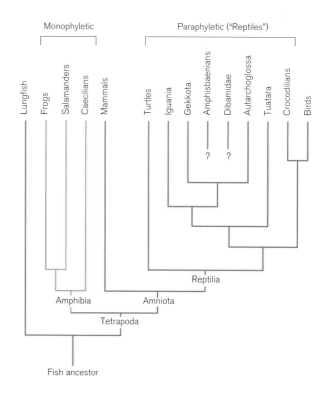

Figure 1.1 Evolutionary relationships among vertebrates. Extant Amphibia is a monophyletic group (clade) because it contains an ancestor and all its descendants. "Reptiles" is a paraphyletic group because birds, which historically have not been considered reptiles, descended from the same common ancestor. This problem is easily solved by including birds within Reptilia. Relationships of dibamids, amphisbaenians, and snakes remain unclear, but snakes likely fall within Varanoidea (see fig. 1.4).

hypotheses. Phylogeneticists reconstruct phylogenies by assembling data on states of various characters (fig. 1.2). Presence or absence, for example, might be two states for the character "cementum" at the base of teeth. Although ancestral states do not provide a phylogenetic signal, shared derived (synapomorphic) character states do contain information that allows phylogenetic inference. If all lizards except teiids lack cementum, nothing can be said about the relationships of those lizards to each other based on that character. But because cementum arose in the ancestor of teiids, all teiids have it and it represents a shared (among all teiids) derived (shows up first in an ancestor to teiids) character state or trait. Character states that arise independently (convergences)—such as the three horns in males of Jackson's chameleon and *Tricer-*

atops dinosaurs—are misleading because they suggest relatedness where there is none. The challenge for phylogenetic systematists is to identify shared derived character states and to exclude convergent ones.

Comparison with appropriate related "outgroups" allows systematists to identify ancestral traits and to "polarize" character state changes of probable shared derived characteristics. These can then be used to construct a phylogenetic tree. Using a closely related outgroup, a tree can be "rooted" to produce a "cladogram" that shows probable genealogical relationships among members of the group (clade) concerned. Past history can be recovered, at least to some extent, from current character states. Exploitation of these techniques to their fullest potential obviously requires examination of as many related taxa as possible—extinctions of existing species truly become "lost pages" in the unread and rapidly vanishing book of life.

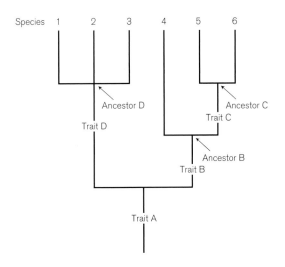

Figure 1.2 "Phylogenetic systematics" showing trait evolution. Each new trait defines a monophyletic group or clade. For example, the ancestor of species 4–6 had trait B; all of its descendants (species 4–6) share that trait. Trait C defines another monophyletic group, whose members (species 5–6) have not only trait B from their common ancestor with species 4 but also the new trait C. Because trait C did not occur in ancestor B, it is called a derived trait. All six species have trait A because the ancestor of them all did too. Species 1–3 share a common ancestor with trait D, but no traits yet discovered tell us the evolutionary relationships among these three species; they will remain an *unresolved trichotomy* until a trait is discovered that ties two of the species together. Traits can also be lost during evolution along a particular lineage, erasing phylogenetic signals to give a misleading perspective on relatedness.

Modern molecular techniques, particularly DNA amplification and sequencing, now allow reconstruction of probable phylogenies of diverse monophyletic groups of organisms (Hillis et al. 1996). Such molecular approaches combined with increasingly rigorous statistical procedures have greatly increased our confidence in particular "resolved" phylogenetic trees. DNA consists of strands of genetic material made up of genes. An individual's evolutionary history is coded in its gene sequences, which determine how development produces the correct set of events and entities that make an individual what it is. As a result, the individual develops into the correct species. Its recent family history is also coded in gene sequences, also reflected in its final morphology. This is why we develop into humans rather than lizards during embryogenesis and why we resemble our parents rather than our neighbor's parents (hopefully!). DNA sequences are really no different from morphological characters used by many past and present systematists. Advantages center on our ability to examine a lot of characters (genes) simultaneously. With morphological traits (e.g., the shape of a particular bone), each trait has a lot of power in an analysis because many genes ultimately contributed to its structure. The disadvantage centers on our inability to determine exactly which gene sequences result in that structure. Concordances between morphologically and molecularly derived phylogenies strengthen our confidence in estimates of evolutionary relationships.

SQUAMATE EVOLUTION

Lizards and snakes form a monophyletic assemblage, a clade, known as Squamata. Understanding squamate evolution requires consideration of their closest relatives, their sister group Rhynchocephalia, which contains some fossils and two species of tuatara, genus *Sphenodon,* found only on offshore islands of New Zealand. The two species of *Sphenodon* are truly "living fossils," as they are the only living descendants of an ancient group about 250 million years old. Because the fossil record of lizards and their closest relatives is scant, we must make educated guesses at relationships and evolutionary scenarios. Most sets of closely related species tend to exhibit niche conservatism, seldom changing much in evolutionary time (Holt 1996). If tuatara haven't changed much during the past 250 million years, they could provide valuable insights into what early ancestors of squamates might have been like (of course, such

The tuatara of New Zealand, *Sphenodon punctatus,* is one of two living representatives of Rhynchocephalia, the sister taxon to all squamate reptiles. The other species, also of New Zealand, is *S. guntheri.* (Paddy Ryan)

reasoning is rather speculative and could be quite erroneous if evolutionary changes *have* occurred). *Sphenodon* has elliptical pupils and low active body temperatures (11°C). These reptiles forage at dawn and dusk, eating large beetles, big orthopterans, and nocturnal geckos (also seabirds). During the day tuatara bask, their body temperatures reaching as high as 28°C or more (Gans 1983). Tuatara skulls are diapsid, having two apertures on each side. (Lizards are phylogenetically diapsids, but have lost the lower temporal arch and so retain only a single upper aperture. This frees the quadrate to move and results in an efficient new hanging jaw mechanism known as streptostyly.) The top of a tuatara's head is covered with small scales. Males have a striking crest of scales running down the back of their neck and the middle of their back. Teeth are not set in sockets but are simply serrations of the jawbones with a layer of hard enamel; they do not appear to be replaced. Tuatara have a fleshy tongue, used when feeding on small prey (Schwenk and Throckmorton 1989).

Tuatara have a conspicuous pineal organ or "third eye" in the center on top of their heads and no external ear openings (although they can hear). Male tuatara have no copulatory organ but do have two shallow paired outpockets at the end of their cloaca, probably antecedents to the paired hemipenes that characterize all lizards and snakes (Arnold 1984a). Tuataras lay eggs, grow slowly, and are quite long lived (up to sixty years). Some of these traits could be novel adaptations evolved in response to the cold New Zealand environment (Gans 1983), making it somewhat risky to consider them ancestral traits of

Squamata. However, we have no other way of inferring ancestral states.

The origin and diversification of lizards must be considered in the context of continental drift. For many years, biogeographers assumed some permanence in locations of continents. As a result, interpretations of faunal similarities between them often relied on hypothetical mechanisms of transport from one continent to another, such as improbable land bridges or "rafting" of organisms across huge water gaps. Such long-distance dispersal events are exceedingly improbable, although some have clearly occurred. Land tortoises, for example, must have rafted to remote volcanic islands such as the Galápagos and Aldabra. Cattle egrets made a successful trans-Atlantic crossing from Africa to South America without human assistance during recorded history late in the eighteenth century. Skinks and geckos drifted to most Pacific islands, even remote ones. However, such events are not the rule, and much classical biogeography is still being reinterpreted in light of the relatively recent discovery that continents are drifting.

Continents are formed of light "plates" of siliceous, largely granitic, rocks about 30 km thick, which float on heavier mantlelike basaltic blocks. Ocean floors are composed of a relatively thin altered top of the earth's mantle. A mountain range on the seafloor in the mid-Atlantic represents a region of upwelling of the mantle. As upwelling proceeds, seafloors spread and continents move apart.

Positions of paleomagnetic anomalies (periodic polarity reversals) in the seafloor allow geologists to calculate the velocity of lateral motion of the ocean floors, which

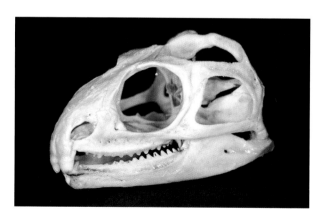

Sphenodon punctatus skull showing two openings that define the diapsid skull. (R.W. Murphy)

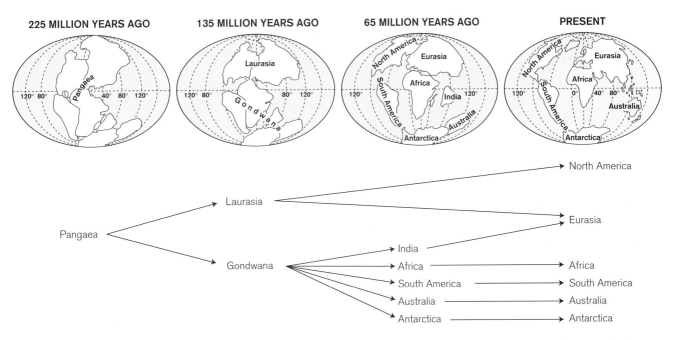

Figure 1.3 Continental drift. During the long history of continental drift, the supercontinent Pangaea first split into two large continents, Laurasia in the north and Gondwana in the south. Much of current North America and Eurasia (except India) is derived from Laurasia, whereas Africa, South America, Australia, and Antarctica are derived from Gondwana. These events can be represented graphically as maps *(top)* or as area cladograms *(bottom)*.

correspond to comparable estimates for landmasses. Except for the Pacific (which is shrinking), the world's oceans are growing, with very young ocean floors in mid-ocean and progressively older seafloors toward the continents. Other evidence, such as ages of islands and depths of sediments, nicely corroborates these conclusions.

Biogeographical evidence for continental drift is ample. Ancient groups of plants, freshwater lungfishes, amphibians, and insects that had arisen and spread before the breakup of the continents now occur on several continents, whereas many other more recently evolved animals, such as some groups of mammals and birds, are restricted to particular biogeographic regions. Cold-adapted fossil plants in the genus *Glossopteris* are found from the Triassic (about 200 million years before present [M.Y.B.P.]) on South America, Australia, Africa, and Antarctica. Similarly, present-day remnants of ancient southern beech forests *(Nothofagus)* are found in southern Chile, Australia, and Tasmania and occur as fossil pollens in Antarctica.

The great southern landmass of Pangaea, consisting of all continents connected together, began to break apart during the Triassic, about 200 M.Y.B.P. (fig. 1.3). A large block known as Laurasia, which later separated into North America and Eurasia, broke away in the early Mesozoic (about 180 M.Y.B.P.) and began to drift north. The remaining southern landmass, Gondwana, consisted of the three southern continents, Africa, South America, and Australia, plus the tectonic plate destined to become India, all of which remained fused together and connected to Antarctica. Earth was much warmer then, and both Gondwana and Laurasia were populated with dinosaurs.

About 100 million years ago, during the Mesozoic, Gondwana began to fragment into several tectonic plates. India broke off and, carrying its flora and fauna with it, sailed like a giant Noah's ark northward, where its collision with Eurasia formed the Himalayas. Two competing scenarios differ in how far away Laurasia was from Gondwana. Under one scenario (Smith and Bryden 1977) Laurasia was thousands of kilometers to the north, and the Indian plate was isolated for a considerable length of time as it moved rapidly, by geological standards, northward through the oceanic region now known as the Indian Ocean. Under the other scenario (Owen 1976), eastern Laurasia (now Southeast Asia) remained in contact with Gondwana during the Mesozoic, allowing considerably greater faunal interchange to occur. South America plus Africa, remaining connected to each other, began

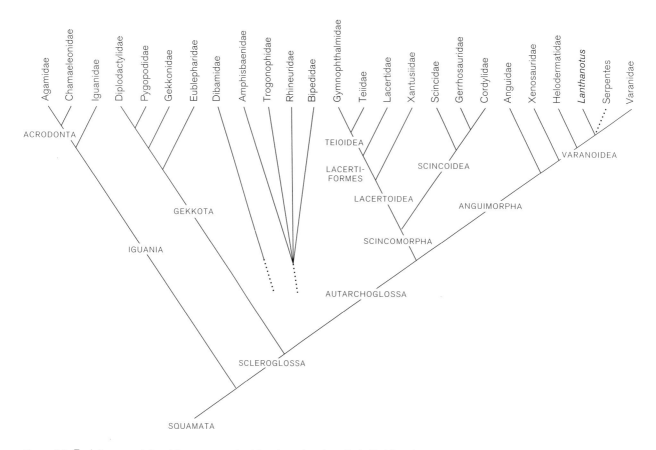

Figure 1.4 Evolutionary relationships among extant lizards and snakes. Note that lizards as a group are paraphyletic because snakes—Serpentes—are nested within them. Slightly different familial relationships are suggested by recent molecular studies, which place dibamids, amphisbaenians, and xantusiids closer to gekkotans. For consistency, we follow the phylogeny shown here throughout this book.

to pull away from Australia-Antarctica; finally, about 100 M.Y.B.P., Africa broke free, forming the Atlantic Ocean, and began to drift northward. By the late Mesozoic (65 M.Y.B.P.), the Indian plate along with its biota was midway across the ocean that is now known as the Indian Ocean. South America, Australia, and Antarctica remained attached until the early Cenozoic.

Imagine the effects of changing climates on plants and animals as tectonic plates and continents drift through different latitudes. About 50 million years ago, Australia finally broke off and began its own slow drift northward. It entered the tropics about 20 M.Y.B.P., still remaining fairly wet. Aridification of Australia began about 10–15 M.Y.B.P., as forests gave way first to grasslands and then to deserts. The Australian continent has been more or less in its present position for the last several million years. As Australia continues to move toward the equator, its climate will gradually become wetter and increasingly more and more tropical.

Among the oldest and most primitive possible relatives of lizards are the paliguanids (Carrol 1988). One fossil, *Paliguana,* is known from an isolated skull from the Upper Permian and Lower Triassic (about 250 M.Y.B.P.) of southern Africa. Another paliguanid from the Lower Triassic of Australia is *Kudnu mackinleyi,* which has been considered by some researchers close to the ancestor of all later squamates. Although relationships of paliguanids to squamates remain uncertain, they are clearly diapsids and likely squamates. The occurrence of paliguanid fossils in South Africa and Australia suggests that lizards arose in Gondwana (Molnar 1985).

An extinct Upper Triassic group, the Kuehneosauridae, is considered the sister group to Lepidosauria, which includes Rhynchocephalia and Squamata (Gauthier et al. 1989; Rieppel 1994)—though other interpretations of their relationships also exist (M. Caldwell 1996; deBraga and Rieppel 1997). Kuehneosaurs were gliding forms (Wuethrich 1997; Frey et al. 1997) with subdermal bones in winglike appendages. True lizards did not arise until the Late Jurassic (about 150 M.Y.B.P.), following a 50-million-year gap in the fossil record (Estes 1983). Most lizard fossils are considerably more recent. Richard Estes (1983) argues that much of early lizard evolution must have occurred during the Jurassic and Cretaceous on the supercontinent of Pangaea.

Some lizard families appear to be of Gondwanan origin, but others presumably arose in the Northern Hemisphere and subsequently dispersed to southern continents. The snakelike pygopodids are thought to have arisen within Australia from diplodactylid geckos (which themselves probably evolved from another gekkotan group within Australia). The four lizard clades that reached Australia underwent extensive adaptive radiations on the island continent, with some, such as varanids, becoming more species rich there than they are today in their probable source areas.

When the northern Laurasian plate separated from the southern Gondwanan plate in the Middle Jurassic, two isolated landmasses were formed. Gondwana presumably had primitive iguanians (Estes 1983) and geckos (Kluge 1987), whereas Laurasia must have contained ancestral eublepharid geckos, scincomorphans, and anguimorphans (Estes 1983). When Gondwana broke apart, its iguanians and gekkotans became isolated on the three large southern landmasses, Africa (agamids and gekkonids), South America (iguanids and sphaerodactylines), and the Australian region (agamids and diplodactylids).

Gekkonids and skinks dispersed widely and became virtually cosmopolitan, while other groups either remained confined to the landmass of their origin (cordylids, corytophanines, crotaphytines, diplodactylids, gymnophthalmids, hoplocercines, iguanines, lanthanotids, leiocephalines, leiosaurines, liolaemines, oplurines, phrynosomatines, polychrotines, pygopodids, sphaerodactylines, tropidurines, and xantusiids) or exhibited a more limited dispersal (agamids, anguids, chamaeleonids, helodermatids, iguanids, lacertids, teiids, and varanids).

Early in the evolutionary history of lizards, two distinct groups arose (fig. 1.4; table 1.1): the Iguania, which retain a primitive fleshy tongue used in feeding (and shared with the tuatara), and the Scleroglossa (meaning "hard tongue"), which use their jaws rather than their tongues when feeding, thus freeing up their tongues for chemoreception (Schwenk and Throckmorton 1989). Iguanians include the following families and subfamilies: Agamas

(Agamidae), chameleons (Chamaeleonidae), and iguanas (Iguanidae). Iguanids are a diverse group containing eleven subfamilies: the herbivorous iguanas (Iguaninae); spiny-tailed lizards or Madagascar swifts (Oplurinae); swifts, horned lizards, and their allies (Phrynosomatinae); lava lizards and their allies (Tropidurinae); South American swifts (Liolaeminae); curly-tailed lizards (Leiocephalinae); flathead anoles and their allies (Leiosaurinae);

TABLE 1.1

Squamate classification

IGUANIA	SCLEROGLOSSA
Iguanidae	Gekkota
Iguaninae	Eublepharidae
Crotaphytinae	Diplodactylidae
Phrynosomatinae	Pygopodidae
Hoplocercinae	Gekkonidae
Liolaeminae	Sphaerodactylinae
Polychrotinae	Gekkoninae
Corytophaninae	*Incertae sedis*
Leiocephalinae	Amphisbaenia
Tropidurinae	Amphisbaenidae
Leisaurinae	Bipedidae
Oplurinae	Rhineuridae
Agamidae	Trogonophidae
Agaminae	*Incertae sedis*
Leiolepidinae	Dibamidae
Chamaeleonidae	Autarchoglossa
	Scincomorpha
	Lacertidae
	Teiidae
	Teiinae
	Tupinambinae
	Gymnophthalmidae
	Xantusiidae
	Scincidae
	Scincinae
	Acontinae
	Feyliniinae
	Lygosominae
	Cordylidae
	Gerrhosauridae
	Anguimorpha
	Xenosauridae
	Anguidae
	Varanoidea
	Helodermatidae
	Lanthanotidae
	Varanidae
	Serpentes *[embedded here]*

anoles and their allies (Polychrotinae); dwarf iguanas and weapon tails (Hoplocercinae); collared and leopard lizards (Crotaphytinae); and helmeted iguanas and basilisks (Corytophaninae). All iguanians are ambush predators, foraging by sitting and waiting for their prey, relying mostly on visual cues. All other lizards (21 families) are scleroglossans, most of which are active foragers, typically using chemosensory cues both to locate prey and to discriminate among prey types. Scleroglossans include the large clade Gekkota, consisting of three families of geckos (Eublepharidae, Gekkonidae, and Diplodactylidae) plus the flap-footed lizards (Pygopodidae). Other clades of scleroglossans are worm lizards (Amphisbaenidae, Bipedidae, Rhineuridae, and Trogonophidae); blind lizards (Dibamidae); racerunners, ameivas, and tegus (Teiidae); microteiids (Gymnophthalmidae); wall lizards and their allies (Lacertidae); night lizards (Xantusiidae); skinks (Scincidae); girdled lizards (Cordylidae); plated lizards (Gerrhosauridae); alligator lizards and glass lizards (Anguidae); knob-scaled lizards (Xenosauridae); earless mon-itors (Lanthanotidae); beaded lizards and Gila monsters (Helodermatidae); and monitor lizards (Varanidae). Of course, snakes are also scleroglossans. Lizard genera are summarized in the appendix.

As a result of these differences in foraging mode and chemosensory capacities, iguanians and scleroglossans exhibit numerous fundamental differences in behavior, natural history, and ecology. For example, bearing of live young (viviparity) has arisen only a few times in Iguania but is widespread in Scleroglossa. Scleroglossans can find sedentary prey items like insect larvae and termites using olfactory cues, whereas iguanians must hunt visually. Indeed, acquisition of heightened chemosensory ability may have facilitated the shift to active foraging. Both have allowed scleroglossan lizards to exploit a new resource base and to occupy a previously unexploited adaptive zone. These factors, in turn, allowed Scleroglossa to undergo a massive adaptive radiation and gain a cosmopolitan distribution (over 3,000 species of lizards in 21 families plus nearly 3,000 species of snakes in 15 families).

CHAPTER 2

GETTING AROUND IN A COMPLEX WORLD

**Tracks of the thorny devil,
Moloch horridus, as it makes its way
across the sand.** (Eric Pianka)

From a lizard's perspective, the world is a complex set of landscapes. Physical structure of the environment is one landscape and includes trees, shrubs, rocks, and the consistency of ground or water. Lizards perform their daily activities within constraints of the particular habitat where they live. Moreover, over evolutionary time this structure has had a lasting impact on morphology of individual species, resulting in the diversity we see today. Depending on morphology and circumstances, lizards run, jump, swim, slither, burrow, and even glide. The shapes of their bodies, tails, limbs, and heads as well as some of their spectacular ornamentation facilitate or impede locomotion within their particular three-dimensional habitat.

Physical landscapes change relatively slowly, but changes can be great given enough time. Because lizards cannot regulate their body temperatures physiologically like mammals and birds, the thermal landscape is of particular importance; indeed, it may be, in many respects, more complex than the physical landscape itself. The thermal landscape determines time of activity during any given day and, because of seasonal shifts in available sunlight (the source of heat), directly impacts when and where lizards can be active. Either directly or indirectly, temperature influences virtually everything a lizard does, including running speed, alertness, and ability to capture prey, to mention just a few. Moreover, temperature influences growth, timing and frequency of reproduction, and other life history characteristics. Water availability is another landscape. Although influenced greatly by latitude and position with respect to mountain ranges, coasts, and major bodies of water, water available to individual lizards is tied largely to temperature. Life cannot exist without water. Nevertheless, lizards have done quite well in some of the driest habitats on Earth. Lizard survival reflects a long evolutionary history of complex trade-offs between physical characteristics of the external environ-

ment, corresponding physiological responses within the lizard, locomotor abilities, and a plethora of intra- and interspecific interactions (fig. 2.1).

LOCOMOTION AND MORPHOLOGY

How lizards move around in their respective habitats depends to a large extent on their morphology, the structure of their bodies and appendages. The vision most people have of a lizard is that of a relatively small, elongate, long-tailed, scaly animal with relatively long hind limbs and shorter forelimbs that scurries across flat surfaces. Lizards, however, come in a wide variety of shapes and sizes. Indeed, several groups within lizards are totally limbless, including snakes, most amphisbaenians, many skinks, and some anguids, including California legless lizards *(Anniella)*. Others have such tiny limbs that for all practical purposes they are limbless (pygopodids, some cordylids, and many skinks). Many are small, with some, like the tiny leaf-litter gecko *Coleodactylus amazonicus,* weighing less than 0.5 g as adults. Others, like the Komodo dragon *(Varanus komodoensis),* are huge—more than 3 m in length and weighing as much as 250 kg! Luckily for us, an ancestor of the Komodo dragon, *Megalania prisca,* is extinct. It exceeded 5.5 m in length and likely weighed nearly

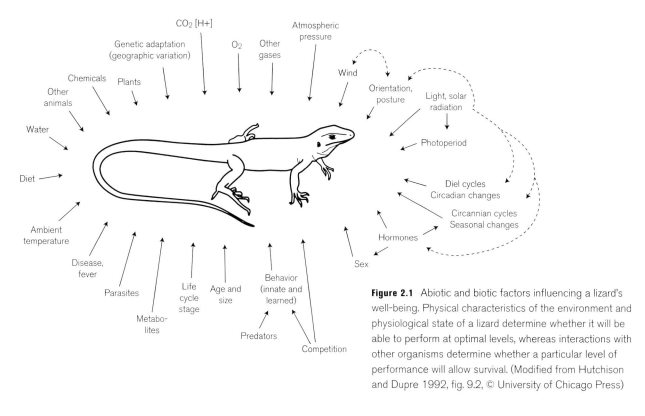

Figure 2.1 Abiotic and biotic factors influencing a lizard's well-being. Physical characteristics of the environment and physiological state of a lizard determine whether it will be able to perform at optimal levels, whereas interactions with other organisms determine whether a particular level of performance will allow survival. (Modified from Hutchison and Dupre 1992, fig. 9.2, © University of Chicago Press)

The spider gecko, *Agamura persica,* from Southwest Asia gets its common name from the way it moves. (R.D. Bartlett)

680 kg! Although lizards may appear to move the way they do *because* of their body form (which is true in a functional context), their various body forms represent a long evolutionary history of the impact of the physical structure of their respective habitats on their morphology. In other words, their bodies are the way they are *because* of historical effects of microhabitats and the biotic environment (e.g., competition and predation) that have molded evolution of their bodies.

Many lizards are long and thin with long tails; others are short and robust with short tails. Some have strongly prehensile tails used as a fifth leg in climbing. Others use their tails as whips to defend themselves. Among lizards with limbs, some have long limbs, others have short limbs, and some have only two limbs. A few lizards have webbed feet and/or skin flaps that can be expanded and used as partial parachutes when jumping from arboreal perches. Many species have laterally compressed tails used in swimming. Still others have adhesive pads on their toes allowing them to scale vertical smooth surfaces; a few even have adhesive pads on the tip of the tail providing a fifth point of secure contact. Heads of lizards vary from flat and wide in some crevice-dwelling species to long and thin in species like the pygopodid *Lialis* that feeds on other lizards. A wide variety of ornamentation exists as well, some for crypsis (camouflage) and some having to do with social interactions. Mentioned here only in passing, these are dealt with in greater detail in chapters 4 and 8.

The thin-bodied lacertid *Takydromus smaragdinus* moves easily through vegetation. (Christopher Austin)

Chameleon tails, as in this *Chamaeleo pardalis,* are the most prehensile among living lizards. (Chris Mattison)

Chameleons, such as *Chamaeleo pardalis*, have zygodactylous toes allowing them to grasp thin branches. (Chris Mattison)

To exemplify the diversity of morphologies in lizards and how they relate to getting around in natural environments, we provide examples particularly interesting to us. Similar and equally interesting descriptions could be presented for virtually every living lizard species.

LIFE ON FOUR LEGS

Most lizards have four legs and can move rapidly across horizontal, and in some instances vertical, surfaces. Movement across solid horizontal surfaces does not require drastic modifications in toe, foot, or limb morphology; as a consequence, lizards of many shapes and sizes move easily across flat surfaces. A vast majority of lizards with legs use a sprawling type of locomotion. Because legs are positioned to the sides of the body rather than underneath it, lizards spend a great deal of time with their ventral surfaces pressed against the substrate. Sprawling locomotion is most easily observed in slow-moving lizards, such as the Gila monster *(Heloderma suspectum)*. A Gila monster's body bends from side to side (laterally), enhancing sprawling locomotion. With the head, neck, and chest lifted off the ground, the right foot stretches forward out and away from the body. The foot pushes the front (anterior) part

of the body forward and slightly away from the foot. The lateral position of the leg results in forces in two directions, sideways and forward. The opposite hind leg is then moved forward and, after coming in contact with the substrate, pushes the rear (posterior) part of the lizard forward and sideways. The opposite front leg is then moved forward and the process repeated. The overall effect is that the lizard moves along in a stiff, snakelike fashion but generally forward. Its tail swings back and forth as a result of the lateral movement of its body.

Numerous specializations in the vertebral and appendicular skeleton facilitate this kind of movement. Movable vertebrae accommodate lateral bending of the body, and this, along with lateral positioning of the limbs, confers a relatively long stride length. Lizard hind legs are typically much longer than forelegs; thus much of the forward propulsion results from hind leg movement. Lizards such as *Crotaphytus* and *Basiliscus,* which run considerable distances bipedally, make maximum use of their extremely long hind limbs, achieving higher running speeds than they would on all four limbs (Snyder 1952). Considering relative length of hind limbs and lateral forces resulting from sprawling locomotion, lizards might be expected to run with little apparent forward thrust. Obviously, however, this is not the case. Stephen Reilly (1995) solved this apparent dilemma when he examined films of locomotion in Clark's spiny lizard *(Sceloporus clarki)*. The sprawling gait of the hind limbs does in fact generate lateral forces pushing the midline away from the limb, but the reaction force generated by each limb is countered by the reaction force from the other limb during rapid locomotion. The result is that most of the motion is forward. Strong muscles (the caudofemoralis) that connect the tail with the femur pull the limb toward the pelvic girdle (Reilly and DeLancey 1997).

Vertical surfaces present special challenges, as do surfaces that offer reduced resistance such as sand or water. Among arboreal lizards, some have sharply curved claws modified to grasp specific substrates, some have adhesive toe pads to facilitate sticking to smooth substrates, and some have fused toes (zygodactyly; see chapter 8) to accommodate grasping. Fringed toes are effective in moving across sand or water.

Claws

All lizards, except for a few gekkonids, have claws. For some species, claws are effective for locomotion on vertical surfaces, particularly those with some three-dimensional structure such as tree trunks and limbs and rough-surfaced rocks. Arboreal skinks ascend tree trunks at rapid rates even though most have no adhesive toe pads. Large-bodied *Iguana* and *Ctenosaura* in New World tropics climb tree trunks with ease, as do Amazonian tropidurines. *Plica plica, P. umbra,* and *Uracentron flaviceps* scale vertical tree trunks at rates that superficially resemble those of *Tropidurus hispidus* running across horizontal rock surfaces. Throughout the world, lizards in most families are capable of varying degrees of climbing, often using claws to grip surfaces. Interestingly, the single claw feature contributing most to clinging ability on rough surfaces is claw height, the vertical thickness of a claw at its base (Zani 2000). Lizard species that climb rough, vertical surfaces often have claws that appear more curved than those using other surfaces, but this feature does not stand out as a factor contributing to climbing performance.

In addition to scaling vertical surfaces, claws are used to dig nests in which to deposit eggs, to make burrows, and to search through debris for prey. A vast majority of field-collected or observed lizards have worn-down claws, attesting to their use in locomotion and other activities. Lizards in captivity often grow claws much longer than those ever observed in nature.

Adhesive Toe Pads

Many geckos, anoles, and some skinks have adhesive toe pads used for arboreal or scansorial (climbing) locomotion. In most, the toe undersurface contains striking numbers of microscopic setae (hairlike structures) arranged on lamellae (enlarged scales) (see chapter 9). Setae are not all the same. Some are hooks. Some, such as those in anoles, have spatulate ends, whereas some in geckos are branched. Thickness of the setal stalk and overall size of setae vary considerably as well. Among skinks, *Prasinohaema virens* has setae similar to anoles and geckos, while a close relative, *Lipinia leptosoma,* lacks lamellae with setae but nevertheless can climb vertical surfaces (Williams and Peterson 1982). *Lipinia* instead has a ruffle-fold scale architecture on the underside of its toes, which functionally accomplishes the same thing: maintenance of an adhesive grip on smooth surfaces. In a general sense, adhesive toe pads function primarily for locomotion on vertical surfaces: most lizards with adhesive toe pads are arboreal or live on rock or tree surfaces. Many geckos have moved in with humans where they often scurry up surfaces as

smooth as glass—they can even walk upside down on ceilings. Although a vast majority of terrestrial lizards do not have adhesive toe pads, terrestrial anoles do. In most if not all terrestrial anoles, adhesive pads are reduced and their function, if any (perhaps they are just vestigial remnants from arboreal ancestors?), remains unknown (see Peterson and Williams 1981). Clinging ability is associated with relative size of adhesive toe pads, and lizards with varying pad structure (anoles, geckos, and skinks) exhibit similar clinging abilities, though geckos are probably superior (Irschick et al. 1996). Nevertheless, larger lizards are better at clinging to surfaces than predicted on the basis of pad size alone, so other factors also influence locomotion on vertical surfaces.

Fringed Toes

Many lizards have fringes of enlarged scales on their toes that increase surface area exposed to substrate. Toe fringes have evolved in more than 150 lizard species, cutting across half of all lizard families (see, e.g., Luke 1986). Varying markedly in shape, fringed toes are a textbook example of convergent evolution, the independent acquisition of similar traits. Fringes can be triangular *(Uma)*, rectangular *(Kentropyx)*, conical *(Teratoscincus scincus)*, or projectional *(Scincus scincus)*. Fringes that appear made up of flat lateral scales triangular in shape from above are considered triangular fringes. Rectangular fringes are made up of rectangular-shaped lateral scales. Conical fringes superficially resemble triangular fringes from above but are circular in cross section. Projectional fringes are made up of dorsal or ventral scales on toes that project laterally over the toe's edges (Luke 1986). Lizards with rectangular-shaped fringes usually have a tongue-in-groove type of connection between scales that comprise fringes, effectively making them into flaps. Lizards with other types of fringes do not have the connections typical of lizards with rectangular fringes. Although each fringe type is associated to at least some degree with locomotion in specific kinds of habitats (e.g., sand, water), an appreciation for the diversity of morphological adaptations associated with different substrates is best achieved by examining several species within each of several important habitat types. Additional morphological adaptations are sometimes associated with locomotion in a particular medium.

Across Water | Lizards in several families (e.g., Iguanidae, Agamidae, Teiidae) can run across the surface of water for varying distances. Presumably, running across water provides escape from predators approaching from shore and also provides access to habitat patches interrupted by expanses of water. It may sometimes provide escape from aquatic predators as well. The basilisk, *Basiliscus plumifrons,* for example, has a linear distribution along streams and shorelines of ponds and lakes in Central America and northwestern South America. These moderate-sized lizards bask on tree limbs adjacent to or overhanging water. When disturbed, they jump into the water, often disappearing under the surface. If they have a running start, they raise their head, chest, and front limbs and run bipedally across the surface, propelling themselves primarily with their hind feet. Rectangular toe fringes greatly increase the surface area of toes contacting the water's surface, providing lift to hold the lizard on top while being propelled forward. As the hind foot pushes against the water, the foot does not submerge, and on the return stroke the foot passes through an air pocket above the foot, minimizing resistance (Glasheen and McMahon 1991). The tail is raised above the water as a counterbalance for the uplifted head and body (Snyder 1952). Other lizards,

WATER, SAND, OR TREES: WHAT DO FRINGES REALLY DO?

Over the past fifteen years I have conducted field studies on five of the eight described species of the teiid genus *Kentropyx.* Some *Kentropyx* use their rectangular-shaped toe fringes to run across the surface of water. In an interesting research article from 1986, when little was known about *Kentropyx* ecology, Claudia Luke, then a graduate student at the University of California, Berkeley, argued that evolution of rectangular toe fringes likely represented an adaptation specifically for propulsion across water. Yet contrary to popular belief, only a single species of *Kentropyx, K. altamazonica,* has a riparian distribution, and it is arboreal most of the time. During the wet season, when rivers and streams flood into the forest, these lizards move up into the vegetation and are agile climbers even on small limbs and across the leafy canopy. When forced into water, they usually swim but occasionally skitter short distances across the surface in a fashion somewhat similar to the iguanid *Basiliscus* (see also Gallagher and Dixon 1991). Although they achieve bipedality, their head and thoracic regions do not lift off the water as much as in *Basiliscus.* Body size and shape differences may partly explain differences in locomotory posture. *Basiliscus* are much larger and more streamlined than *Kentropyx.*

Other Amazonian species, *K. calcarata, K. striata,* and *K. pelviceps,* are also arboreal on low vegetation, particularly as juveniles. Although their distribution is not riparian, individuals often occur along streams, where they gain access to direct sun for thermoregulation. Likely, the misconception that these species are riparian stems from a false impression based on access: the easiest way to observe wildlife in Amazonia is to travel by boat and search areas near rivers. In fact, however, *K. pelviceps* is most common near treefalls in forest; *K. calcarata* is common in all types of edges, including treefalls, trails, the interface between forest and Amazon savanna, and the edge associated with streams; and *K. striata* is an Amazon savanna species, common in grasslands. These lizards are not arboreal in the same sense as *Basiliscus* and *Uranoscodon,* both of which use arboreal perches primarily as basking sites. *Kentropyx* move rapidly through vegetation searching branches, small limbs, and upper and lower surfaces of leaves for food. Their foraging behavior is very much like leaf gleaning observed in birds. Even much of their social behavior occurs in vegetation. Although these lizards can run short distances across the surface of water, they more often swim when forced into water.

Kentropyx vanzoi is a small-bodied species restricted to cerrado grasslands of central Brazil. It lives in patches of habitat with fine-grain sandy soil, foraging on the ground and within clumps of vegetation, including grass, moving rapidly through vegetation. I studied this species in southern Rondônia in habitat with no streams. Clearly, it is not a riparian species. Other species—*K. paulensis, K. borckiana,* and *K. viridistriga*—are found in open habitats as well, seeking refuge in vegetation rather than running across water (Gallagher and Dixon 1991).

Taken together, these observations suggest that evolution of rectangular toe fringes from a group of lizards (teiid ancestors of *Kentropyx*) more likely is associated with a shift to arboreality rather than to locomotion across water. This is consistent, moreover, with the likelihood that *Kentropyx* is derived from an Amazonian *Ameiva*-like ancestor (Gallagher and Dixon 1991). A shift to arboreality, or at least ability to climb and move about in vegetation, would be greatly enhanced by development of a fringe, particularly for juveniles that move rapidly across leaf surfaces. Feeding and escaping into vegetation also offsets predation by and competition with the widespread larger-bodied teiid *Ameiva ameiva* and including *Hydrosaurus* and *Physignathus* (Agamidae), also appear to use rectangular fringes the same way.

The Amazonian lizard *Uranoscodon superciliosus* also has rectangular fringes. These lizards use vertical and horizontal vines and narrow tree trunks and limbs as perches along and in watercourses. When approached, they typically remain motionless, relying on cryptic morphology and coloration to escape detection. However, when approached too closely, like basilisks they jump into water and scurry across the surface with the front part of their body lifted off the water surface and hind limbs providing propulsion. Laterally compressed tails of *Uranoscodon* also appear to provide forward thrust, moving back and forth in the water as the lizard runs. *Uranoscodon* usually run to another vertical vine or trunk and disappear into the foliage.

Large-bodied lizards that run across water have received the most press because of their size and easily observed body ornamentation, which bring them into the arena of wildlife worth watching. At the small end of the size scale are tiny geckos in the genera *Coleodactylus, Pseudogonatodes,* and *Lepidoblepharus.* These New World

species of *Tupinambis*, both of which lack toe fringes and are poor climbers. Further enhancing their climbing abilities are highly keeled ventral scales, which, among teiids, occur only in *Kentropyx*. Once present, rectangular fringes would allow some of these lizards to run across water surfaces for short distances. Interestingly, nearly all, if not all, lizards capable of running across water are also arboreal.

Evolution of rectangular toe fringes must have involved some intermediate steps. This transition during shifts to ever greater arboreality appears straightforward: increased surface of fringes allows better access to vegetation. An intermediate-sized fringe might allow only short-distance runs across water, which would render the lizards vulnerable to aquatic predators—making this explanation again less likely. The step from not being able to maintain support on water while running to maintaining it for distances necessary to facilitate escape is a big one; in contrast, running across water with fringes already in place for another reason—arboreality—is a relatively small step.

When size is considered, it makes even less sense to have rectangular fringes developing first for locomotion across water. Many *Kentropyx* rarely if ever run across water, but *Kentropyx*

are small relative to the size of adult *Basiliscus* or even *Uranoscodon*. Simple physics dictates that mass increases as length cubed, while surface area increases as the square of length, so larger lizards like *Basiliscus* and *Uranoscodon* would require relatively larger fringes than smaller lizards to maintain locomotion across water. Moreover, because surface tension of water is a constant, anything heavier is more likely to break the water surface. Once again, a stretch of the imagination is required to develop a scenario in which rectangular fringes evolved specifically for locomotion across water in lizards that may have had large ancestors (all known corytophanines are relatively large).

This example reaffirms the importance of detailed natural history information for species used in comparative studies of adaptation. Furthermore, given that all *Kentropyx* have fringes, two closely related "corytophanines" have them, and but a single tropidurine has them, a single evolutionary change even within each group is sufficient to explain independent evolution of rectangular fringes. The corytophanine most closely related to *Basiliscus* is supposedly *Corytophanes*, a large-bodied arboreal species. The teiid most closely related to *Kentropyx* is likely *Ameiva*, a

terrestrial genus without fringes, and the tropidurines most closely related to *Uranoscodon* are terrestrial, rock dwelling, and arboreal. Other lizards with rectangular fringes are also arboreal (e.g., *Hydrosaurus* and *Physignathus*). Square fringes appear to have evolved independently at least five times: in corytophanines, tropidurines, agamids, lacertids, and teiids (Luke 1986). The direct connection to water just doesn't appear to be there. The bottom line is that the search for an evolutionary cause of rectangular fringes in lizards remains open.

I would propose the following hypothesis for the independent evolution of rectangular fringes in these lizards. First, ancestors in each group with slightly enlarged scales on lateral toe surfaces could climb about in vegetation better than those with smaller scales; they gained access to portions of arboreal habitat previously unavailable, giving them an advantage in escaping predators and finding prey. Through selection, fringes became more enlarged. Once fringes were in place, they functioned well for rapid bipedal locomotion across sand and into water. Consequently, the evolutionary shift accounting for running across water was behavioral; the morphology was already in place for another reason. *(VITT)* geckos live in lowland tropical forest, usually in leaf litter on the forest floor. They have no toe fringes, and they have no expanded adhesive toe pads or webbing between toes that might accommodate locomotion across water. Their tiny claws are hidden within a sheath of scales. As adults, these lizards weigh less than a gram. Their leaf litter microhabitats can be severely flooded during rainstorms, and in many parts of their geographic ranges the forest becomes flooded during the wet season. Because they are so small, not only can they walk across water, but they can stand perfectly still on the surface without breaking surface tension. Moreover, when forced under the surface, an envelope of air surrounds them so they never really come into direct contact with water! The skin microstructure of many geckos includes what look like fields of tiny villi that hold the surface film of water at bay, forming this envelope of air. Two slightly larger sphaerodactyline geckos, *Gonatodes humeralis* and *G. hasemani,* cannot stand or walk on water. However, when forced under the surface, an envelope of air forms around them as well.

Through Water | A number of both small- and large-bodied lizards regularly swim through water, including the Galápagos marine iguana *(Amblyrhynchus),* numerous gymnophthalmids (e.g., *Alopoglossus, Neusticurus*), some agamids *(Physignathus* and *Hydrosaurus),* many skinks (e.g., *Eulamprus tympanum* and *E. quoyii),* the crocodile lizard *(Crocodilurus),* the caiman lizard *(Dracaena),* and some monitors *(Varanus).* In nearly all of these, the body and tail are streamlined, and the tail is laterally compressed with one or two dorsal crests. Unlike aquatic frogs, turtles, and some mammals, lizards that swim do not have webbing between toes, which may seem paradoxical considering that geckos appear to have had no problem evolving webbing for locomotion across sand or through the air. Yet the reason is simple. While swimming, lizards fold their limbs against their sides to minimize resistance, and swim in a serpentine fashion to maximize forward thrust. Because tails of vertebrates in general are simply extensions of the historically primitive locomotory apparatus (caudal fins in fishes and their ancestors), tetrapod vertebrates with large tails and a flexible axial skeleton are preadapted for swimming. Moreover, the musculature associated with the axial skeleton and nervous control accommodating locomotory movements of the body and tail were already in place through-

out vertebrate history. This is why most desert lizards can swim when thrown into water—though admittedly some are better than others! Nevertheless, on a global level, aquatic tetrapod vertebrates with short, robust bodies have webbed feet, whereas, with few exceptions, elongate aquatic tetrapods do not. Most of the latter use lateral undulation locomotion with the tail as the primary source of thrust. Crocodilians are the main exception, but their tails remain the primary power source.

While swimming in backwaters of the Amazon River, *Crocodilurus* typically meander through the water with their heads just above the surface. Their bodies move gracefully back and forth with the tail following, providing thrust necessary for forward locomotion. They easily dive to the bottom and often walk on it. Galápagos marine iguanas dive into the ocean from their perches on lava rocks and gracefully swim to the bottom, where they graze on marine algae. Gymnophthalmids, such as *Neusticurus ecpleopus,* move rapidly in a serpentine manner while in water, often disappearing to the bottom or up underneath the stream margin. Their locomotion appears salamanderlike. In all these lizards, limbs appear to be used primarily for stabilization during slow movement. Even typically nonswimming lizards, such as the desert lizards *Callisaurus draconoides* and *Cnemidophorus tigris,* when forced into water swim with rapid undulatory motions of the body, with limbs playing no apparent role. *Amblyrhynchus,* and likely other truly swimming lizards, can easily maintain position by "sculling." Sculling involves undulatory motion of the tail, similar to that seen in swimming but at a lower rate. The tail is usually at a sharp angle to the water surface, and little energy is expended because the tail is moved slowly. Lateral compression of the tail increases its surface area, much like rubber fins used by swimmers.

LIFE WITHOUT LEGS

Many anguids, including *Ophisaurus, Anguis, Anniella,* and *Ophioides,* nearly all amphisbaenians, many scincids, some gymophthalmids and cordylids, pygopodids, and dibamids either have no limbs or only tiny vestigial remnants of limbs. Limblessness or near limblessness has often evolved independently and in most cases appears associated with burrowing in one form or another or moving through grass. Limb loss is associated with elongation of the body, but locomotion varies considerably among limbless lizards. Limbless anguids (glass lizards) move in a serpentine fashion using lateral undulation in which portions of their bodies are pushed against the substrate. As in snakes, force of the body against the substrate produces equal and opposite forces, a portion of which results in forward locomotion. In terms of locomotion, pygopodids are, for all practical purposes, snakes. Many are terrestrial, using lateral undulation to move about. Others, such as *Aprasia* and *Ophidiocephalus,* are burrowers. Some species, such as *Aclys* and most species of *Delma,* are excellent climbers, moving rapidly through shrubs. Even though *Delma* are snakelike, their locomotion is quite different. In emergencies they saltate, jumping off the ground repeatedly with rapid forward thrust, appearing to spring from place to place. In vegetation, saltatory movements of their powerful tails followed by straightening of the body thrust them through vegetation like arrows! The only other lizard observed using saltation in vegetation is the lacertid *Takydromus.* One key observation is that elongation in terrestrial limbless or reduced-limbed lizards (e.g., *Ophisaurus, Ophiodes,* and *Chamaesaura*) emphasizes increases in tail length: most of these have tails more than twice their snout-vent length. Elongation in subterranean limbless or reduced-limb lizards (e.g., *Bachia, Typhlosaurus,* and *Anniella*) involves increased trunk length but decreased relative tail length.

Into Windblown Sand

An open sandy desert poses severe challenges for its inhabitants: (1) windblown sands are always loose and provide little traction; (2) surface temperatures at midday rise to lethal levels; and (3) open sandy areas offer little food,

The Australian pygopodid *Ophidocephalus taeniatus,* showing head scalation. (Dr. Hal Cogger)

Like most fossorial lizards, *Anniella pulchra* has a countersunk lower jaw. (L. Lee Grismer)

shade, or cover for evading predators. Even so, evolutionarily, lizards have adapted well to sandy desert conditions. Subterranean lizards simply bypass most problems by staying underground, and actually benefit from the loose sand since underground locomotion is made easier. In some parts of the world unrelated lizards have evolved several, and in some cases all, of the following morphological specializations for movement through sand: secondary loss of limbs, shovel-shaped heads and elongate snakelike bodies, countersunk lower jaws, loss of external ears, nasal plugs, and specialized scales capping the eyes to form a "brille," or spectacle. They spend their lives virtually swimming through sand. Along the Pacific coast of southern California and Baja California, all members of the small subfamily Anniellinae (Anguidae) are completely subterranean. Although most common in coastal dunes, they also occur in fine-grained soil in some interior valleys. Near Riverside, California, for example, *Anniella pulchra* can be found commonly in soil under a canopy of juniper trees where a layer of leaf debris combined with loose soil provides a substrate rich with termite prey. In South Africa, skinks in the genus *Typhlosaurus* occur in fine-grained sand of deserts; another species occurs in the soil of montane grasslands (Broadley 1968). Like *Anniella, Typhlosaurus* spend most of their time under sand, only occasionally moving across the surface. Also like *Anniella, Typhlosaurus* are common in sand under vegetation producing mats of leaf litter where their termite prey are common. Most activity occurs at night and individuals appear to move relatively little (Huey et al. 1974). *Anniella* and *Typhlosaurus* even give live birth while buried in the sand. In both, embryos are oriented parallel with the female's body. However, in *Typhlosaurus* only the tail is folded back on the embryo's body, whereas *Anniella* embryos are usually folded two to three times (Miller 1944). As partial compensation for the relatively larger amount of female body space taken up by folded embryos, female *Anniella* produce offspring from only one ovary. In both cases, orientation and positioning of elongate embryos minimizes the effect of carrying embryos on female locomotion. In viviparous lizards with legs, embryos are generally coiled such that the overall shape of an embryo is spheroidal, like a large reptilian egg.

In a set of relictual sand dunes located along the Rio São Francisco in the state of Bahia, Brazil, gymnophthalmid lizards in the genus *Calyptommatus* have evolved a set of morphological and ecological traits for existence in fine-grained sand similar to those found in *Typhlosaurus* and *Anniella*. These sand-dwelling lizards have elongate bodies, countersunk lower jaws, shovel-shaped heads, no external ear opening, eyes covered by an ocular scale, and hind limbs reduced to small spikelike structures with no toes—and they lack forelimbs (Rodrigues 1991). Unlike most gymnophthalmids (see chapter 10), they are most active just before and just after dark. As in other snakelike sand-dwelling lizards, *Calyptommatus* leave trails in sand when active near the surface. Although they eat a variety of insects, as in *Anniella* and *Typhlosaurus,* termites are important in their diets (Moraes 1993).

Additional lizards throughout the world have converged on subsand morphology and ecology described above, as have many snake genera including *Chilomeniscus, Chionactis,* and *Simoselaps*. Independent evolutionary origins of extreme morphologies associated with similar ecological traits (fossoriality, or living underground in loose soil) underscore the power specialized microhabitats exert on evolution of morphology. All legless lizards were derived from terrestrial ancestors with legs. The exact set of circumstances that led to this specialization in different parts of the world will never be known for certain, but similar habitat shifts over evolutionary time doubtlessly played an important role. A likely scenario is that well-developed chemical sensory abilities allowed lizards to detect prey in burrows and under surface objects, which, in turn, favored body elongation. As bodies elongated and became more effective at serpentine locomotion, limbs became reduced or were lost because their presence interfered with such locomotion. An interme-

diate morphology might be that found in gymnophthalmids like *Bachia* in South America or skinks like *Larutia* in Malaysia. These move through soft soil pushing themselves slowly along with tiny legs. Their long tails are effective for escaping predation via autotomy (reflex separation of the tail) in shallow substrates where they can easily be exposed. Some also use their long tails to catapult themselves to a different position within leaf litter for escape. As lizards move deeper in granular substrates like sand where exposure to predators is minimized, advantages of long tails for predator escape are outweighed by advantages of short tails in locomotion driven by undulatory movements of the body. This may explain why legless subterranean lizards living in sand (e.g., *Anniella* and *Typhlosaurus*) have short tails whereas legless lizards living on the surface (e.g., *Ophisaurus* and *Ophiodes*) have long tails. Long, muscular tails in elongate, legless terrestrial lizards facilitate locomotion and escape via autotomy.

Limbless locomotion in amphisbaenians is quite different. While on the surface, species such as *Amphisbaena alba* move in an accordionlike manner, akin to rectilinear locomotion in snakes but in many ways exaggerated. Portions of the body lift slightly off the ground and are pulled forward by the portion of the body in front, which is securely in contact with the ground. Once pulled forward, that portion pulls the next part. Muscles originating in the skin pull the vertebral column directly forward. A repeated series of such movements results in forward locomotion. Forward movement on the surface is relatively slow, even though rectilinear movements occur at a fast rate. When disturbed on the surface, some amphisbaenians convert to a stiffened body flexing that results in rather inefficient lateral undulation, causing limited forward locomotion. Such rapid surface movements function more to distract or surprise a potential predator than to facilitate movement from one place to another.

Amphisbaenians spend most of their time under leaf litter or burrowing through the soil. While burrowing, they move headfirst to produce passages through which their body follows (Gans 1974). Their rectilinear locomotion, with the body oriented linearly in a tunnel, facilitates movement through such burrows. Exactly how headfirst burrowing occurs varies among amphisbaenians and is associated with head shape. The head is thrust forward into the substrate. In species with dorsoventrally (top-to-bottom) compressed heads, such as *Leposternon infraorbitale,* heads are moved up and down and thrust forward, thus packing upper and lower burrow walls and expanding the forward part of the burrow. Other amphisbaenians with laterally compressed heads, such as *Ancyclocranium,* move their heads back and forth, accomplishing the same thing on lateral burrow walls. Species with rounded heads, such as *Amphisbaena alba,* rotate their heads back and forth while moving them slightly up and down. Not all amphisbaenians are limbless. *Bipes* use their molelike front limbs in burrowing, and do not thrust their rounded head forward into the substrate as do other amphisbaenians.

Spatulate head of the South American worm lizard, *Leposternon infraorbitale.* (Dante Fenolio)

Across Sand

Some lizards are inactive under sand but active on the surface. Shortly after sunrise, but before sand temperatures climb too high, diurnal sand-dwelling lizards scurry about above ground. These sand-specialized lizards provide yet another striking example of convergent evolution and ecological equivalence. Lizards scattered throughout the world's deserts have found a similar solution for getting better traction on loose sand: enlarged scales on toes, or lamellae, forming fringes have evolved independently in six different families of lizards, Scincidae, Lacertidae, Iguanidae, Agamidae, Gekkonidae, and Gerrhosauridae. A skink, appropriately dubbed the "sand fish," literally swims through sandy seas in search of insect food in the Sahara and other eastern deserts. These sandy desert regions also support lacertids (*Acanthodactylus*) with fringed toes and shovel snouts. Far away in the southern hemisphere, on windblown dunes in the Namib Desert of southwestern Africa, independent lineages of lacertids (*Meroles [Aporosaura] anchietae*) and gerrhosaurids (*Angolosaurus skoogi*) have evolved similar life forms.

North American *Uma* (Iguanidae) have also adopted this body form, foraging by waiting in the open. These fringe-toed lizards eat various insects, such as sand roaches, beetle larvae, and other burrowing arthropods. They also listen intently for insects moving buried in the sand, and dig them up. Sometimes they dash, dig, and paw through a patch of sand and then watch the disturbed area for movements. When seasonal flowers become abundant, they switch to partial herbivory (Durtsche 1992). When disturbed, fringe-toed lizards run across dunes, disappearing over the top and seeming to disappear altogether. However, careful examination of the sand reveals tracks that lead directly to the lizard, which has buried itself after losing sight of its pursuer!

Amphisbaenia alba **is the largest of South American worm lizards.** (Laurie Vitt)

All these activities require lizards to move rapidly and effectively over the sand surface. Their tracks consist of small depressions where each foot has landed. Fringe-toed lizard tracks are easily distinguished from tracks of other lizards because all other lizards lack fringes and leave toe marks. To test the effectiveness of toe fringes on locomotion across loose sand, John Carothers, then of the University of California, Berkeley, performed a simple but insightful experiment. After removing fringes on a sample of lizards, he ran lizards with and without fringes on a rubber racetrack. As expected, absence or presence of fringes did not influence a lizard's ability to run on substrates that were not loose. When the experiment was repeated on sand, however, opposite results were obtained: lizards lacking fringes performed more poorly than lizards with fringes (Carothers 1986). Fringes provided the extra support necessary for lizards to achieve high running speeds on sand.

All these lizards have flattened, duckbill-like, shovel-nosed snouts, which enable them to make remarkable "dives" into sand even while running at full speed. They then wriggle along under the surface for a meter or more. One must see such sand diving to appreciate its effectiveness as a disappearing act.

Some Namib Desert lizards found another solution to gain traction on powdery sands: froglike webbing between the toes, as seen in the geckos *Palmatogecko* and *Kaokogecko*. Webbed feet work the same way snowshoes do, increasing effective surface area to better distribute an individual's weight. Webbing between toes also serves other functions, such as digging (Russell and Bauer 1991).

THROUGH AIR:
JUMPING, PARACHUTING, AND GLIDING

Considering the great success birds (feathered reptiles) and bats have had taking to the skies and the apparent success of pterosaurs during the age of dinosaurs, lack of flying forms within lizards is surprising. The structure of limbs, pectoral girdles, and possibly bones of lizard ancestors likely placed evolutionary constraints on evolution of morphological traits necessary for flight. Considering that no extant ectothermic vertebrates fly, thermal physiology also must have constrained evolution of flight within lizards. Although tropical and desert species could achieve body temperatures high enough to sustain high metabolic rates necessary for true flight, they likely would not be able to maintain those temperatures once in flight.

The web-footed gecko, *Palmatogecko rangeri*, moves across sand in the Namib Desert just as a human moves across snow with snowshoes. (Eric Pianka)

Whatever the historical reasons are, it just hasn't happened. Nevertheless, many lizards jump from elevated surfaces, some parachute, and others simply glide.

The simplest aerial locomotion is to drop from perches to the ground or into water. Green iguanas *(Iguana iguana)* throughout Central and South America forage in the canopy of large trees overhanging streams, particularly large rivers. In Belize, for example, one of the truly exciting experiences for ecotourists is a riverboat ride on which hundreds of large iguanas can be seen basking and foraging in bordering trees. When approached too closely, iguanas drop into the river and either swim on the surface to shore or disappear under water. In relatively clear streams, lizards can be observed walking on the bottom after submerging. No doubt, dropping from tree limbs into the water is an effective means of escaping predators approaching from shore or from within a tree. Iguanas are not the only lizards that jump or drop from arboreal perches into water to escape potential predators. *Basiliscus, Hydrosaurus, Physignathus, Varanus,* and many snakes also do this.

Many lizards drop from perches to the ground for escape as well. Most arboreal iguanian lizards sleep on branches, usually near ends of thin ones. The North American desert lizard *Urosaurus graciosus,* for example, sleeps on ends of branches of smoke trees *(Dalea spinosa),* mesquites *(Prosopis),* and creosote bushes *(Larrea divaricata).* Many species of *Anolis* sleep on the ends of thin branches or tall blades of grass. In the Amazon rain forest, *A. fuscoauratus* sleep on ends of thin twigs 1–4 m above the ground, whereas *A. transversalis* sleep on ends of tree branches more than 5 m above the ground (perhaps as high as 45 m up!). When branches are disturbed, lizards jump, disappearing into the darkness, though many anoles simply jump to the ground and freeze. The Amazonian tropidurine, *Plica umbra,* which sleeps on horizontal limbs, jumps to the ground and runs at high speed, disappearing into the forest. Because this is a diurnal lizard, one has to wonder how many have bashed their heads against tree trunks as they sped away in the dark. Slow-moving relatives of anoles, *Polychrus,* totally arboreal, also sleep on ends of branches. When they drop after their branches are disturbed at night, they typically pull their legs under their bodies and curl up, remaining motionless in the leaf litter.

In many ways, dropping from arboreal perches doesn't really fall into the category of locomotion because the liz-

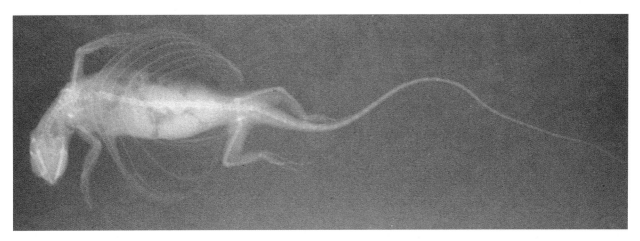

X-ray of a flying lizard, *Draco,* showing folding ribs that support the "wings." (C.H. Diong)

ards are not actively doing anything: gravity does the work for them. Many arboreal lizards, however, do leap from perch to perch during normal activity, and most are very good at it. The tiny diurnal gecko *Lygodactylus klugei,* which is arboreal in semiarid caatinga of northeastern Brazil, lives on small tree trunks and limbs as well as on man-made fences constructed of limbs. They easily jump from branch to branch during social interactions and can do so in rapid succession, appearing to nearly fly through the air. In addition to having adhesive pads on their feet typical of gekkonines, an adhesive pad on the tip of their tail enables them to "stick" more easily to the limbs and branches through which they move. Adhesive pads on tips of gecko tails are not restricted to *Lygodactylus* or even Gekkonidae. Diplodactylid geckos in the genera *Naultinus, Hoplodactylus, Bavayia, Eurydactylodes, Rhacodactylus,* and *Pseudothecadactylus* also have adhesive pads on their tails (Bauer 1998).

High in the largest trees within the Amazon forest, the tropidurine *Uracentron flaviceps* runs across horizontal limbs, frequently making short jumps to other limbs while foraging and during social interactions. Lizards living on rock surfaces, such as *Tropidurus hispidus* and *T. semitaeniatus* in northeastern Brazil and *Sceloporus merriami* in west Texas, frequently jump from one surface of a rock to another. Jumping appears to be as much a part of standard locomotion for many lizards as walking or running.

A few lizards glide or parachute. Parachuting occurs if the angle of descent is less than 45° from vertical, and gliding occurs if the angle of descent is flatter than 45°. The geckos *Ptychozoon* and *Thecadactylus* expand webbing be-

An expert hang glider, *Draco cornutus* floats on air between trees in a forest in Borneo. (Tim Laman)

tween their toes, fringes on posterior leg surfaces, and skin along lateral surfaces of their bodies to produce an airfoil used to parachute when they leap from arboreal perches. Because these lizards move horizontally (out) as well as vertically (down), they clearly are gliding, and because their webbing serves as an airfoil, they are parachuting as well. The best-known and most spectacular gliding lizards are Malaysian flying dragons, *Draco* (Russell 1979). Rather than having simple flaps of skin along lateral body surfaces, they have pataglia, flaps of skin fully supported by movable ribs. When not in use, pataglia are folded against the body, much like a folded fan. When the lizards take to the air, however, the pataglia are erected by muscle contractions, pulling the first two ribs forward, followed by the remaining ribs, which are attached to the first two by ligaments. Although neither of us has observed gliding by *Draco,* descriptions indicate that it should be on the life list of observations to make for every serious saurologist. Reportedly (J. McGuire, pers. comm.), these lizards leap from trees falling at a steep angle with their pataglia held against the body and tail raised at a sharp angle. As a lizard drops, its tail is pulled downward and its pataglia are erected, catching the air and sending the lizard out at an angle of 15–20°, resulting in near horizontal gliding. When a *Draco* approaches a landing site, its rate of gliding slows considerably as it raises its tail, with consequent repositioning of the airfoil. The lizard then lands relatively gently, usually on tree trunks facing upward, and scampers up and away.

THERMOREGULATION

Because lizards cannot produce heat internally (physiologically), temperature—or more precisely, heat availability—is a master limiting factor. This becomes rather obvious when you think about the geographical distribution of lizards in the world. Number of species drops off rapidly with increasing latitude and elevation. Only a single species, *Lacerta vivipara,* enters the Arctic Circle, and but a few occur on the highest mountains. For example, some high Andean gymnophthalmids and a New Zealand gecko, *Hoplodactylus kahoutarae,* are active on sunny days when ice or snow is on the ground. Availability of heat limits reproductive options as well. The proportion of lizards that lay eggs drops off rapidly with latitude and elevation. As its name implies, *Lacerta vivipara* produces live offspring, even though all of its lower-latitude lacertid relatives lay eggs. Thus, not only does heat availability directly influence activities of individual lizards, it also influences their ability to get offspring into future generations by limiting reproductive options.

Animals that, like lizards, cannot produce heat internally are called ectotherms, because they receive their bodily heat from external sources. The term "cold-blooded" is frequently applied to lizards, but this antiquated term misrepresents the thermal biology of lizards. Indeed, many lizards are active at body temperatures higher than those of humans, so "cold-blooded" is quite inappropriate! Another common misconception is that lizard temperatures are always the same as those of surrounding environments. Some lizards are quite good at maintaining body temperatures within relatively narrow ranges as long as heat sources are available.

Why is temperature—or more accurately, maintenance of relatively high body temperatures—so important to many lizards? The answer is simple: all physiological functions (and thus behavioral functions) of these animals are temperature dependent, from prey capture, predator evasion, and courting of mates to such physiological processes as digestion, sperm production, conversion of lipids into eggs, and growth. At low temperatures, these processes occur at slow rates, whereas at high temperatures they occur at higher rates. Ultimately, temperature determines how successful one individual is compared to all others in its population at capturing and processing food and converting that food into new lizard tissue (eggs or offspring). All of these can easily be clustered into a single currency: performance (fig. 2.2). Performance determines which individuals survive to produce offspring and which do not. As a result, natural selection has had a major impact on temperature relations in lizards.

Interest in thermoregulation in lizards has grown enormously since early studies by two pioneers in the field, Raymond B. Cowles and Charles M. Bogert. Among their most important findings was that desert lizards maintain relatively constant body temperatures while active even though the temperature of the overall environment changes throughout the day. Desert lizards regulate their body temperatures by behavioral adjustments; when they begin to cool, they bask in sun to gain heat, and when they begin to overheat, they seek shade and lose heat to the environment (Cowles and Bogert 1944). As simple minded as these discoveries might appear today, they had a profound effect on the way we think about so-called cold-blooded animals. These lizards were homeotherms—able to regulate their temperature—at least while active. The primary difference between lizards and animals that we typically think of as homeothermic (mammals and birds) is that lizards gain and lose heat directly or indirectly from and to the external environment (they are ectothermic), whereas mammals and birds produce their own heat internally (they are endothermic). So, desert lizards, as well as many others, can be considered ectothermic homeotherms. Of course, not all lizards regulate their body temperatures as well as some desert lizards, and body temperatures of some lizards such as nocturnal geckos do seem to vary directly with environmental temperatures (mak-

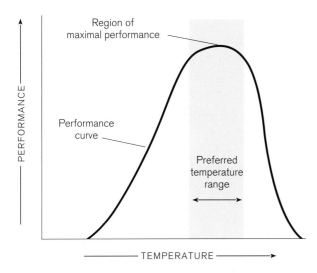

Figure 2.2 Relationship between temperature and performance. Optimal performance usually occurs within a relatively narrow range of temperatures. Performance can include almost anything that a lizard might do, including physiological processes such as digestion and respiration.

ing them poikilothermic ectotherms). Such passive thermoregulators are also known as thermoconformers.

Endothermy, as seen in mammals and birds, has some advantages. A warm animal can perform well all of the time; it can respond to predators and react quickly; it can move about to find food and mates. The cost of endothermy is high, however, because metabolic processes that generate heat require a large intake of energy. Taking in a lot of energy requires spending more time searching for prey, which increases the risk of predation. Because of continual heat production, endotherms cannot survive long without food—they are poorly buffered against even relatively short-term environmental changes.

Advantages of ectothermy center on the fact that none of the energy acquired by a lizard is used to generate heat: it can all be directed into growth, social behavior, and reproduction. If conditions are too cold for activity, lizards can remain inactive. Their low body temperatures during periods of inactivity result in very little energy being used metabolically. The downside of ectothermy is that when heat sources are not available, activity must cease, and while inactive a lizard is less able to escape predation or any other hazards. At a population level, the same amount of resources can support a much larger number of lizards than mammals or birds of similar size because

of these differences in metabolic costs of thermoregulation. Moreover, an environment with very low resources or with resources that fluctuate could support a lizard population even if it would not support a single bird or mammal.

As recognized by Cowles and Bogert, lizards are active within a relatively narrow range of temperatures, cease activity above and below those temperatures, and at extreme temperatures (hot or cold) appear to lose the ability to save themselves from either more extreme exposure or predation attempts.

SOURCES OF HEAT GAIN AND LOSS

The ultimate source of heat for lizards is the sun, and many lizards take advantage of sunlight by seeking out patches of direct sun exposure in their habitat and basking to gain heat. Such animals are called heliotherms. The color of lizard skin, amount of surface exposed, intensity of the sun, and compounding factors such as air movement are all important in determining heat gain. Dark skin absorbs more heat than light-colored skin, while skin with efficient capillary beds to transport heat away from the source and into the body aids in warming a lizard rapidly. Some lizards, such as horned lizards *(Phrynosoma)*, the greater earless lizard *(Cophosaurus texanus),* and the zebra-tailed lizard *(Callisaurus draconoides),* stick their head out of the sand in early morning, exposing it to direct sunlight. Capillary beds in their head heat up and warm blood moves into their bodies, to be replaced by cooler blood. The process continues until a lizard is warm enough for activity. A lizard therefore enters its aboveground terrestrial environment at a body temperature allowing optimal performance. Whereas a cold lizard on the surface would have difficulty escaping a predator, a warm lizard is prepared to use appropriate escape behaviors should an encounter occur.

Heliothermic lizards occur in nearly all habitats, even some of the warmest ones. Heat gain is not by any means restricted to that associated with direct sunlight. All lizards gain some heat by conduction, which is the transport of heat from one object to another (fig. 2.3). A lizard perched on a rock in the sun gains heat directly from the sun, directly from the rock it sits on by conduction, and indirectly from other objects as heat moves through the air (convection). Subterranean lizards gain heat directly from their microhabitat and can regulate heat gain and

Like other lizards standing on hot surfaces, *Crotaphytus collaris* curls its toes up to cool them off. (Adam D. Leaché)

loss simply by moving up or down in the thermal gradients of their respective media. Some species, such as the Galápagos marine iguana, *Amblyrhynchus cristatus,* gain heat by exposure to the sun and black lava rocks in the terrestrial environment but maintain high active body temperatures while diving in the cold ocean by restricting blood flow to the peripheral circulatory system (Bartholomew and Lasiewski 1965).

On western beaches of Central America, whiptail lizards of the species *Cnemidophorus deppii* face a complex mosaic of temperature change in their preferred habitat. In early morning, when the habitat is relatively cool, they lie in direct sunlight with ventral surfaces pressed against warming sand to gain heat. Once their body temperature reaches about 40°C, which occurs by 9 A.M., they begin normal activities, foraging for food and seeking mates, spending considerable time in sun or at least partial sun (Vitt et al. 1993). By 10 A.M, the sand surface exposed to direct sun exceeds 45°C, and by noon it exceeds 55°C. Full shade is rare because very little vegetation exists on the beaches and existing vegetation is sparse. Even in full shade, surface temperatures reach about 35°C by mid-afternoon. During midday, when sand surfaces are hot, lizards seek small patches of partial shade under branches of shrubs. They orient their bodies parallel to branches to gain maximum exposure to shade and move their bodies from side to side, burying themselves partially in cooler soil and thereby dumping excess heat. Once they have cooled down sufficiently, lizards begin to forage in and out of the sun. As soon as they begin to overheat, the process is repeated. Late in the day, when environmental temperatures fall below 40°C, lizards remaining active again commence basking in sun against the warm sand for heat gain. For these lizards, gaining heat is relatively easy, whereas losing excess heat is more problematic because the overall environment reaches temperatures much higher than lizard body temperatures during much of the day. Throughout the day, *C. deppii* maintain body temperatures near 40°C by shifting their behavior. Body temperatures slightly higher than 40°C are lethal for lizards, so *C. deppii* operate at the top end of their preferred temperature range over their entire period of activity.

Students of thermal physiology have often noted an apparent upper thermal limit of about 40°C for most of Earth's eukaryotic creatures (most plants, invertebrates, and vertebrates). This thermal "lid" has frequently been used as evidence for an extremely archaic and inflexible fundamental physiological process (perhaps some enzyme fundamental to all life processes, such as a dehydrogenase, denatures). An intriguing hypothesis ("max-

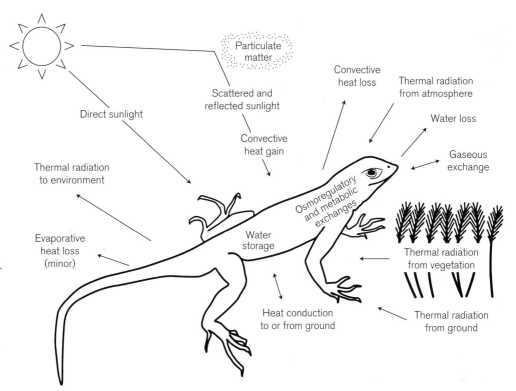

Figure 2.3 Avenues of heat gain and loss for a diurnal basking lizard. Skin is modified in many species so that heat is picked up by the circulatory system and rapidly distributed while the lizard basks.

Labels in figure: Direct sunlight; Particulate matter; Scattered and reflected sunlight; Convective heat gain; Convective heat loss; Thermal radiation from atmosphere; Water loss; Gaseous exchange; Thermal radiation to environment; Osmoregulatory and metabolic exchanges; Evaporative heat loss (minor); Water storage; Heat conduction to or from ground; Thermal radiation from vegetation; Thermal radiation from ground

ithermy") for the evolution of homeothermy suggests that it is a by-product of advantages gained from maintaining maximum body temperatures in the face of such an innate physiological ceiling (Hamilton 1973). Many lizards operate close to their upper lethal thermal limits to maximize their performance levels and ultimately their lifetime reproductive success.

A lizard distantly related (but in the same family, Teiidae) to *C. deppii, Kentropyx calcarata,* lives in lowland forest of the Amazon River basin. Like *C. deppii, K. calcarata* is a heliotherm that seeks microhabitats receiving direct sunlight in the morning to gain heat. Herein lies a problem, however, for patches of direct sun are limited in lowland rain forest. About 45 m off the ground, a continuous canopy formed by rain forest trees shuts off most direct sunlight that floods treetops. At ground level, only filtered light reaches the forest floor, except in areas along edges or in treefalls. Even on sunny days clouds frequently block the sun, further limiting availability of sunlight for heat gain. On cloudy days, *K. calcarata* are not active. On sunny days, they bask in direct sunlight at forest edges and on surfaces of logs in treefalls. After reaching active body temperatures of about 34.5°C, *K. calcarata* forage along the forest edge, often entering the forest short distances to capture prey. When their body temperatures drop, they

return to perch sites exposed to sun to gain heat. Despite these impediments, *K. calcarata* body temperatures are always above those of any forest microhabitats, which average less than 32°C (Vitt, Zani, and Lima 1997). Dark dorsal coloration and an ability to climb up on perches providing access to sun contribute to the ability of these forest lizards to gain sufficient heat to maintain the active lifestyle typical of teiid lizards. Like *C. deppii, K. calcarata* maintain relatively stable body temperatures during the day by adjusting behavior appropriately. However, because these lizards live in an environment in which gaining heat is the primary challenge, they operate at the lower end of their preferred body temperature range. In such dark forests, lizards that rely on patches of sun as heat sources must expose themselves to predators that learn to find lizards by looking for patches of light.

Another frequent misconception about the thermal ecology of lizards is that the physical environment of major habitats is the primary determinant of lizard body temperatures. If this were true, we would expect all desert lizards to have high body temperatures and all tropical forest lizards to have relatively low body temperatures. Moreover, we might expect all desert lizard species to have about the same body temperatures and all tropical forest lizards to have about the same body temperatures. Nei-

The Brazilian lizard *Kentropyx vanzoi* has keeled ventral scales, a feature that sets this genus off from other teiids. (Laurie Vitt)

ther of these is true, even when comparisons are restricted to diurnal lizards (nocturnal lizards in general have lower body temperatures regardless of habitat). Among North American desert lizards, for example, *Cnemidophorus tigris* averages 39.5°C when active, while in the same habitat *Uta stansburiana* averages only 35.3°C. *Dipsosaurus dorsalis* averages 40.0°C, whereas *Sceloporus magister* averages only 34.8°C. In the Kalahari Desert, *Nucras tessellata* averages 39.3°C, whereas *Mabuya spilogaster* averages only 34.5°C. Body temperatures of diurnal lizards in Australian deserts vary considerably as well (Pianka 1986). In the wet lowland tropical forest at Cuyabeno, Ecuador, two species of teiid lizards, *Kentropyx pelviceps* and *Tupinambis teguixin,* are active with body temperatures of about 34°C, whereas two diurnal gekkonids, *Gonatodes humeralis* and *G. concinnatus,* are active with body temperatures between 27°C and 28°C (Vitt and Zani 1996a). Each lizard species uses different microhabitats, has its own unique history, and interacts with its thermal environment differently.

Degree of similarity in thermal ecology of lizard species occurring in any habitat has a historical component that has been largely overlooked by physiologists. All teiid lizards, for example, tend to have higher body temperatures than diurnal gekkonids, regardless of where they occur. All xantusiids have relatively low body temperatures compared to those of most other lizards. This "phylogenetic" component of lizard body temperatures was clear in early summaries of lizard body temperature data assembled more than thirty-five years ago (Brattstrom 1965).

Ecologically optimal temperatures do not always coincide with physiological optima (Huey and Slatkin 1976), particularly in nocturnal lizards. The observation that nocturnal lizards are active at lower body temperatures than most diurnal species is by no means surprising: heat sources available during the day, such as the sun or sun-heated surfaces, are absent at night (although some surfaces, such as large rocks, retain substantial amounts of heat well after dark, so nocturnal lizards do have limited access to heat sources). Nocturnal lizards also tend to have more variable body temperatures than diurnal lizards, for similar reasons.

Because nocturnal lizards are active at lower body temperatures than most diurnal lizards, an obvious expectation is that their physiology should shift such that performance is optimized at the lower body temperatures. However, this does not appear to be the case. An insightful set of experiments conducted by Raymond B. Huey and colleagues (1989) revealed that several species of nocturnal geckos *(Hemidactylus turcicus, H. frenatus, Lepidodactylus lugubris, Coleonyx brevis,* and *C. variegatus)* not only perform better in sprint performance trials at temperatures considerably above those experienced while active at night, but when given a choice, they actually select temperatures higher than their typical active body temperatures as recorded in the field. Temperatures selected by geckos in the laboratory are similar to those selected by some diurnal lizards and, on average, about 6°C higher than body temperatures of active geckos in their natural habitats. Their critical thermal minima and maxima are also similar to those of diurnal lizards. As expected, sprint speeds increase with temperature but, rather than leveling off at a temperature near the active field body temperatures, continue to increase, leveling off only at 35°C, which corresponds to preferred temperatures recorded in the laboratory.

Most likely a combination of factors explains this seemingly nonintuitive result. These geckos live in warm environments and seek retreats during the day when temperatures are highest. They likely experience higher temperatures during the day while in retreats. Because sprinting usually reflects response of other physiological traits to temperature, performance may in fact be optimal at a time—daytime—when evolutionarily important events are most common. Diurnal predators that capture geckos while in their retreats may be a more significant source of mortality than nocturnal predators (which are also cold) that attempt to capture active geckos, thus maintaining selection for optimal performance at daytime temperatures rather than at nighttime temperatures. Social behavior may also be most prominent while geckos are in their diurnal retreats, adding to benefits associated with optimal performance at temperatures higher than those experienced by geckos at night. Low body temperatures tolerated by geckos at night can be viewed as necessary costs paid to avoid competition and predation during the day. Moreover, nocturnality offers access to resources such as nocturnal insects that are unavailable during the day.

ACTIVITY CYCLES

Lizards are not active all the time. Both daily and seasonal cycles of activity are tightly tied to temperature, one of many consequences of ectothermy. The fact that some species are diurnal and others nocturnal also suggests basic differences in how lizards respond to light and dark. In a general way, most desert lizards active during daytime are not active during night and most desert lizards active during the night are not active during the day, or at least not active in the same way as they are at night. For example, among twelve lizards from a Sonoran Desert site, ten are strictly diurnal, one is strictly nocturnal, and one is quasi-nocturnal (see chapter 10). Among tropical lizards, those in open habitats (savannas, cerrado, caatinga, etc.) appear similar to desert lizards, with a distinct diurnal/nocturnal dichotomy. For example, among the thirteen lizard species at a single site in caatinga of northeastern Brazil, three are strictly nocturnal and the remainder strictly diurnal. Nocturnal species at both sites and many others throughout the world are in one of three gecko families, Gekkonidae, Eublepharidae, and Diplodactylidae. Exceptions do occur. Not all geckos are nocturnal, and in fact, some of the most abundant ones are strictly diurnal (e.g., *Coleodactylus, Lygodactylus, Phelsuma*). Some Australian desert skinks and pygopodids are nocturnal. Helodermatids can be either diurnal or nocturnal depending on season. Consequently, the nocturnal versus diurnal dichotomy can be attributed partially to historical effects because some evolutionary groups are mostly diurnal or nocturnal. Effects of thermophysiology on behavior or species interactions may also offset activity periods between potentially competing species pairs.

The diurnal versus nocturnal dichotomy is further complicated by daily and seasonal patterns of activity that themselves are influenced by historical effects and species interactions. The places where each species lives—or more succinctly, their microhabitats—also influence body temperatures and activity, sometimes in not very obvious ways. The relationship between thermoregulation and activity varies among species as well, and nearly every possible exception exists. For example, desert lizards are active at midday in spring and fall, shifting to morning and late afternoon during summer. (One North American desert lizard exception is *Dipsosaurus dorsalis,* which is active at midday in summer, apparently to avoid predators.) In tropical forests, many lizards are active during midday in wet seasons but shift to morning activity in dry seasons.

Differences in temporal patterns of activity, use of space, and body temperature relationships are hardly independent. Rather, they complexly constrain one another, sometimes in intricate and obscure ways. For example, thermal conditions associated with particular microhabitats change temporally in characteristic ways; a choice basking site at one time of day becomes an inhospitable hot spot at another time. Perches of arboreal lizards receive full sun early and late in the day when ambient air temperatures tend to be low and basking is therefore desirable, but these same tree trunks are shady and cool during the heat of midday when heat-avoidance behavior becomes necessary. In contrast, the fraction of the ground's surface in the sun is low early and late in the day, when it is cold and when shadows are long, but reaches a maximum at midday. Terrestrial heliothermic lizards may thus experience a shortage of suitable basking sites early and late in the day, while during the heat of the day their movements through relatively extensive patches of open sun can be severely curtailed. Hence, ground-dwelling lizards encounter fundamentally different and more difficult thermal challenges than do climbing species (Huey and Pianka 1977c).

Hot, arid regions typically support rich lizard faunas, whereas cooler forested areas have considerably fewer lizard species and individuals. In deserts, lizards can enjoy the benefits of a high metabolic rate during the relatively brief periods when conditions are appropriate for activity, becoming inactive during adverse conditions. By facilitating metabolic inactivity on both a daily and a seasonal basis, ectothermy thus allows lizards to capitalize on unpredictable food supplies. Ectotherms are low-energy animals; one day's food supply for a small bird will last a lizard of the same body mass for a full month! Most endothermic diurnal birds and mammals must wait out the hot midday period at considerable metabolic cost, whereas lizards can effectively reduce temporal heterogeneity by retreating underground, becoming inactive, and lowering their metabolic rate during harsh periods (though some desert rodents do estivate—enter a state of summer torpor—when food and/or water is in short supply). Ectothermy may well contribute to the apparent relative success of lizards over birds and mammals in arid regions. Temperate forests and grasslands are probably simply too shady and too cold for ectothermic lizards to be very successful because these animals depend on basking to reach body temperatures high enough for activity. In contrast, birds and mammals can do quite well in such areas provided sufficient energy resources are available, partly because of their endothermy.

WATER BALANCE

Most organisms are 65–75 percent water. All life processes depend on water but also require specific chemicals involved in the many chemical reactions supporting metabolic processes. More specifically, life depends on critical balances between water and the chemical makeup of cells. Loss of water increases concentrations of solutes in organisms, whereas flooding of water in organisms reduces solute concentrations. Unlike amphibians, which lose and gain water at a rapid rate because of high skin permeability, reptiles lose water rather slowly and gain most water by drinking or from water contained in their food (preformed water). The reason many amphibians can afford to lose water is that they usually live in microhabitats where water is available, so they can easily regain it. Many reptiles live in environments where water may not be available for extended periods of time. The skin of reptiles thus protects them to a large degree from excessive or rapid water loss. Most water, in fact, is lost through respiration. Unlike mammals, moreover, whose urea must be dissolved in water, lizards (and birds) excrete nitrogenous wastes as uric acid, a dry white paste; hence mammals lose more water than lizards or birds do.

Water loss generally decreases with increasing body size in lizards (Mautz 1982; Nagy 1982). The primary reason for this is reduced effective surface area because surface-to-volume ratio decreases with increasing body size. Thus a 110-g *Amphisbaena alba,* for example, loses water at 2.68 mg per gram of body weight per hour, whereas a 5.4-g *A. caeca* loses water at 40.69 mg per gram of body weight per hour (Gans et al. 1968). However, the relationship between body size and water loss varies considerably among lizard species, independent of size. Thus a 0.67-g *Sphaerodactylus klauberi* in a mesic habitat loses water at 16.7 mg per gram of body weight per hour, whereas a 0.3-g *Sphaerodactylus nicholsi* in a semiarid habitat loses water at only 2.0 mg per gram of body weight

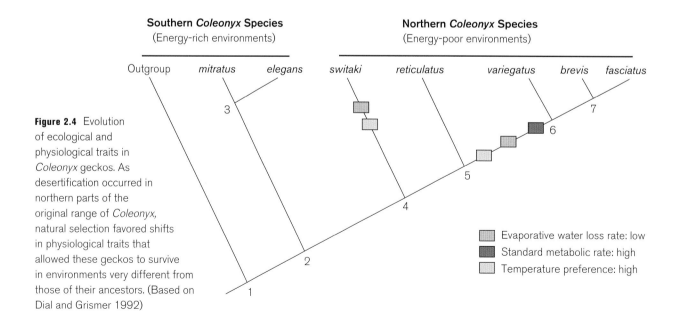

Figure 2.4 Evolution of ecological and physiological traits in *Coleonyx* geckos. As desertification occurred in northern parts of the original range of *Coleonyx,* natural selection favored shifts in physiological traits that allowed these geckos to survive in environments very different from those of their ancestors. (Based on Dial and Grismer 1992)

South Texas is home to this photogenic eublepharid gecko, _Coleonyx brevis._ (Laurie Vitt)

Coleonyx mitratus is a subtropical eublepharid gecko that probably experienced energy rich environments during its evolutionary history. (Louis Porras)

per hour (Heatwole and Vernon 1977). Because _S. nicholsi_ is smaller than _S. klauberi,_ it should lose water at a greater rate based on body size alone. A portion of the variation in water loss rates in lizards results from adaptation to differing habitats: lizards in xeric environments lose water more slowly than lizards in mesic environments. The most plausible explanation is that lizards in mesic habitats have a long evolutionary history of having water available and have adapted accordingly, whereas those in xeric habitats have not.

As innocuous as eyes might seem in the general scheme of water flux in lizards, water loss is substantial across eye surfaces of some species. In juvenile monitor lizards, _Varanus gouldii,_ water loss across eye surfaces accounts for 65 percent of all water loss (Green 1969). Lizards with eyelids lose about twice as much water from their head region as species with a spectacle (brille) (Mautz 1982).

In addition to losing water more slowly, some lizards in xeric environments tolerate remarkable increases in solute concentrations within their bodies. Several species of _Ctenophorus_ living in arid habitats in Australia, for example, experience long periods of drought during summer, becoming more and more dehydrated. Because they have no salt glands for excreting extra salts, some salts—sodium in particular—build up to high levels (called hypernatremia) in body tissues. These lizards usually feed on ants, which themselves contain large amounts of sodium. During drought, the lizards stop feeding on ants, become inactive, and withstand high solute concentrations of sodium until rainfall allows them to drink and excrete the salts (Minnich 1982). Some agamids, iguanids,

Coleonyx switaki is a desert eublepharid gecko that probably experienced energy poor environments during its evolutionary history. (L. Lee Grismer)

lacertids, scincids, xantusiids, teiids, and varanids excrete salts through nasal salt glands.

Water intake from prey is critical in desert lizards, even on a microgeographic level. Two closely related iguanids, _Urosaurus ornatus_ and _U. graciosus,_ live nearly side by side in riparian areas of the central Sonoran Desert. _U. ornatus_ lives on trunks and large limbs of large, well-shaded mesquite trees _(Prosopis)_ near rivers, and _U. graciosus_ lives in the canopy of smaller mesquite trees some distance away from rivers as well as on creosote and other shrubs. The thermal environment is much warmer for _U. graciosus_ than for _U. ornatus,_ and water loss rates in _U. graciosus_ are considerably higher than in _U. ornatus._ Partly to balance its excessive loss of water, *U. graciosus* eats more insects and a greater total volume of insects than *U. ornatus* (Congdon et al. 1982).

EVOLUTION OF PHYSIOLOGICAL TRAITS

Just as morphological traits vary geographically and among closely related species, physiological traits vary as well and are usually tied with identifiable variation in lizard environments. Although one can easily observe that diurnal desert lizards are active at higher body temperatures and might have lower rates of water loss than diurnal mesic forest lizards, cause-and-effect relationships are confounded by numerous uncontrollable variables. For example, if the diurnal desert lizards are whiptails *(Cnemidophorus)* and the mesic forest lizards are anoles *(Anolis)*, any differences may reflect evolutionary differences between lizard genera rather than adaptation to different environmental conditions. To investigate this problem, Ben Dial and Lee Grismer (1992, 1994) examined physiological traits within the eublepharid genus *Coleonyx* in the perspective of their evolutionary history and the history of climate change throughout their historical and present-day distribution in the New World (fig. 2.4). Historically, these lizards were widespread in mesic tropical forests in what is now North America. Thus, during the early Cenozoic or late Cretaceous ancestors of present-day *Coleonyx* were mesic adapted, with high rates of water loss, relatively low preferred body temperatures, and rel-

atively low standard metabolic rates. Drying conditions during the Eocene pushed mesic forests southward, and some *Coleonyx* moved with them (southern *Coleonyx*). Others remained isolated in the north under drying conditions (northern *Coleonyx*). Because environmental conditions for the southern group were relatively similar to those of their ancestors, they retained low preferred body temperatures, high rates of water loss, and low standard metabolic rates. The northern group was split into two, one isolated on Baja California, which, until relatively recently, experienced mesic conditions, and the other on the mainland, which experienced a transition to xeric conditions. Species in the mainland xeric group evolved low water loss rates, higher preferred temperatures, and higher standard metabolic rates. The single species in the more recently xeric Baja California group, *C. switaki*, also developed some xeric physiological traits, including lower water loss rates and higher preferred body temperatures, but not to the extreme that those on the mainland did.

For lizards, getting around in the real world involves moving through often complex structural habitats, gaining enough heat to maintain body temperatures optimal for performance, and retaining critical balances between water and solute levels within their bodies. The many challenges that structural and thermal landscapes present throughout the world have resulted in evolution of a diversity of lizard forms and functions within constraints set by the evolutionary history of each lizard family.

CHAPTER 3

LIZARDS AS PREDATORS

African chameleons, like this *Chamaeleo parsoni,* remain nearly motionless as they prepare to send their tongues catapulting toward unsuspecting insects. (Bill Love)

Although a few lizards are herbivorous, most feed on other animals and swallow them whole. Insectivorous lizards attack dozens if not hundreds of insects each day, and no doubt from the perspective of an insect these are spectacular events. To place such an event in human perspective, consider the Komodo monitor (also called the Komodo dragon), the only extant lizard large enough to kill and devour humans. To paraphrase Walter Auffenberg (1981): On the tropical island of Flores, a few hundred kilometers north of Australia, a large male Komodo monitor lizard waited, hidden in tall grass alongside a game trail. A few weeks earlier, several hundred meters away from this spot, he had ambushed and killed a small deer. He had eaten most of it (about 25 kg), but that meal had now been digested away: this gigantic lizard was hungry and hunting again. He saw (and smelled) a large mammal coming down the trail, and he crouched lower, like a cat, ready to lunge when his prey came within striking distance. This time it was a large wild boar, ambling along the trail; the huge lizard waited, preparing for ambush. The Komodo actually allowed the pig to pass, but when the boar was about a meter away, the lizard attacked from the rear, grabbing the boar's right hind leg with his sharp, serrated teeth, quickly severing tendons. The hamstrung boar squealed and turned, trying to defend itself, but the lizard hung on and pulled back hard on his bite, tearing the pig's flesh and arteries. The boar began bleeding profusely and fell over on its side. Immediately, the Komodo monitor released the pig's leg and bit into the soft flesh of its belly. Blood was gushing from both wounds and the boar's intestines were falling out. The entire attack was over within seconds.

WHAT LIZARDS EAT

Striking variation exists among lizards in the kinds of things they eat. The common misconception that lizards eat whatever is available stems partly from a lack of dietary data for most lizard species and the observation that most lizards available through the pet trade will feed, and often do well, on crickets and mealworms (*Tenebrio* beetle larvae). Even herbivorous lizards like green iguanas will eat mealworms in captivity. To set the stage for the next section, where we discuss why lizards eat what they eat, we offer a minitour through the diets of lizards, starting within Iguania and moving across the lizard phylogeny. Among the most obvious patterns emerging is that many lizards in the Iguania (excluding iguanines) rely to a large degree on ants; very few Scleroglossa regularly eat ants. We will revisit this point. Of course, any examination of lizard diets requires data from lizards in their natural habitats. Lizards are ideal for dietary studies because most swallow their prey whole, making identification of prey easy. Table 3.1 represents a typical summary table for a lizard diet.

One subfamily of iguanids, Iguaninae, and one subfamily of agamids, Leiolepidinae, are herbivorous, as are the agamid *Hydrosaurus* and a few other iguanians. These are the only iguanians that appear capable of discriminating food based on chemical cues, apparently to avoid some plant chemicals. *Dipsosaurus dorsalis, Sauromalus obesus, Ctenosaura pectinata, Phymaturus punae, Pogona vitticeps,* and *Uromastyx aegyptius* have been shown experimentally to discriminate food chemicals (Cooper and Alberts 1990; Cooper, pers. comm.). Within Iguaninae, adults of all species feed primarily on plant parts including fruits, flowers, and leaves. Desert iguanas, *Dipsosaurus dorsalis,* feed on a variety of annual and perennial plants (Mautz and Nagy 1987; Norris 1953), which make up well over 90 percent of their diet (Pianka 1971a). Diet composition changes seasonally as plant availability changes. During spring, they feed to a large extent on yellow flowers, including those of the creosote bush *(Larrea).* Although desert iguanas, even juveniles, rarely eat arthropods in nature, they will gorge themselves on mealworms in captivity, and it is relatively easy to feed mealworms to free-ranging wild individuals in the field. The yellow color of mealworms may be enough to elicit a feeding response, considering that these lizards will also eat pieces of yellow plastic flagging tape!

In central Mexico, the spiny-tailed iguana, *Ctenosaura pectinata,* feeds predominantly on a variety of plants in the families Fabaceae, Bombacaceae, Convolucaceae, and Euphorbiaceae (Durtsche 2000), with diet composition changing seasonally as different plants dominate available forage. The most interesting aspect of its diet, however, is that it undergoes an ontogenetic shift: although juveniles can eat plants, they are primarily insectivorous, eating beetles, grasshoppers, and beetle larvae, shifting to complete herbivory only as adults. A similar dietary shift occurs in *C. similis* (Van Devender 1982). An insect diet may provide growing lizards more protein than plant food. An ontogenetic shift occurs in food chemical discrimination as well: juveniles respond to insect chemicals but not plant chemicals, whereas adults respond to both (Cooper, pers. comm.). Distinct ontogenetic dietary shifts do not occur in other studied iguanines, although insects are included at low frequencies in some juvenile diets, such as Galápagos marine iguanas, *Amblyrhynchus cristatus* (Nagy and Shoemaker 1984), which otherwise dive into the ocean to feed on a wide variety of algae (Carpenter 1966; Trillmich 1979). While much feeding takes place on shallow reefs, marine iguanas can dive as much as 10 m and often remain underwater for half an hour or more,

TABLE 3.1

Representative diet of a Neotropical teiid, *Ameiva ameiva*

PREY TYPE	NUMBER OF PREY ITEMS	PERCENT OF TOTAL NUMBER	VOLUME (MM³)	PERCENT OF TOTAL VOLUME	NO. OF LIZARDS CONTAINING PREY TYPE
Odonates	2	0.04	152.69	0.02	2
Grasshoppers and crickets	398	8.89	85,434.2	12.92	215
Roaches	261	5.83	130,449.23	19.72	143
Mantids and phasmids	8	0.18	1,908.08	0.29	7
Beetles	376	8.4	99,921.77	15.11	171
Hemipterans	101	2.26	22,783.66	3.44	59
Homopterans	21	0.47	1,681.41	0.25	19
Termites	1,542	34.44	20,961.63	3.17	80
Flies	14	0.31	1,065.63	0.16	12
Hymenopterans (non-ant)	27	0.6	2,417.94	0.37	27
Ants	186	4.15	6,168.83	0.93	65
Lepidopterans	24	0.54	15,390.36	2.33	6
Springtails	2	0.04	0.38	0	2
Earwigs	9	0.2	295.95	0.04	5
Thysanopterans and thysanurans	2	0.04	28.67	0	2
Insect larvae, eggs, pupae	729	16.28	95,328.36	14.41	207
Miscellaneous unidentified insects	12	0.27	5,826.68	0.88	11
All vertebrate material	31	0.69	47,769.56	7.22	27
Spiders	345	7.7	69,552.55	10.52	194
Mites	39	0.87	3.21	0	18
Opiliones	32	0.71	1,464.87	0.22	23
Scorpions	16	0.36	13,816.66	2.09	14
Centipedes	31	0.69	9,018.89	1.36	26
Millipedes	64	1.43	3,868.69	0.58	40
Isopods	37	0.83	3,729.54	0.56	14
Mollusks	67	1.5	3,963.12	0.6	50
Earthworms	46	1.03	11,388.8	1.72	29
Plant material	56	1.25	6,963.76	1.05	18
Totals	4,478	100.00	661,355.12	100.00	—
Niche breadths*		5.79		8.37	

Note: Based on stomach contents of 445 *A. ameiva* from a variety of Amazonian localities.
*Niche breadths are estimated dietary diversities (see chapter 7).

scraping their jaws across lava rocks on reefs and tearing off pieces of algae. Similar to other iguanines, marine iguana juveniles acquire the intestinal fauna necessary for digestion of plant material by eating feces of adults. Iguanines have specialized valves in their colons slowing the passage rate of digested plant food, facilitating endosymbiotes that produce cellulases necessary to process plant food (Iverson 1982).

Remaining iguanids eat primarily arthropods, with striking differences among species. Phrynosomatines and tropidurines have varied diets, but ants constitute significant proportions of the diets of most species, with some (e.g., *Phrynosoma*) specializing on them. Numerically, ants are the most common prey taken by three arboreal phrynosomatine lizards—*Urosaurus ornatus, U. graciosus,* and *Sceloporus magister*—in the Sonoran Desert in both wet and dry years (Vitt et al. 1981). By volume, ants make up about 50 percent of these species' diet in a dry year but only about 30 percent in a wet year, suggesting that these lizards switch to other prey as they become more abun-

The only herbivorous lizard that feeds in the ocean, *Amblyrhynchus cristatus* consumes marine algae in the Galápagos. (Eric Pianka)

dant. Four Amazonian species, *Plica plica, P. umbra, Uracentron azureus,* and *U. flaviceps* (Avila-Pires 1995; Vitt 1991, 1993; Vitt and Zani 1996a,b; Vitt, Zani, and Avila-Pires 1997a), likewise specialize on ants. However, the sister taxon to all other tropidurines, the strange riparian lizard *Uranoscodon superciliosus,* rarely feeds on ants, taking most of its prey from debris washed up on shores of streams and rivers (Howland et al. 1990). *Callisaurus draconoides,* a terrestrial desert phrynosomatine, feeds mostly on grasshoppers, crickets, beetles, and insect larvae, but it does take some plant food (Pianka and Parker 1972; Pianka 1986). Herbivory shows up sporadically within the tropidurines, and many *Tropidurus* eat some plant material. The southernmost lizard in the world, *Liolaemus magellanicus* (Liolaeminae) appears to be nearly totally herbivorous (Jaksíc and Schwenk 1983).

A few iguanids eat relatively large prey. *Gambelia wislizeni* (Crotaphytinae) feed on other lizards and large or-

thopterans, and the truly bizarre *Corytophanes cristatus* (Corytophaninae) eats large lepidopteran and beetle larvae, grasshoppers, katydids, and cicadas (Andrews 1979). Diets of most agamid lizards mirror those of nonherbivorous iguanid lizards, with many species feeding on large numbers of ants. Most Australian *Ctenophorus* eat large numbers of ants, and *Moloch horridus* is an ant specialist (Pianka 1986; Pianka et al. 1998). Even *Ctenophorus isolepis,* a species living on open spaces in the Western Australian desert, feeds mostly on ants (Pianka 1971b). Larger species of agamids appear to eat fewer ants, and some are herbivorous. For example, the diet of *Agama hispida* in South Africa includes beetles, lepidopteran larvae, and centipedes, but termites, ants, and even some vegetation are eaten as well (FitzSimons 1943). In the Kalahari, the diet of *Agama hispida* is 49 percent ants by volume. Spiny-tailed agamids or mastigures *(Uromastyx),* large-bodied lizards ecologically somewhat similar to chuckwallas (*Sau-*

Appropriately named, leopard lizards *(Gambelia wislizeni)* feed on other lizards such as this unlucky *Urosaurus graciosus.* (Cecil Schwalbe)

romalus of the desert southwest of North America), are completely herbivorous and have specialized front teeth for cutting vegetation (P. Robinson 1976).

Chamaeleonids eat a diversity of insects, which they capture with their long projectile tongues. Watching a chameleon eat has to be among the best natural comedies known. The lizards perch on branches moving only their eyes—each one independently. When an insect lands or moves within striking distance, both eyes fix on the insect and the lizard moves at what seems to be less than a snail's pace. Its head and body slowly orient toward the insect; the lizard then goes through a painfully slow stretching sequence in which it extends its head and neck in the direction of the prey, its mouth opens slowly, its tongue slowly lifts ever so slightly from the floor of its mouth, and in a millisecond its tongue flies through the air twice the distance of the chameleon's body length, sticks to the insect, and is rapidly hauled back in with insect attached.

Like other western collared lizards, *Crotaphytus insularis* frequently eats side-blotched lizards *(Uta stansburiana).*

(Cecil Schwalbe)

LIZARDS AS PREDATORS

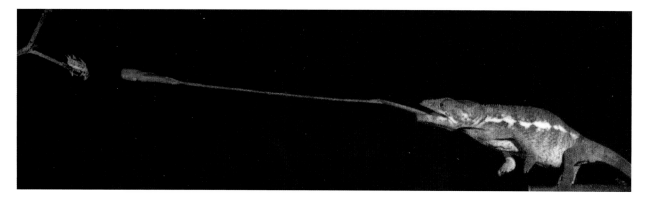

This *Chamaeleo pardalis* shoots its long ballistic tongue at an unsuspecting insect. (M. Vences and F. Rauschenbach)

One has to wonder if maybe the insect simply fails to respond because it experiences neural shock at the sight of the sticky blob hurtling at it from some unknown source in the vegetation! In South Africa, prey of *Chamaeleo dilepis* include grasshoppers, flying insects, beetles, reproductive winged (alate) termites, and spiders, whereas the bizarre *C. namaquensis,* which lives in low shrubs and on the ground, feeds mostly on beetles (FitzSimons 1943). Other South African chameleons, such as *Bradypodion* species, feed on a diversity of insects, mostly those that fly or hop in vegetation. Dwarf chameleons *(Brookesia)* of Madagascar have projectile tongues that exceed twice their body length, which they use to capture flies and crickets from vegetation and the forest floor (Raxworthy 1991).

Among gekkotans, diets are quite varied, but ants drop out nearly completely. Within Diplodactylidae, some species such as *Diplodactylus ciliaris, D. damaeus,* and *D. elderi* eat a combination of spiders, crickets, beetles, and other insects, whereas other species, like *D. conspicillatus, D. pulcher,* and *Rhynchoedura ornata* are termite specialists (Pianka and Pianka 1976; Pianka 1986). Among gekkonids, the large nocturnal Amazonian geckos *Thecadactylus rapicauda* and *Hemidactylus palaichthus* eat mostly roaches and crickets (Vitt and Zani 1997). However, the introduced gecko, *H. mabouia,* feeds mostly on moths, which it captures near lights on houses. In the Brazilian caatinga, the tiny diurnal gekkonid *Lygodactylus klugei* eats worker termites in the genus *Nasutitermes.* It actually breaks open the soft mud-encased termite trails, feeding on workers until soldiers swarm the area. Most geckos in the subfamily Sphaerodactylinae feed on a wide variety of insects. For example, Amazonian *Gonatodes humeralis* eat insect larvae, grasshoppers and crickets, roaches, hymenopter-

ans, and even earwigs among other things (Vitt, Caldwell et al. 1997). *Gonatodes hasemani* eats spiders, insect larvae, roaches, and homopterans (Vitt, Souza et al. 2000). The tiny leaf-litter *Coleodactylus amazonicus* eats mostly tiny springtails, insect larvae, and homopterans, and *Pseudogonatodes guianensis* eats mostly tiny insect larvae and grasshoppers. Several gecko genera, including *Hoplodactylus, Phelsuma,* and *Rhacodactylus,* contain species that are omnivorous, feeding on a variety of insects and fruits.

The most unusual gekkotan diets are found in the snakelike pygopodids. *Lialis burtonis* eats only other lizards, which it swallows whole using flexion of the skull to force prey down its throat (Pianka 1986; Patchell and Shine 1986). *Delma butleri* eats mostly spiders, whereas *Pygopus nigriceps* eats spiders and scorpions (Pianka 1986).

Autarchoglossan families, if anything, have shifted away from ants (at least adult ants) and tend, with one exception, to eat fairly large prey. The exception is termites, which many autarchoglossans eat. Some even specialize on them. In a sense, termites are like large prey because a large mass of these juicy social insects can usually be found in one place (high payoff with little searching). Insect larvae, mollusks, earthworms, and a diversity of other prey items generally not encountered on the surface appear commonly in autarchoglossan diets as well. Although some autarchoglossans eat substantial numbers of ants, these insects remain relatively unimportant volumetrically. Possible exceptions are Mediterranean *Podarcis.*

Diets of amphisbaenians and dibamids are poorly known, but both appear to feed primarily under the surface. Numerically, the large-bodied South American *Amphisbaena alba* eats mostly termites (43 percent) and ants (29 percent), but termites, grasshoppers, beetles, and insect larvae are most important volumetrically, with ants

Phelsuma **eating a cockroach.** (Bill Love)

Chamaeleo rhinoceratus **stretches out using its hind legs and prehensile tail to hold on as it positions itself to capture an insect.** (Steve Wilson)

Many lizards, like this gecko, *Diplodactylus taenicauda,* slowly stretch their bodies out as they prepare to capture insects. (Steve Wilson)

comprising only a meager 8.6 percent (Colli and Zamboni 1999). Other amphisbaenians including *A. gonavensis* and the strange *Bipes* of Mexico eat a variety of prey as well (Cusumano and Powell 1991; Papenfuss 1982, pers. comm.). Amphisbaenians have well-developed chemosensory systems and powerful jaws that aid them in finding, identifying, and manipulating prey underground (López and Salvador 1992).

Small teiids and lacertids are insectivorous, with some, particularly *Cnemidophorus* in open habitats, feeding on large numbers of termites. Typical prey include orthopterans, spiders, roaches, and insect larvae. For example, the Amazonian teiid *Cnemidophorus lemniscatus* eats grasshoppers, beetles, and spiders (Vitt, Zani, Caldwell et al. 1997). *Kentropyx pelviceps* and *K. altamazonica* in the western Amazon feed primarily on grasshoppers, crickets, and spiders (Vitt, Sartorius et al. 2000), whereas *K. striata*, living in open habitats of Amazonian savanna, feeds mostly on frogs, insect larvae, and spiders (Vitt and Carvalho 1992). The larger teiid *Ameiva ameiva* eats roaches, beetles, insect larvae, grasshoppers, and spiders, but because of its larger size it eats larger invertebrates. *Ameiva* also eats vertebrates, including frogs and lizards. The largest teiids, *Tupinambis, Crocodilurus,* and *Dracaena,* eat not only larger prey for the most part, but also different prey. *Tupinambis* frequently eat frogs, with vertebrates constituting nearly 75 percent of the diet in *T. teguixin.* The natural diet of some *Tupinambis,* especially *T. rufescens* and *T. merianae,* is varied and includes enough plant material (mostly fruits) that they could be considered omnivorous. *Dracaena* is the most specialized large teiid, feeding on mollusks, crushing their shells with its powerful jaws and large posterior teeth.

Among the most striking dietary shifts in teiid lizards is the shift to herbivory in several endemic island species, including *Cnemidophorus arubensis* and *C. murinus* (Schall 1990; Schall and Ressel 1991; Dearing and Schall 1992). Many plants available to both species contain toxins, and as a result, these lizards avoid some plants (especially those containing phenols, saponin, and alkaloids). Use of plants by these two species may represent a single origin of omnivory with a secondary reduction in arthropod intake. The closely related *Cnemidophorus lemniscatus* eats some fruits as well (Vitt, Zani, Caldwell et al. 1997; Markezich et al. 1997).

Among lacertids, diets are also variable. In Tunisia, not only does *Mesalina olivieri* feed on a wide variety of insects and other arthropods, but its diet changes both spatially and temporally, suggesting that it is an opportunistic species (Nouira and Mou 1982). In the Namib Desert, *Meroles cuneirostris,* which lives in vegetated valleys between sand dunes, eats a wide variety of arthropods (mostly insects), only rarely eating plant material. *Meroles anchietae,* however, which lives on sand dunes with little vegetation, switches to grass seeds and seeds of the desert plant *Trianthema* (Aizoaceae) when insect abundance is low (M. Robinson and Cunningham 1978). In the Ebro Delta area of northeastern Spain, *Acanthodactylus erythrurus* and *Psammodromus algirus* occur together in a dune area (Carretero and Llorente 1993). Whereas *A. erythrurus* eats mostly fly and beetle pupae and spiders, *P. algirus* eats mainly beetle pupae, bugs, ants, and lepidopteran larvae. Some lacertids, including *Gallotia galloti* in the Canary Islands, feed on vegetation. *Gallotia* disperse seeds as a result of feeding on fruits (Valido and Nogales 1994). In the Kalahari, four to five species of lacertids occur together at most sites (Pianka 1971c, 1986; Pianka et al. 1979). One species, *Nucras tessellata,* specializes on large scorpions, which it digs up in their diurnal retreats. All others eat a fair number of termites, and two species, *Heliobolus lugubris* and *Ichnotropis squamulosa,* are termite specialists. Other Kalahari lacertids feed on a wider variety of insects, including beetles and grasshoppers.

A number of lacertids appear to eat large numbers of ants. Volumetric contribution of ants to their diets is unknown, because most dietary studies of these lizards are based on either numerical data from stomach contents or counts of remains in fecal pellets, neither of which provides estimates of prey volumes. Nevertheless, high frequency of ants in diets of *Acanthodactylus* and some other lacertids suggests that some species do not avoid ants (e.g., Gil et al. 1993).

Because of their small size, gymnophthalmids are restricted to small insects and spiders. However, considerable variation exists among species in what they eat. Many that occur in leaf litter, such as *Iphisa elegans, Cercosaura ocellata,* and *Prionodactylus eigenmanni,* eat small individuals of the same kinds of prey as small-bodied teiids— orthopterans, roaches, and spiders. Semiaquatic species such as *Neusticurus ecpleopus* eat small crickets, earthworms, beetles, and insect larvae, many of which are larvae of aquatic insects (Vitt et al. 1998). Among the most specialized gymnophthalmids are fossorial *Calyptommatus* living in sand dunes associated with the Rio São Francisco in northeastern Brazil (Moraes 1993). Diets of *Calyptommatus sinebrachiatus, C. leiolepis,* and *C. nicterus* are dominated by termites and beetle larvae, which they presumably capture underground.

Because xantusiids live under decaying Joshua trees *(Xantusia vigilis),* in rock crevices *(X. henshawi),* under and within dead tropical tree trunks *(Lepidophyma),* or under rocks *(Cricosaura),* their prey reflects what is available in those microhabitats. In the Mojave Desert, *Xantusia vigilis* feeds primarily on beetles and ants but also eats insect larvae, spiders, roaches, termites, flies, and moths (Pianka 1986). In Cuba, *Cricosaura* eats house crickets, moths, and insect larvae (Schwartz and Henderson 1991). In lowland forest of Nicaragua, *Lepidophyma flavimaculatum* eats mostly termites and crickets (Vitt and Zani 1998a). However, *L. smithi* eats figs that fall into the caves where it lives in Guerrero, Mexico (Mautz and Lopez-Forment 1978). The large island night lizard *Xantusia riversiana* is apparently omnivorous.

Because of high ecological, morphological, and species diversity, skink diets range from insects to plants. The smallest skinks eat insects, including many termites. Among eleven species of Philippine skinks studied by Walter and Troy Auffenberg (1988), termites are the most common prey for six species, the second most common for two species, and the third most common for the other three. Crabs dominate the diet of *Emoia atrocostata,* a species that lives on the beach, and earthworms dominate the diet of *Otosaurus cumingii,* a large species that spends most of its time in crevices or holes. High species diversity of Australian skinks and a plethora of dietary data offer a unique opportunity to examine variation in diets within this large family. In the genus *Ctenotus* alone, most species prey on some termites, but termites constitute more than 75 percent of the diet for four of fourteen species (Pianka 1986). One species, *C. piankai,* doesn't eat termites, but feeds primarily on spiders, bugs, and grasshoppers. In the Kalahari, termites comprise more than 90 percent of the diet of subterranean *Typhlosaurus,* but terrestrial *Mabuya* eat beetles and termites. In Amazon rain forest, the semiarboreal skink *Mabuya nigropunctata* feeds on grasshoppers, roaches, bugs, and spiders (Vitt, Zani, and Lima 1997). The largest skinks, including the tree skinks of the Solomon Islands *(Corucia)* and the Australian blue-tongues *(Tiliqua),* have shifted to eating fruits and flowers, although insects are still taken. Some *Egernia* are omnivorous.

Most cordylids are insectivorous. Sungazers, or girdle-tailed lizards *(Cordylus),* for instance, feed on beetles and other large insects captured in or at the edge of crevices where they live. A few cordylids are facultative (or opportunistic) herbivores as well. The Cape flat lizard, *Platysaurus capensis,* typically feeds on insects, particularly simulid flies. When figs become available after falling to the ground, *P. capensis* switch to them, often congregating at unusually high densities under fig trees (Whiting and Greeff 1997). The sister group to cordylids, gerrhosaurids, especially larger species, eat a combination of large insects, termites, lizards, small mammals, and fruits.

Most anguids have powerful jaws that are used to subdue large prey. *Ophisaurus apodus,* the largest "glass lizard," feeds on snails, lizards, and mice (Honders 1975; Schmidt and Inger 1957). The North American glass lizard, *O. attenuatus,* eats large insects, mostly grasshoppers, katydids, crickets, and spiders (Fitch 1989). *Diploglossus lessonae* of northeastern Brazil eats mostly spiders and scarab beetle pupae (Vitt 1985). The large alligator lizard *Elgaria multicarinata* eats a diversity of invertebrates—beetles, orthopterans, moths and their larvae, and spiders being the most common—and some vertebrates. Of the latter, five species of lizards (including *E. multicarinata),* small mammals, and fledgling birds have been reported from *E. multicarinata* stomachs (Cunningham 1956).

Like anguids, xenosaurids have powerful jaws that can subdue a variety of prey. *Xenosaurus grandis* in Veracruz, Mexico, feed mostly on lepidopteran larvae and various orthopterans (Ballinger et al. 1995). Because the lizards rarely leave their rock crevices, prey are presumably captured either within or at the edge of same.

Among varanoid lizards, diets have shifted toward larger invertebrates and vertebrates, which are swallowed whole or torn into large pieces and swallowed. Chemoreception of prey odors is well developed (Cooper 1989). The large Gila monsters and beaded lizards swallow very large prey, killing with their powerful jaws rather than with venom. Gila monsters *(Heloderma suspectum)* eat young mammals, including ground squirrels, rock squirrels, cottontails (Daniel Beck [1990] saw an adult eat four 40-g cottontails), jackrabbits, kangaroo rats, and likely desert mice and packrats, all of which are swallowed whole. Reptile eggs—particularly those of tortoises—and bird eggs are also taken. The natural diet of beaded lizards *(H. horridum)* includes mammals, fledgling birds, bird eggs, and reptile eggs (Bogert and Martín del Campo

1993). Gila monsters and beaded lizards are agile climbers and take bird eggs and fledglings from nests in thorn forest trees (Beck and Lowe 1991; Campbell and Lamar 1989). In Jalisco, Mexico, beaded lizards are particularly fond of the eggs of spiny-tailed iguanas *(Ctenosaura)*, which they dig up during the spring nesting season.

The varanoid *Lanthanotus* eats squid and small bits of fish and liver in captivity, but their natural diet remains largely unknown, as do most aspects of their natural history. Wild-caught *Lanthanotus* stomachs contained earthworm setae (Greene 1986).

Among varanid lizards, prey vary from invertebrates to vertebrates much larger than the lizards themselves. The tiny (by *Varanus* standards) *Varanus brevicauda* eats a combination of large insects and reptile eggs (Pianka 1970b). The small-bodied pygmy monitors *V. caudolineatus* and *V. gilleni* eat insects (mostly orthopterans) and small lizards (tails of geckos are eaten), whereas the slightly larger pygmy monitor *V. eremius* eats mostly lizards (Pianka 1968, 1969a, 1994a). *Varanus tristis* preys on bird eggs and baby birds as well as various other species of lizards; *V. gouldii* eats many other species of lizards. The only om-

nivore among Anguimorpha, *V. olivaceus*, eats fruits when in season. Prey types of larger monitors are at least partly a function of the habitat they occupy. In the Alligator River area of northern Australia, *Varanus panoptes* eats a wide variety of invertebrates and vertebrates, though by volume, vertebrates—fish, frogs, lizards, snakes, birds, and carrion (portions of an echidna!)—constitute most of the diet (Shine 1986a). The arboreal emerald monitor, *V. prasinus,* eats mostly insects, only occasionally taking small vertebrates (Greene 1986). At the extreme are Komodo monitors, *V. komodoensis,* which can kill prey at least as large as water buffalo (Auffenberg 1978). After eating enormous meals, they fast, similar to snakes. Visceral contents of dead prey or decomposition odors attract these huge lizards from distances exceeding 2 km. Many lizards often converge on a decaying prey item, tearing the carcass apart and rapidly swallowing large pieces. Walter Auffenberg (1981) watched a 50-kg female consume an entire 31-kg boar in a mere seventeen minutes!

Clearly, lizard diets are quite diverse. Most species eat a variety of invertebrates, with the particular mix of prey varying from species to species. Larger lizards feed on

A number of lizards, such as this Australian skink *(Cyclodomorphus),* **include snails in their diets when available.** (Steve Wilson)

LIZARD LIFESTYLES

Having captured a large lizard, this Australian monitor, *Varanus panoptes*, swallows it whole. (Steve Wilson)

larger prey, and many take vertebrates as well as invertebrates. Herbivory occurs in several lizard families, the most spectacular development of herbivory being in the Iguaninae. Specialization on ants has occurred repeatedly in Iguania but not in Scleroglossa, whereas specialization on termites is common in Scleroglossa but rare in Iguania. We now examine in some detail why lizards eat what they do.

WHY LIZARDS EAT WHAT THEY EAT

At some point in the history of the evolution of lizard diets, species interactions (competition, predation, parasitism) played a major role in determining what lizards eat, either directly, by determining the "menu" of available prey, or indirectly, by influencing shifts in microhabitat use, behavior, and morphology. The observation that lizard assemblages throughout the world are structured with respect to niche axes (food, place, and time) confirms this (see chapter 7). Species interactions no doubt are important as ongoing processes, easy to detect in systems where resources are limited but subtler in systems where resources appear unlimited. Each species car-

A BELATED BONUS BACK HOME

Once, while tracking a medium-sized thorny devil in the Great Victoria Desert of Western Australia, I encountered a sand goanna, *Varanus gouldii,* which I collected and preserved for later analysis in the lab. I gave up on finding that thorny devil: its tracks just vanished into thin air. Six months later, back in the United States, I set about dissecting the monitor lizard and found that thorny devil in its stomach! Sitting at my lab bench in Texas, I was taken right back to that day in Australia. Even today I remain amazed that the thorny devil's spines didn't puncture the monitor's thin stomach wall. I also found another small thorny devil in the stomach of a racehorse monitor lizard, *Varanus tristis.* Raptors and bustards are also probable predators of thorny devils. *(PIANKA)*

ries its own unique evolutionary history, which influences its ability to successfully interact with others. We return to species interactions in chapter 7. Here we focus on some of the more obvious correlates of dietary patterns among lizards.

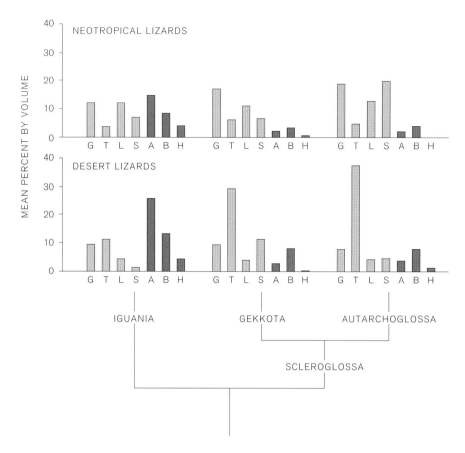

Figure 3.1 Mean percent utilization of the seven most important prey categories by Neotropical *(top)* and desert lizards *(bottom)*. Proportions of palatable prey types are shown in green. Proportions of ants and other noxious insects, shown in brown, decrease from high values in the Iguania to low values in the Scleroglossa. Ants are replaced by a combination of grasshoppers, insect larvae, and spiders in Neotropical lizards, and by termites and spiders in desert lizards. Ants, some beetles, and other hymenopterans are known to contain numerous chemicals for defense.

G = grasshoppers and crickets
T = termites
L = insect larvae, pupae, and eggs
S = spiders
A = ants
B = beetles
H = non-ant hymenopterans

Determinants of diets for lizard species are complex but involve the interplay of evolutionary history, body size, microhabitat specialization (or lack thereof), and prey availability (past and present). Iguanians, for example, are sit-and-wait foragers (except iguanines) and thus predestined to feed on mobile prey, which they detect by vision. In contrast, scleroglossans are active foragers (with exceptions) with keen chemosensory systems and thus predestined to add nonmobile prey to their diets. Use of chemical cues to discriminate prey also allows them to avoid noxious prey items. All species in the North American genus *Phrynosoma* are ant specialists, indicating that ant specialization evolved early in the evolutionary history of this clade and was carried through to all present-day descendants. Similarly, all iguanines are obligate herbivores as adults, reflecting herbivory in their common ancestor.

These are all historical effects. To examine some of the most obvious historical patterns further, we provide an ex-

ample from Neotropical and desert lizard diets, asking: How different are diets of iguanians, gekkotans, and autarchoglossans? To answer this question, we generated dietary summaries for 83 Neotropical species and 92 desert species. We then calculated average percentages of prey use across species within each of these three clades. This provides a mean value as well as a measure of variation. Results are striking (fig. 3.1). Certain dietary characteristics appear to have a phylogenetic basis. Ants are common in Neotropical and desert iguanians but relatively rare in gekkotans and autarchoglossans. Ants, particularly those in the subfamily Myrmicinae, produce a variety of noxious chemicals that may cause gekkotans and autarchoglossans to avoid them. Other insects that produce noxious chemicals (alkaloids and others), including other hymenopterans, hemipterans, homopterans, and beetles, are less common in autarchoglossans than in iguanians, adding support to this hypothesis. Some insects are staples in diets of all lizards. Grasshoppers and crickets make up a large portion of the diets for iguanians, gekkotans, and autarchoglossans in Neotropical forests and deserts, but the proportions in desert lizards are much lower. Termites are eaten by most lizards but form a larger portion of diets of desert lizards, especially in the Kalahari. Spiders and roaches form a larger portion of Neotropical lizard diets. These differences likely reflect habitat-specific activity and abundance patterns of specific prey. Diurnal and nocturnal roaches, for example, are common in the Neotropics but uncommon in deserts. Similarly, nocturnal termite alate releases are encountered by a large number of nocturnal lizards in open habitats in the Kalahari and Australian deserts, whereas only one or two nocturnal Neotropical lizards encounter such swarms. Termites sometimes swarm during the day as well in the Kalahari and Australia, when they are eaten by many diurnal species of lizards; termites released during the day in Neotropical forest, however, would be less exposed than in deserts because of the high habitat structural diversity and thus less available to lizards.

FORAGING MODES

Foraging mode is also a historical factor, being deeply rooted in the evolutionary history of lizards. Many lizards, like horned lizards *(Phrynosoma)*, ambush their prey, sitting and waiting for potential prey to wander past. Other lizards actively search out and pursue their prey. These two foraging modes are known as "sit-and-wait" versus "widely foraging" tactics (Pianka 1966; Huey and Pianka 1981; Perry and Pianka 1997). Although this dichotomy is somewhat artificial, numerous animal groups do fall into either one category or the other. Thus iguanids, agamids, chamaeleonids, and most gekkotans primarily sit and wait for their prey, whereas teiids, scincids, and most varanids usually forage widely. Chameleons are very efficient sit-and-wait predators, relying on camouflage and using their long projectile sticky tongues to catch all sorts of insects, including some extremely active ones like flies. Chameleons could be viewed as the ultimate sit-and-wait predators because they continue to sit even during prey capture, minimizing the possibility that their crypsis will be compromised. Because a few lizards, particularly iguanines, are herbivorous, eating flowers and leaves, they don't fit neatly into sit-and-wait or widely foraging; rather, they are essentially grazers.

This evident natural dichotomy in foraging modes has

The helmeted iguana, *Corytophanes cristatus,* often remains motionless to take advantage of its cryptic morphology.
(Janalee Caldwell)

impacted theories of optimal diets and competitive relationships among species (Huey and Pianka 1981; Magnusson et al. 1985; McLaughlin 1989; Perry et al. 1990; Perry 1999; Pietruszka 1986; Perry and Pianka 1997). The realization that differences in foraging behavior might influence other aspects of lizard biology has had a cascading effect on research in behavior, physiology, reproductive biology, and ecology. Initial studies simply compared species known to be sit-and-wait foragers with species known to be active or wide foragers. We now know that a major phylogenetic component exists such that differences in foraging behavior did not evolve independently in each species of lizard: rather, shifts in foraging mode had a historical basis and were carried through major clades (fig. 3.2). Moreover, sensory system evolution occurred simultaneously and cannot be easily separated from behavioral components of foraging.

Lizard ancestors appear to have been sit-and-wait foragers that depended primarily on visual cues for detecting prey. They had rudimentary chemosensory systems

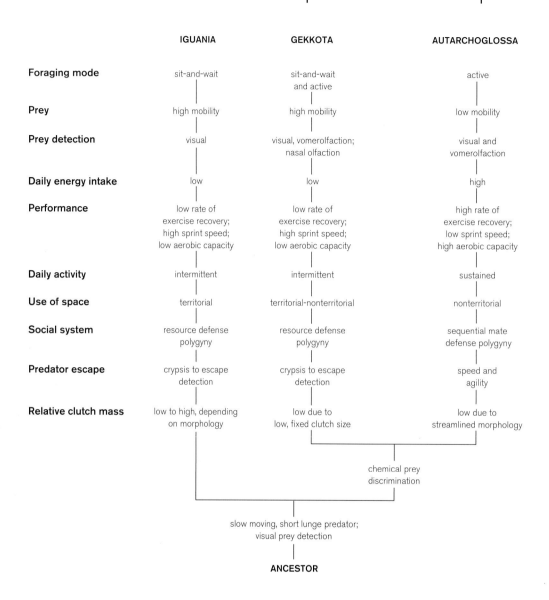

Figure 3.2 Attributes of Iguania versus Scleroglossa. Numerous aspects of morphology, behavior, physiology, and ecology are associated with the dramatic shift that occurred when scleroglossans split from iguanians. (Evolutionary reversals, discussed elsewhere, are not shown in this diagram.)

(see chapter 6) used primarily in a social context. These characteristics were carried through to the Iguania. Sit-and-wait foraging requires remaining motionless waiting for prey to pass nearby, and crypsis offsets detection. Thus prey most commonly eaten by sit-and-wait lizards are mobile, active on the surface, and visually conspicuous. Development of nasal olfaction and vomeronasal systems for prey detection and discrimination in scleroglos-

sans opened up an adaptive zone untapped by iguanians and lizard ancestors. A great diversity of insects, spiders, and other invertebrates were relatively undetectable by sit-and-wait foragers because they were either cryptically colored and didn't move much or remained hidden under surface objects, within vegetation, or in crevices or holes. Chemical detection of prey changed all this and no doubt led to the remarkable diversification that produced scleroglossan lizards, including the highly successful snakes.

Foraging and escape behavior are closely linked, and are correlated with numerous physiological, life history, and ecological traits as well. Because sit-and-wait lizards rely on crypsis for escaping detection by predators, natural selection in most cases has not favored physiological traits allowing long escape runs. Nevertheless, most are capable of rapid bursts of speed for short distances, allowing them to capture prey and providing rapid access to retreats.

Likewise, sit-and-wait foraging has favored territoriality because individuals move around very little. Good perches for prey detection and capture likely provided the impetus for defending space as well as social benefits. No doubt, territoriality was in place in lizard ancestors.

Crypsis has many benefits as long as it works for escape from detection by predators. Morphological diversification, moreover, can be virtually unlimited as long as morphology does not interfere with crypsis. Tanklike bodies of horned lizards *(Phrynosoma)* are cryptic against the ground, leaflike morphology of *Stenocercus* renders them nearly invisible on the forest floor, leaflike morphology of chameleons makes them nearly invisible in vegetation or on leaf litter (depending upon species), and long, thin bodies of anoles and many other lizards become cryptic on thin branches.

Reproductive consequences of morphological diversification are also remarkable. Tanklike bodies allow for voluminous clutches; *Phrynosoma,* for example, can produce clutches weighing nearly as much as an adult female. A voluminous clutch in a cryptically colored lizard that moves very little presumably does not incur a great risk of predation. Among active foragers, continual movement while foraging has an energetic cost that likely feeds back on clutch production. More important, their overall morphology is streamlined to facilitate rapid escape, and this alone constrains reproductive investment at any one time. Active lizards must have low clutch volumes.

Community-level consequences of foraging mode exist as well. For example, sit-and-wait species tend to be microhabitat specialists. *Plica plica* is nearly always found on trunks of large Amazonian trees, *Urosaurus graciosus* are nearly always found on thin limbs of Sonoran Desert shrubs and small trees, and *Phyllopezus pollicaris* in northeastern Brazil are nearly always found in crevices in boulder fields. Within lizard assemblages, too, sit-and-wait

Hidden by its morphology and color pattern, the Madagascan tropical gecko *Uroplatus sikorae* can remain in one position for hours. (Steve Wilson)

species tend to separate by microhabitat. The spectacular evolution of ecomorphs within the genus *Anolis* provides a prime example (see chapter 7).

Among active foragers, the situation is very different. Active or wide foraging requires movement, which itself renders these lizards easy to detect by visually oriented predators. Alert behavior allows them to detect the presence of many predators, and high performance levels, especially associated with long-distance running, allows them to escape predator attacks. Foraging over long distances precludes defense of particular pieces of real estate (thus they are nonterritorial), but more important, it means that they traverse many microhabitats in their

The tiny chameleon *Rhampholeon robecchi* could easily be mistaken for a twig. (Rolf Leptien; Chimaira Edition)

search for prey. Not all active foragers remain on the surface, however; many have gone underground, where they can move about with reduced risk from surface predators. At a community level, active foraging species tend to segregate on the basis of body size because differences in body size translate into differences in prey size, which reflect to some degree differences in prey types.

Certain dietary differences are associated with foraging tactics as well. Ambush predators rely largely on moving prey, whereas widely foraging predators encounter and consume nonmoving types of prey items more frequently. For the sit-and-wait tactic to pay off, prey must be relatively mobile and prey density must be high (or predator energy requirements low). During periods of prey scarcity, therefore, the sit-and-wait tactic is less prevalent than the wide-foraging method. The success of the wide-foraging tactic is also influenced by prey mobility and prey density, as well as by a predator's energetic requirements (which are usually higher than those of ambush predators), but searching abilities of a predator and the spatial distribution of its prey assume substantial importance. Desert sites in North America and Australia support similar numbers of species of sit-and-wait foragers, whereas fewer species use this foraging mode in the Kalahari (Pianka 1986). By the same token, markedly fewer species forage widely in western North America (only one species, the teiid *Cnemidophorus tigris*) and in the Kalahari (an average of four species per site) than in the Australian deserts (mean number of widely foraging species per area is 10.1, mostly skinks in the genus *Ctenotus*). History of colonization may explain these differences.

Neotropical lizard assemblages contain mixtures of sit-and-wait and widely foraging species. Among caatinga lizards of northeastern Brazil, six are wide foragers and seven are sit-and-wait foragers (Vitt 1995). In lowland rain forest of Amazonian Ecuador, twelve species are wide foragers and twelve are sit-and-wait foragers (Vitt and Zani 1996a). In a lowland rain forest of Nicaragua (Vitt and Zani 1998a), five species are wide foragers, fourteen are sit-and-wait foragers, and one is a grazer. For the most part, sit-and-wait species in these lizard assemblages are microhabitat specialists, whereas widely foraging species use a wide variety of microhabitats. Furthermore, while widely foraging species are predominately terrestrial and home ranges of different species overlap considerably, most sit-and-wait foragers use some kind of elevated perches or are nocturnal (see chapter 14). Similarities in microhabitat use among widely foraging species are offset by differences in body size, which ultimately result in differences in prey size (see below). Widely foraging species appear to constitute an ordered progression of different body sizes (Vitt, Sartorius et al. 2000).

Although foraging mode is generally similar among species within each lizard family, exceptions exist. Two species of Kalahari lacertids, *Meroles suborbitalis* and *Pedioplanis lineo-ocellata,* sit and wait for prey, whereas two other sympatric lacertid species, *Heliobolus lugubris* and *Pedioplanis namaquensis,* forage widely for their food (Pianka et al. 1979; Huey and Pianka 1981; Huey et al. 1984). Time budgets of these lacertids reflect their modes of foraging (fig. 3.3; Pianka 1986). Overall energy budgets of widely foraging species are nearly double those of sit-and-wait species (Huey and Pianka 1981). Compared with sit-and-wait species, widely foraging Kalahari lacertids eat more termites (sedentary, spatially and temporally unpredictable but clumped prey). One widely foraging lacertid, *Nucras tessellata,* specializes on scorpions that it digs from burrows.

Because of their more or less continual movements, widely foraging Kalahari lizards expose themselves and tend to be highly visible to predators. As a result, they are more vulnerable to attack than sit-and-wait foragers, which are better protected by camouflage. Widely foraging species fall prey to lizard predators that hunt by ambush, whereas sit-and-wait lizard species tend to be eaten by predators that forage actively, generating "crossovers" in foraging modes between trophic levels. Widely foraging lizard species are also more streamlined and have longer tails than sit-and-wait species, features that presumably enhance escape ability (Huey and Pianka 1981).

Even more striking are shifts in foraging mode within species. When winds are blowing, the Namibian lacertid *Meroles anchiete* sits and waits for seeds and insects to blow by, but during calm periods the same individual forages actively, searching out more sedentary insect prey items.

Another important spinoff of foraging mode involves reproductive tactics. Clutch volumes and often number of eggs of widely foraging species are smaller than those of sit-and-wait species, probably because the former can less afford to weigh themselves down with eggs as can the latter (Vitt and Congdon 1978). Hence foraging style constrains reproductive prospects (as well as vice versa). Huey and Pianka (1981) summarize many of these ecological correlates of foraging mode.

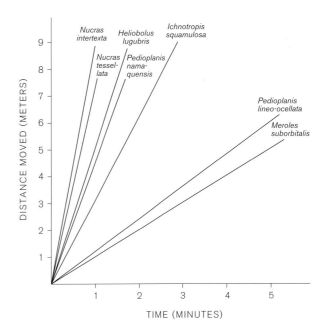

Figure 3.3 Sit-and-wait foraging versus active foraging. Variation in movement rates among lacertid lizards demonstrate that the dichotomy between sit-and-wait and wide foraging is not necessarily dramatic. Nevertheless, *Pedioplanis lineo-ocellata* and *Meroles suborbitalis* move considerably more than iguanian sit-and-wait foragers. (Redrawn from Pianka 1986 by permission of Princeton University Press)

SPECIALISTS VERSUS GENERALISTS

Some lizards are dietary specialists, whereas others are generalists. Each mode has advantages. Among dietary specialists, body size does not appear to constrain prey size to as large an extent as in dietary generalists. This is particularly true for specialists that eat social insects like ants and termites. Among nonspecialist lizards that swallow their prey whole, in contrast, body size does limit sizes of prey (fig. 3.4): small lizards simply cannot eat large prey that larger lizards eat (though larger individuals do also eat small prey). Extremes in relative prey size are seen in the pygopodid *Lialis,* which swallows other lizards whole; helodermatids, which swallow mammals and birds; and varanids, which both swallow large prey and maim even larger prey, returning to dismember them later. In the Komodo example at the beginning of this chapter, the boar was eaten on the spot. With larger prey, Komodos sheer through muscle of the victim's leg, whereupon it runs off and bleeds to death. When the carcass begins to rot, the smell attracts the lizards, which then dismember the animal and swallow large pieces.

Dietary specialization can also reflect superabundant localized distribution of prey—a concentrated food supply. For example, the Australian agamid *Moloch horridus* eats essentially nothing but ants, mostly of a single species of *Iridomyrmex.* Convergent North American horned lizards, genus *Phrynosoma,* and Amazonian *Plica umbra, P. plica, Uracentron flaviceps,* and *U. azureum* specialize on ants that are small compared to the lizards' head and gape size. Even though they could eat larger prey, they usually do not. Perhaps wide heads facilitate rapid tongue extrusion. The Kalahari lizards *Heliobolus lugubris* and *Typhlosaurus,* some diurnal Australian *Ctenotus* species, and the Australian nocturnal geckos *Diplodactylus conspicillatus* and *Rhynchoedura ornata* consume virtually nothing but termites, even though other species in the same habitat never eat termites as prey. Food specialization on ants and termites is economically feasible because these social insects normally are clumped and hence constitute a concentrated food supply.

Other lizard species, while not so specialized, also have narrow diets. For example, both the Kalahari lacertid *Nucras tessellata* and the Australian pygopodid *Pygopus nigriceps* consume many more scorpions than other lizard species (Pianka 1986). *Nucras* forages widely by day to capture these large arachnids in their diurnal retreats, whereas the nocturnal *Pygopus* sits and waits for scorpions and large spiders to move by at night. In North America, a small sand-swimming snake, *Chionactis occipitalis,* has usurped this ecological role. Evolution of dietary reliance on scorpions has been facilitated by the fact that these prey items, though solitary, are extremely large and nutritious. For similar reasons, specialization on other lizards as food items has evolved in North American leopard lizards *(Gambelia wislizeni),* a South American teiid *(Callopistes),* as well as most Australian varanids.

Dietary specialization is often associated with specialized living habits, and in many cases convergence in diets reflects convergence in microhabitat specialization and morphology. Lizards in several scleroglossan families, for example, have evolved elongate bodies, limb reduction or loss, countersunk lower jaws, and additional morphological characteristics to take advantage of total or nearly total subterranean habits (table 3.2). Life in loose soil provides easy access to subterranean invertebrates. As a result, most fossorial lizards have specialized on termites, a prey type that dominates the underground world in many habitats, particularly deserts and tropics.

Less specialized lizard species have much more catholic diets, eating a considerably wider variety of foods. Dietary niche breadth also varies within species from time to time and place to place as dietary composition changes in response to fluctuating prey abundances and availabilities (Pianka 1970b, 1986). Generally speaking, however, lizard diets are remarkably consistent, suggesting a profound impact of microhabitat utilization, foraging mode, as well as various anatomical and behavioral constraints imposed by phylogeny.

FEAST OR FAMINE?

Diverse issues in ecology, physiology, and evolution are concerned with energy balance. To determine whether lizards are generally in positive energy balance, we explored a massive data set (N = 18,223), representing 127 species of lizards from nine families distributed on four continents, primarily in Temperate Zone deserts but also in the Neotropics (Huey et al. 2001), assessing the proportion of individual lizards with empty stomachs. Across all species, average percentage of individuals with empty stomachs was low (13.2 percent), even among desert lizards, suggesting that most lizards are in positive energy balance most of the time. Species varied substantially in the percentage of individuals with empty stomachs, from

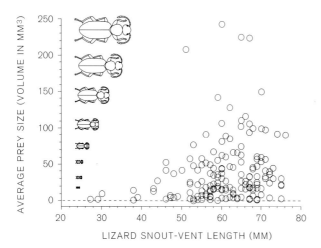

Figure 3.4 Prey size versus lizard size in the teiid *Cnemidophorus lemniscatus.* In many lizards, especially those that eat a variety of prey types such as *C. lemniscatus,* prey size and lizard size are correlated. However, the correlation rests on the fact that larger lizards eat some larger prey, while smaller lizards do not. Large lizards continue to eat small prey as well.

TABLE 3.2

Lizards that have converged on morphology (elongation), microhabitats (subterranean, living in loose soil such as sand), and diet (feeding predominantly on termites)

GENUS	FAMILY	GEOGRAPHIC REGION
Acontius (all species)	Scincidae	Southern Africa
Anomalopus (all species)	Scincidae	Australia
Larutia (all species)	Scincidae	Southeast Asia
Lerista (many species)	Scincidae	Australia
Neoseps reynoldsi	Scincidae	Florida
Ophioscincus (all species)	Scincidae	Australia
*Scelotes (all species)	Scincidae	Southern Africa
*Typhlacontias ngamiensis	Scincidae	Southern Africa
Typhlosaurus (all species)	Scincidae	Southern Africa
*Bachia (some species)	Gymnophthalmidae	South America
Calyptommatus (all species)	Gymnophthalmidae	Northeastern Brazil
Anniella (all species)	Anguidae	Southern California, Baja California
Aprasia (all species)	Pygopodidae	Southern Australia

Sources: Chan-Ard et al. 1999; Cogger 1992; Rodrigues 1991; FitzSimons 1943.

*Although dietary studies do not exist for these genera, termites likely form the diet of some species.

o to 66 percent. Among species with unusually high frequencies of empty stomachs, several patterns were detectable. Nocturnal lizards, on average, "run on empty" more often (24.1 percent) than do diurnal species (10.5 percent). This pattern holds even for nocturnal geckos (21.2 percent) versus diurnal geckos (7.2 percent). Top predators generally, but not always, have empty stomachs more often than do species that feed at lower trophic levels.

Dietary niche breadth and body size are not related to frequency of empty stomachs. Relative to sit-and-wait species, widely foraging species sometimes have a higher frequency of empty stomachs, but patterns vary among continents and are confounded by phylogeny and trophic level. Ant-eating and diurnal termite specialists display uniformly low frequencies of empty stomachs, but nocturnal termite specialists exhibit very high frequencies. Lizards from certain families—Agamidae, Iguanidae, Lacertidae, Scincidae, Teiidae, Gymnophthalmidae, and Varanidae—are less likely to have empty stomachs than are those of other families (gekkotans).

Herbivory in iguanines, scincids, teiids, agamids, gerrhosaurids, and xantusiids raises a number of interesting questions about evolution of lizard diets. Why would any lizard choose low-energy plant material over high-energy animal material, especially when so many plants contain toxic secondary compounds? In some instances, the answer is simple: animal food is rare, either temporally or spatially. Thus the xantusiid *Lepidophyma smithi,* which lives in caves with low invertebrate availability, feeds on energy-rich figs that fall into the caves, just as the gerrhosaurid *Angolosaurus skoogi* switches to high-energy seeds when insects become unavailable. Nevertheless, although there should be no cost to eating digestible plant parts lacking or low in cellulose, especially if they are energy rich like fruits and some flowers, this does not explain the switch to herbivory among species that eat leaves. Processing leaves—which, because they contain cellulose, cannot be broken down by lizard digestive systems—requires an appropriate gut fauna, intake of large amounts of plant material, and maintenance of high enough body temperature to digest it. Essentially, a fermentation chamber is a requisite. Iguanines, two genera of agamids *(Uromastyx* and *Hydrosaurus),* and the large-bodied Solomon Island skink *Corucia zebrata* maintain symbiotic gut faunas that break down cellulose, making it available to the lizard (Pough et al. 1998).

Among scleroglossans that have switched to plant foods (fruits, flowers, and leaves), the transition ought to have been relatively easy evolutionarily because a well-developed chemosensory system for prey discrimination was already in place. Just as scleroglossans appear to avoid ants, many of which produce alkaloids and other noxious chemicals, scleroglossans should have no problem discriminating "good" and "bad" plants. Experimental studies by Joe Schall and his students at the University of Vermont clearly demonstrate this ability in endemic island species of *Cnemidophorus* (Schall 1990; Schall and Ressel 1991).

Within Iguania, herbivory has evolved several times, once in the ancestor of Iguaninae, once in the ancestor of *Leiolepidinae,* and once in the ancestor of *Hydrosaurus.* None of these are in clades in which chemosensory discrimination of prey occurs in nonherbivorous species. Consequently, evolution of chemosensory prey discrimination in herbivorous lizards may well have occurred along with the switch to herbivory and acquisition of the symbiotic gut fauna. Exactly what drove this remains a mystery, but herbivory has one clear benefit: in many environments where temperatures are warm, plants appear to be a virtually unlimited resource. Even in deserts with low productivity, plants are unlimited for short time periods, and lizards are capable of long-term fasting. So the impetus may simply have been the superabundance of plants relative to lizard abundance.

For chuckwallas *(Sauromalus),* the entire desert floor is a virtual banquet during good seasons, though they typically do not move far from rocks to forage. Their restriction to rocks and rock crevices likely reflects a high risk associated with harvesting plants far from the boulder fields where they live, and the same is likely true for some *Uromastyx.*

How did lizards acquire the gut microfauna necessary for the breakdown of cellulose? One possibility is that omnivorous lizards, while feeding on fruits, also feed on insects feeding on feces of herbivorous animals that contain gut microorganisms capable of living in lizard digestive tracts. The fact that desert iguanas feed on feces of herbivorous mammals, particularly in spring (Norris 1953), is suggestive. To explore this possibility further, *Dipsosaurus* endosymbionts should be compared with those of mammals.

MAKING DIETARY CHOICES

Without doubt, prey availability, past or present, influences what lizards eat. Prey availability is not simply total biomass or number of invertebrates in a particular environment. Rather, it is what a lizard actually has access to. A lizard has to find, capture, subdue, and eat each prey item. An environment with a high density of subterranean insect larvae might have high insect biomass, but none of these prey items would be available to a strict sit-and-wait predator. Prey availability must therefore be considered in the context of the unique ecology and behavior of each species. Given this constraint, and using what has become popularized as "optimal foraging theory," some interesting predictions can be made about how lizards should respond to variation in prey availability.

In an environment with a scant food supply, because expectation of prey encounter is low and mean search time per item encountered long, a consumer cannot afford to bypass many inferior prey items (MacArthur and Pianka 1966). In such an environment, a broad diet maximizes returns per unit of expenditure, favoring generalization. In a food-rich environment, however, a foraging animal encounters numerous potential prey items and search time per item is low. Here, substandard prey can be bypassed economically because expectation of finding a superior item in the near future is high. Hence, rich food supplies favor selective foraging and lead to selection of a few relatively profitable prey items. Because food is one of the three primary niche dimensions (food, time, and place) in lizards (Pianka 1986, 1993), specialization would result in narrow food niche breadths (indicating the relative use of available resources). These arguments are supported by the North American teiid lizard *Cnemidophorus tigris,* which eats a greater diversity of foods in drier-than-average years (presumably times of low food availability) but, like most lizards, contracts its diet during periods of prey abundance (Pianka 1970a, 1986).

A more extreme example occurs in deserts after heavy summer rains when termites send out millions of alates and virtually every species of lizard eats them exclusively—even lizard species that normally never consume termites. During such fleeting moments of extraordinary prey abundance, competition for food is negligible and dietary overlap among lizards can be nearly complete.

Lizards also make choices about the size of prey they eat. Most lizards eat prey considerably smaller than themselves and swallow them whole. Some, however, eat extraordinarily large prey. How is this possible? Large varanids have evolved serrated teeth, which allow them to cut and tear large prey such as pigs and deer into smaller pieces for ingestion. Deer and pigs are not native to the Sunda Shelf islands of Indonesia where Komodo

In all fields of science, a handful of people with extraordinary insight set the direction of the field for many generations. Robert H. MacArthur was the twentieth-century visionary in ecology, bringing what had been largely a descriptive, mechanistic science into the evolutionary arena.

I was a graduate student at the University of Washington during the early 1960s, an exciting time in evolutionary ecology. Robert MacArthur held center stage then, regularly offering intriguing new ideas. Each time a MacArthur paper appeared, everyone read it with great interest and we discussed his brilliant ideas. I wrote to him and asked if I could do postdoctoral work with him. He accepted me, and I applied for and received a three-year National Institutes of Health (NIH) postdoctoral fellowship. At the time, MacArthur was at the University of Pennsylvania, but shortly before I was to leave Seattle, Princeton lured him away with the offer of a house within walking distance of his office.

When I arrived at Princeton in 1965, I chanced into an ideal situation: Robert's graduate students had stayed behind at Pennsylvania! This brilliant scientist was virtually without any colleagues, extremely approachable, and eager for interaction and intellectual stimulation. Immediately we began to discuss his newest idea—then just a germ—on costs and benefits of various foraging activities. Robert had identified search time per item eaten and pursuit time per item eaten as currencies. Although it is difficult to believe now, foraging theory did not exist then.

The speed with which MacArthur's mind worked, as well as its clarity, was dazzling. Never before had I encountered true genius. It was exhilarating to be part of the two-man brainstorming effort that ensued during the fall and winter of 1965–1966, but it was also a humbling experience. Each evening I went home determined to think of something really neat, but precious little came. In the end, other than acting as a sounding board for MacArthur's fine mind, my major contribution was to propose and outline a table summarizing our results. MacArthur's generosity in making me a coauthor of the resulting paper, "On Optimal Use of a Patchy Environment" (1966), was typical of his dealings with lesser scientists. Quite simply, I was exceedingly fortunate to be in the right place at the right time.

We developed a graphical model of animal feeding activities based on costs versus profits, outlining a forager's optimal diet. Interesting predictions emerged. Because prey abundance affects search time per item eaten, it influences the degree to which a consumer can afford to be selective. Diets should therefore be broad when prey are scarce (long search times), but narrow if food is abundant (short search times), since a consumer can afford to bypass inferior prey only when there is a reasonably high probability of encountering a superior item in the time it would have taken to capture and handle the previous one. As simple and intuitively obvious as these ideas are, this was the first time they had been proposed to ecologists.

Our paper and J. M. Emlen's (1966), published back to back, ushered in the concept of optimal foraging, which has blossomed greatly. Behavioral ecologists embraced foraging theory because it confers rigor and generates testable predictions in an otherwise fairly subjective field. Although optimality models have been savagely attacked, they remain one of the most powerful approaches to adaptation currently available. The theorem that diets contract when food is abundant and expand when food is scarce has proven to be exceedingly robust and now constitutes a basic tenet of evolutionary ecology.

Robert MacArthur's life was cut short by cancer, and his death at age forty-two, in 1972, left a huge void in the field of ecology. (PIANKA)

---

monitors live, but only a few thousand years ago this region supported pygmy elephants, which have since gone extinct. Natural selection could well have favored evolution of large size in *Varanus komodoensis* in response to the availability of such large prey (Auffenberg 1981; Diamond 1987). Fortunately, these spectacular gigantic lizards did not die out when pygmy elephants did, but managed to survive on alternate prey such as megapode birds until humans introduced deer and pigs. Although *V. komodoensis,* like most monitor lizards, forages actively when hunting for small prey such as rodents and snakes, it has secondarily evolved an ambush tactic—yet another adaptation enabling it to capture the large prey on which it once again relies.

Biologically significant variation in utilization of certain relatively minor food categories exists among lizard species as well. For example, diets of climbing species include more hemipterans-homopterans and mantids-phasmids as well as various flying insects (wasps, dipterans, and lepidopterans) than do those of terrestrial lizards. Likewise, geckos consume more nocturnal arthropods (scorpions, crickets, roaches, and moths) than most diurnal species (though some diurnal lizards do capture nocturnal prey in their diurnal retreats). Such prey items are thus indicators of spatial and temporal patterns of activity.

Sensory capabilities of lizard species play important roles in dietary decisions. Visually oriented iguanians can be expected to eat different prey than chemosensory-oriented scleroglossans. Moreover, within scleroglossans, gekkotans might be expected to take different prey because of their reliance on a combination of visual signals and probably nasal olfaction, compared with autarchoglossans, which rely on a combination of visual and vomeronasal signals. Studies on use of airborne chemical cues for prey detection are in their infancy, but observations on Komodo dragons finding carrion (Auffenberg 1981) and lacertids locating fruit (Cooper, pers. comm.) indicate that these capabilities are important in some species as well.

CHAPTER 4

ESCAPING PREDATORS

"Se vende lagartos" (Lizards for sale) is a commonly heard phrase in southern Mexico and Central America, where iguanas and spiny-tailed iguanas are sold for food. (David M. Hillis)

Evolutionary interactions between lizards and their predators can be thought of as a never-ending arms race. Because being faster and more wary should result in higher survival rates than being slow and less wary, natural selection has led to traits that enable successful escape. On the opposing side, because ability to capture meaty prey such as lizards allows energy to be converted into offspring faster, selection has led to ever more successful predators. If lizards as prey cannot stay ahead in this "race," they go extinct.

Selection is strong on prey because failing to escape is certain death: there is no genetic future for a captured lizard. Selection on predators is also strong, but consequences of failing to capture a particular prey item are less severe: a predator doesn't die, it simply misses a meal. It takes a long time to translate lost meals into a reduction in future genetic success (called "fitness" by biologists), whereas lost life reduces fitness immediately. Within the context of this race, population characteristics—morphology, behavior, ecology, and life history traits—reflect the remarkable impact predation has had in molding lizard biology.

In nature, lizard populations remain fairly stable, their size heavily influenced by the absolute amount of resources available (fig. 4.1). If population size is low relative to resources, growth occurs exponentially, leveling off when resources become limiting. Most populations do not reach equilibrium levels based on resources alone, however. Finding and capturing food requires activity, and to capture prey, lizards expose themselves to predators. In other words, resource acquisition is balanced against the risk of acquiring resources. Predation and competition usually result in populations remaining well below their carrying capacity.

Predation can occur during any life history stage, but usually is heaviest on eggs and juveniles. Lizard predators include birds, mammals, snakes, other lizards, fish, spiders, centipedes, scorpions, tailless whip scorpions (amblypygids), and insects, such as praying mantises and ambush bugs (reduviids). The set of potential predators is limited by lizard size relative to predator size. Adult Komodo dragons, for example, have no natural predators (humans being a recent addition to the fauna of the Lesser Sunda Islands). Birds—which because of high metabolic demands associated with heat production eat large amounts of food compared to reptiles and amphibians—are the most important predators on moderate- to small-sized diurnal lizards during periods of activity. During periods of inactivity, mammals (which also have big appetites due to endothermy) and snakes are the most likely predators. A summary of predation data in Mediterranean habitats across the world, for example, identified 21 species of birds, 5 mammals, and 12 snakes feeding on small reptiles (mostly lizards) (Jaksíc et al. 1982). Surprisingly few studies have examined predation on lizards by other vertebrates. Invertebrates, particularly spiders, undoubtedly take large numbers of juvenile lizards and adults of some very small lizard species (e.g., Bauer 1990).

Field observations on predation events, though relatively scarce, provide valuable insight into how lizards deal with predators. The most commonly observed predation events include various species of cuckoos, especially roadrunners, with lizards dangling from their beaks. Similarly, horned lizards, geckos, and other species impaled on agave spines attest to avian (shrikes') abilities at capturing lizards.

The impact of a history of predation on lizards is reflected by a remarkable diversity of predator escape mechanisms. They include, but are not limited to, cryptic coloration; armor; alertness and high running speeds; mimicry; scratching, biting, and aggressive displays; tail autotomy; skin loss; and even bad-tasting blood. Among the subtler escape mechanisms are synchronous hatch-

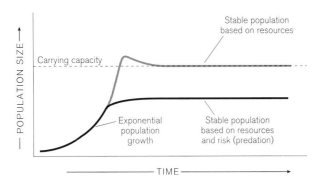

Figure 4.1 Resources, risk, and population growth. In one way or another, growing populations of lizards are always limited by either absolute resource levels or predation.

Tropical vine snakes such as this Amazonian *Oxybelis aeneus* capture geckos *(Gonatodes humeralis)* from their arboreal perches. (William E. Magnusson)

Although "flying" snakes *(Chrysopelea)* of southern Asia do not really fly, they are agile climbers and often parachute from higher to lower branches in pursuit of lizard prey, in this case an agamid *(Calotes).* (S.M.A. Rashid)

Shrikes frequently "crucify" lizards like this Kalahari barking gecko *(Ptenopus garrulus),* returning later to eat them. (Eric Pianka)

ing of eggs and evolution of large clutches of many small offspring in the face of high nonselective juvenile mortality. Some mechanisms, indeed, such as the roles of chemicals and ultraviolet light reception, remain largely unexplored and may be so subtle that they will never be fully understood.

For a predation event to occur, a predator must first detect its prey—let's say, a lizard. This requires sorting out sensory information attributable to the lizard from the sensory signals surrounding it, thus allowing identification. Once a lizard is identified as a potential prey item, the predator must approach, capture, kill, and ingest it. Different predator escape mechanisms come into play at each phase of a predation event (fig. 4.2), and likely approach or exceed the number of living lizard species. Not only do lizard species use a variety of predator escape tactics, these tactics can change with a lizard's age. What works for juveniles may not work for adults. Geographic, particularly latitudinal, variation in predator diversity makes this story even more interesting.

AVOIDING DETECTION

Avoiding detection by predators sometimes involves shifting activity to times when risks are relatively low. During periods of activity, a combination of cryptic col-

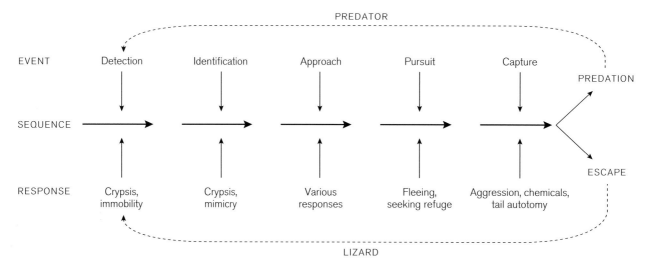

Figure 4.2 Sequence of events in predation. The "evolutionary arms race" between predators and prey results from natural selection operating on (a) the ability of prey to escape predation at any stage of a predation event and (b) a predator's ability to detect, identify, pursue, and capture the prey. Successful predators and prey live to pass on genes that contribute toward successful behaviors.

In the Sonoran Desert of the southwest United States, coachwhips *(Masticophis flagellum)* are major predators of lizards such as this unfortunate desert iguana *(Dipsosaurus dorsalis)*. (Jeff Howland)

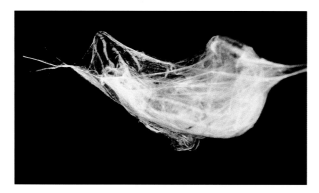

Juvenile lizards like this *Eumeces fasciatus* are subject to predation by many invertebrates, especially spiders. (Laurie Vitt)

oration or morphology and immobility can render a lizard nearly invisible to a predator. The Amazonian lizard *Enyalius leechii,* for example, is active by day on leaf litter or on low vegetation in undisturbed forest. Its color pattern is disruptive such that the lizard's outline is obscured against complex backgrounds. The set of colors making up the disruptive pattern represents a subset of colors of leaves on the forest floor. When approached, this lizard remains absolutely motionless, becoming nearly indistinguishable against the leaf litter. *E. leechii* is found from the eastern Amazon nearly to Bolivia, a linear distance of about 1600 km, yet only about twenty individuals have ever been found!

Cryptic coloration or morphology occurs to some extent in most lizard species. The predominant mechanism of predator avoidance in iguanians and gekkotans is crypsis to avoid detection at the beginning of a predation sequence. Because most species in these taxa are sit-and-wait foragers that remain relatively immobile while waiting for prey to pass by, their presence is not revealed due to movement. Most autarchoglossans, in contrast, are active foragers that move from place to place searching for prey. Movement alone reduces the efficiency of crypsis. Autarchoglossans offset this apparent cost to crypsis with their alert behavior, fast response time, and high running speeds.

If lizard coloration we observe in a particular habitat represents an evolutionary response to predation selecting for crypsis, it would seem that lizard color patterns should vary geographically within a species. Evidence

In leaf litter as well as on branches in low vegetation, the Amazonian leiosaurine *Enyalius leechii* is nearly invisible as long as it doesn't move. (Laurie Vitt)

from a variety of lizard species suggests that this is indeed the case. The short-horned lizard, *Phrynosoma douglassi,* for instance, occurs in a variety of habitats in mid to high elevations from the Sonoran Desert to the western and northern reaches of the Great Basin Desert and beyond. In virtually every place this lizard is found, its dorsal coloration is nearly identical to that of the local sand or soil. The same is true for many other species of lizards. Side-blotched lizards are black on black-colored volcanic rock in the Mojave Desert, while several other species, including *Sceloporus undulatus, Holbrookia maculata,* and *Cnemidophorus inornatus,* are nearly white on the gypsum sands in New Mexico, even though populations in other nearby habitats are colored very differently.

Many lizard habitats contain repeatable structural components that some lizards mimic. The North American round-tailed horned lizard, *Phrynosoma modestum,* and the Australian earless dragon, *Tympanocryptis cephalus,* combine a stout morphology, skin with architecture like a rock surface, and motionless posturing to mimic small rocks scattered over the surface of their desert habitats (Sherbrooke and Montanucci 1988). Various arboreal lizards, including the South American grass anole *Anolis auratus,* the South American gymnophthalmid *Prionodactylus argulus,* and the North American long-tailed brush lizard *Urosaurus graciosus,* remain perfectly still on thin branches that they match both in morphology (long, thin bodies with long tails) and color.

INTERFERING WITH IDENTIFICATION

Once detected, a lizard can still deceive a predator prior to an actual attack. Running a short distance, stopping quickly, and remaining immobile is one option. Running catches the predator's attention, but a quick stop combined with crypsis can make a lizard appear to disappear again,

Shaped like a pebble with a tail different from the body, this Australian agamid, *Tympanocrytis cephalus*, mimics rocks.

especially if a predator is tracking its motion. An effective alternative is to mimic a potentially dangerous organism. This tactic, known as Batesian mimicry, involves a model, a mimic, and a dupe. The model is an organism that is potentially dangerous to the predator and advertises that danger by bright coloration or specific threatening behavior. A mimic is usually not dangerous, but looks or behaves like the dangerous model. The dupe is the predator who is fooled by the mimic. Few mimicry systems have been described in lizards, and the best-supported ones involve mimicry of noxious invertebrates.

In the Kalahari Desert of southern Africa, defenseless juvenile lacertid lizards *(Heliobolus lugubris)* mimic noxious "oogpister" beetles (the Afrikaans translates euphemistically as "eye squirter"), which emit pungent acids, aldehydes, and other chemicals when disturbed (Huey and Pianka 1977a). In stark contrast to adult *H. lugubris,* which are buff-colored and pale red, matching the color of Kalahari sands, juveniles have jet black bodies with white spots (though their tails are red, matching sand color). Moreover, whereas adults walk with a normal

A combination of spines, spots, and color render *Phrynosoma mcalli* well camouflaged against sandy desert substrate. (Paul J. Gier)

tetrapod lizard gait, their backs undulating from side to side, juveniles walk stiff-legged, with backs arched vertically as they hold their reddish tails flat against the ground (making the tail difficult to detect). When pursued, young *H. lugubris* abandon their "beetle walk" and dart rapidly for cover. At about 45–50 mm snout-vent length (SVL) (the size of the largest oogpister beetles), these lizards "metamorphose" into cryptic adult coloration and permanently abandon their stilt walk. Frequency of broken and regenerated tails is lower in juvenile *H. lugubris* than among juveniles of closely related lacertids in the same habitats exposed to common pred-

ators, suggesting that beetle mimicry reduces predatory attacks. Once, while collecting beetles to measure their size, Pianka stepped on a juvenile lizard thinking it was a noxious beetle.

In semiarid caatinga in northeastern Brazil, a poverty-stricken area renowned for its extended droughts, the moderate-sized lizard *Diploglossus lessonae* occurs among and under rocks on boulder-strewn hillsides. As adults, they are rather dull colored dorsally but orange red underneath. Adults inflict impressive bites, so red coloration could warn some predators that they are dangerous. Juveniles, however, are banded with black and white, a pat-

During March 1998 I was working in undisturbed forest at the Parque Estadual Guajará-Mirim in western Rondônia, Brazil, when I observed a spectacular predation event. As I walked along a trail in the forest, I noticed what appeared to be a small branch on a shrub moving back and forth. The reason it caught my eye is that the air was dead calm in the forest: nothing else was moving. As I slowly approached the swaying "vine," it lunged forward, picked a small lizard off of a vine just below, and returned to its original position, discontinuing the swaying movement. I had just seen a vine snake, *Xenoxybelis argenteus,* capture a relatively rare gymnophthalmid, *Prionodactylus argulus!* Although this particular snake is often quite common, finding one during

the day is nearly impossible because they imitate vines so perfectly. At night, however, they can be found sleeping, coiled on top of shrubs—a position that makes them easy to see. *Prionodactylus argulus* are small, thin-bodied lizards with very long tails. Among gymnophthalmids I have studied in Brazil, these are the most difficult to find because they are cryptic, wary, and tend to live in shrubs and on vines. No doubt, they are more common than simple observations would lead one to believe. Stomachs of two species of *Xenoxybelis* commonly contain *P. argulus*. As the snake approaches the lizard prior to striking, its tongue is straight out, with no detectable movement. The swaying motion I observed may have represented the snake's use of multiple ob-

servation points to accurately determine distance of the prey, optimizing binocular vision in a species with eyes placed laterally on the head. Grooves along each side of the head likely aid as well, much like the sight on a rifle. The swaying movement may have been cryptic in purpose, intended to reduce the probability that the lizard would detect the snake (Fleishman 1992). Or it may have been meant to distract the lizard momentarily, causing it to discontinue forward movement in an attempt to avoid visual detection by a potential predator. The visually and chemically oriented snake, however, was not fooled by the lizard's crypsis. It struck accurately, grabbed the lizard in its mouth, and envenomated it with its rear fangs. *(VITT)*

The Amazonian gymnophthalmid *Prionodactylus argulus* is nearly impossible to detect when stretched out on a vine or twig. (Laurie Vitt)

ESCAPING PREDATORS

tern utterly out of place in a habitat with very little structure to match. The explanation comes with the arrival of the wet season, when a moderate-sized millipede, *Rhinocricus albidolimbatus,* becomes superabundant on the surface and under rocks. This millipede produces a wide variety of noxious chemicals, including trans-2-dodecenal, 2-methyl-1, 4-benzoquinone, and 2-methyl-3-methoxy-1, 4-benzoquinone, which are toxic (Eisner et al. 1978). Eggs of *D. lessonae* also hatch early in the wet season. Not only do the millipedes and juvenile lizards occur together in the same microhabitats, but they are also the same size, and even the cruising gait of young lizards is similar to that of millipedes. As the lizards grow, the wet season slowly comes to an end and millipedes disappear. Coloration of juvenile lizards then changes to a light banding pattern, followed by the drab, nonbanded pattern of adults. A similar mimicry system occurs in a close relative, *D. fasciatus,* in southern Brazil, with juvenile lizards resembling the local millipede with its yellow and black banding. Millipedes are among the oldest of all terrestrial animals and have occurred throughout the entire evolutionary history of lizards, providing ample time for mimicry systems to evolve (Vitt 1992b).

No doubt, other lizard mimicry systems exist. The banded pattern of *Eremiascincus richardsoni,* for example, closely matches the banding pattern of the large centipede *Scolopendra mortisans* in the Great Victorian Desert of Australia. The African lacertid *Holaspis guentheri* sleeps on leaves with its body hidden under a leaf and its brightly banded tail exposed in a coiled posture resembling a millipede. Southeast Asian geckos (*Ptychozoon*) also coil their tails, which strongly resemble an armored millipede. Carrying the tail above the body while slowly cruising along the desert floor may represent mimicry of scorpions in the geckos *Coleonyx, Eublepharis,* and *Teratoscincus roborowski,* in addition to serving as a distraction (Autumn and Han 1989).

An Australian flap-footed lizard, *Pygopus nigriceps,* has a black head and brown body very reminiscent of the coloration of juveniles of a sympatric venomous elapid snake, *Pseudonaja nuchalis.* When threatened, *Pygopus* behave like these snakes, rearing up with inflated necks, waving their heads menacingly, hissing and lunging as if they are a threat. Only when they stick out their short blunt geckolike tongues do they give away their true identity. In some pygopodids, even autotomized tails behave like snakes (Greer 1991).

RESPONDING TO APPROACH

When a predator initiates approach, no matter how subtle that approach may appear, a lizard must be prepared to take evasive action. One alternative to simply darting off, which may result in pursuit, is to signal awareness of the predator's presence. This provides advantages to both predator and prey in terms of time and energy saved. From a predator's point of view, a lizard that has already detected it is much more likely to escape, rendering likelihood of success low. Directing energy toward pursuit of alternative prey makes sense. From the lizard's perspective, of course, convincing a predator to move on negates the need to flee as well as immediate chances of being eaten. Such behaviors fall into a general category called "pursuit deterrence."

Waving, upturned, and vibrating tails, short-distance movements, and even performance of portions of social displays can function in pursuit deterrence. When approached or followed by a large animal, the zebra-tailed lizard, *Callisaurus draconoides,* curls its tail up over its hindquarters and back, exposing a bold black and white pattern underneath, and coyly wriggles its tail from side to side. In this case, the signal might be "Here I am, I see you, and you can't possibly catch me." If pursued farther, zebra-tailed lizards resort to extreme speed (estimated at up to 20–30 km/hr) and long zigzag runs (Dial 1986; Hasson et al. 1989). Similar behavior occurs in the closely related genus *Cophosaurus* (Dial 1986). An Australian

A large mantid eats a skink in Australia. (Steve Wilson)

desert skink, *Ctenotus calurus,* lashes and quivers its bright azure blue tail alongside its body continuously as it forages slowly through open spaces between plants. Similarly, tiny *Morethia butleri* juveniles twitch their bright red tails as they move around in litter beneath *Eucalyptus* trees, and *Leiocephalus carinatus* curl their tails over their backs when approached. *Anolis cristatellus* perform social displays when a predator is at the base of their tree, and these displays intensify as the predator moves closer (Leal and Rodriguez-Robles 1997). Explanations for these behaviors not involving pursuit deterrence are not supported (Cooper 2000).

When predators approach or initiate an attack, lizards often respond with threatening behavior, which can vary from simply filling their lungs and expanding their body to actual open-mouth lunges at a predator. Many lizards, such as Australian blue-tongue skinks or the South American *Enyalioides palpebralis,* open their mouths to expose a brilliantly colored interior. Presumably the color warns a predator that the lizard can inflict a significant bite, reducing potential payoff to the predator as a result of increased risk. Others, such as Gila monsters, Mexican beaded lizards, and many large varanid lizards, not only perform open-mouth threats but also bite when attacked. Gila monsters and Mexican beaded lizards are venomous

Open-mouth displays, as in this Amazonian hoplocercine, *Enyalioides palpebralis,* present the impression that a lizard is much more threatening than it actually is. (Laurie Vitt)

and can inflict life-threatening wounds. Many varanid lizards are effectively venomous due to pathogenic bacteria in their mouths.

In northern Australia, the frilled dragon, *Chlamydosaurus kingii,* erects a large fold of skin while opening its mouth when disturbed, giving the false impression that it is a much larger lizard. Bearded dragons, *Pogona barba-*

20/20 HINDSIGHT

In early February 1991 I was bitten by a medium-sized (1.3-m) perentie *(Varanus giganteus)* while helping Daniel Bennett photograph it. Daniel wanted to get a picture of the lizard on the ground while I stood by watching. The lizard, though, dashed behind me and began to climb up the back of my leg. I had visions of it reaching my head and scratching my face with its large, sharp claws. Reflexively, I reached my left arm behind my back to try to get the lizard off. It chomped down on my wrist with its 6–8-mm serrated cutting teeth. Daniel remained motionless as I began to bleed profusely. Then I reached around with my right arm to try to extricate the big lizard, whereupon it clamped down on my right thumb and began to

chew. It hurt! By the time Daniel came to my rescue, I was slashed up and bleeding like a stuck pig. Fortunately, no tendons were severed.

I washed the bites in alcohol and bandaged them to keep the flies out. They healed over, and I thought I had recovered. But a few days later, I was bitten again on the same poor thumb by a much smaller *Varanus tristis* (teeth only about 3–4 mm), which also chewed. My thumb became infected, and then the left wrist went sympathetic and got infected, too. Luckily, we had some antibiotics, and after I had taken them for a couple of days the swelling went down. Again I thought I was going to be okay.

A month later, however, my thumb

swelled up again to twice its normal size. Apparently, blood-borne antibiotics do not easily reach poorly vascularized synovial joint capsules. Five prescriptions of several different antibiotics later, the infection was still not completely cured. (Aussies and Aborigines say that a *Varanus* bite never heals: since they feed on carrion, their mouths harbor an impressive bacterial community.)

There are two morals to this story: (1) don't stand still around a monitor lizard, and (2) if you should ever be so unfortunate as to find a monitor lizard on your backside, whatever you do, don't put your arms behind your back to try to get it off. Instead, lie down on your belly, and let it walk or run away by itself.

(PIANKA)

Australian bearded dragons *(Pogona vitticeps)* perform defensive displays by opening their mouth and expanding a frill on their throat.
(Steve Wilson)

tus, accomplish the same thing by erecting a large flap of skin under the jaw while opening their mouth to expose a bright yellow interior. When cornered, large varanids and teiids inflate their bodies and necks while posturing with the body lifted off the ground, often adding loud hisses. Such displays intimidate predators and prevent attacks.

ESCAPING CAPTURE

Lizards that are sit-and-wait predators normally rely on crypsis for escape from detection and run only short distances to escape predators that approach too closely. Frequently, these short evasive movements are into protective retreats such as under rocks, into crevices, or down burrows. Typically, these lizards can move rapidly but have limited endurance. For example, the Amazonian lizard *Uracentron flaviceps* lives on large limbs of trees, often over water. Males use nests of arboreal termites *(Nasutitermes)* as perch sites, and individuals of both sexes and all ages forage along branches and in portions of the tree canopy. When approached, all individuals quickly disappear into hollow areas within the tree trunks or large limbs. *Tropi-*

durus semitaeniatus, which lives on relatively flat surfaces of rocks in the semiarid caatinga of northeastern Brazil, responds similarly, with all individuals on a rock surface entering tight crevices under cap rocks. Such retreats protect lizards from predation by birds and most large predators. Some species have adaptations that render them difficult to extract from retreats. For example, the chuckwalla, *Sauromalus obesus,* not only enters crevices when approached but then fills its lungs with air such that it becomes wedged between slabs of rock. Chuckwallas can inflate their lungs to four times their normal capacity. The most interesting aspect of defensive inflation is that it does not occur by simple aspiration; rather, the lizard closes its mouth and, with a repeated series of buccal pulses, forces air into its lungs (Deran et al. 1994). Throat movement fills the buccal cavity and movement of the dorsal surface of the throat pushes air into the lungs. When restricted (while inside of a crevice), the maximum pressure averages 260 mm H_2O, double the pressure achieved by nonrestricted buccal pumping. Pulse pumping behavior and subsequent inflation are made more effective by the presence of abrasive scales that increase the lizard's adherence to rock surfaces. Claws also help hold lizards in crevices.

Once lungs are inflated, claws grip, and scales are pressed against both rock surfaces, an adult chuckwalla is nearly impossible to extract from a tight crevice. Native Americans pierced chuckwallas with sharpened sticks to deflate them so that they could be removed and eaten.

Active foraging lizards continually scan the area immediately surrounding them and appear to be aware of all potential threats. When a human approaches, they often move out and away, monitoring the observer's position while continuing to forage at a safe distance. When an observer, or presumably a predator, approaches too closely or too rapidly, lizards may run great distances before resuming foraging behavior. Anyone who has watched a New World teiid or an Old World lacertid or varanid has experienced this behavior. Such lizards seldom seek refuge in retreats because they can easily outrun most terrestrial predators as long as they maintain a distance adequate to allow an appropriate response time.

Some Old World skinks, agamids, varanids, and New World gymnophthalmids and iguanids enter water and swim or run rapidly to the other side or disappear under the surface. Most *Neusticurus* (Gymnophthalmidae), for example, are linearly distributed along streams. When approached, they jump into the water, submerge, and after a few moments poke their heads above the water's surface, scanning the horizon for predators. Others, including basilisks, some anoles, and *Uranoscodon,* run considerable distances across the surface, lifting their head and forelimbs off the water as they run bipedally.

Many lizards escape by simply dropping from perches when approached. This is taken to the extreme by some arboreal lizards that glide or parachute through the air to different perches. The geckos *Ptychozoon* and *Thecadactylus* have webbing between their toes and loose skin along ventrolateral body surfaces that is extended when gliding. Not only can they glide long distances, effectively making it impossible for most predators to follow, but they can control the direction of gliding much like a skydiver. Slight movement of a foot with its miniparachute can result in a substantial change in direction. The most spectacular adaptations for gliding occur in the many species of arboreal Asian *Draco* (Colbert 1967), which jump from arboreal perches and glide considerable distances when disturbed. Unlike geckos discussed above,

RUN FOR YOUR LIFE!

While working along the Río San Juan in Nicaragua in 1993 I became interested in an anole (*Anolis oxylophus*) that is linearly distributed along small forest streams. Although they rely on cryptic coloration to escape detection, when disturbed they jump to the water's surface and run to the opposite shore. My first impression was that the behavior was identical to that of Central American basilisk lizards (*Basiliscus*) or the South American lizard *Uranoscodon,* both of which run across water when approached. In each case the lizard hits the water, lifts its front end while propelling with its hind limbs, and races to the other side. Most anoles I observed had dropped to the ground and were able to enter the water after a running start. Their total time on the water was at most a couple of seconds.

I wondered if they would do as well without a running start, so I captured a few anoles and conducted a crude but, as I discovered, insightful experiment. The first lizard was an adult about 5 cm in SVL. I dropped it in the center of the stream, expecting it to take off running. As the lizard hit the surface, a fish appeared from nowhere and disappeared with the anole in its mouth! A fish predator had *not* been part of my experimental design. I had naively assumed that anoles run across water only to escape terrestrial predators, which of course is partly true, but I had never asked why they run completely across streams. I then repeated the experiment several times and got the same result. When I set anoles on the ground, however, and allowed them to run on the surface, they all made it to the other side. As a result, I had to change my initial idea from "Anole runs to water to es-

cape predator" to "Anole runs to water to escape terrestrial predator and runs rapidly across the surface to the other side to escape fish predators." Over the next several days, I repeated the experiment with a variety of *Anolis* species, none of which could run across the surface without a headstart and all of which were attacked by fish.

As crude as this experiment was, it exemplifies the importance of designing experiments with the natural history of an experimental animal well in hand. Had I brought these anoles into the laboratory, designed a test chamber, and conducted a highly controlled experiment to determine whether the lizards needed to have a running start to get across the stream, I would have totally missed the most interesting part of the story. (*VITT*)

Draco have no membranes between the toes; rather, their ribs are elongated and movable, connected by membranes of skin that form "wings" when ribs are extended. Presumably, gliding is a mechanism to escape predators approaching from within trees, but gliding is also used (more by males) for changing position within their home ranges (Mori and Hikida 1993, 1994). Moreover, sexual color differences on the ventral portion of the "wings," with males typically having the brighter coloration, suggest that they are used in social signaling as well. Females have relatively larger wings than males, which appears to represent an evolutionary mechanism to maintain aerodynamics during flight in females carrying clutches of eggs (Shine et al. 1998).

A TAIL-LASHING EXPERIENCE

I will never forget the first time I attempted to capture a large *Iguana iguana* in the field. It was about 4 m off the ground in a thorny tree in northeastern Brazil. Because it was the first iguana I had ever seen in the field, I simply had to get a closer look—which meant catching it. As I approached the tree it moved out on one of the top limbs, watching me carefully. My first thought was to climb the tree, but I knew that if I did, the iguana would simply jump to the ground and escape. After analyzing the situation for several moments, I decided that I should be able to shake the lizard from the tree and, if I were quick, catch it as it fell. The technique seemed straightforward enough at the time.

When I began shaking the tree, however, the iguana only held on more tightly. I shook and shook, and just as I was about to give up, the iguana made a long jump. I scrambled to get under it and managed grab it about midbody. I was not wearing a shirt. In the few seconds that I held on to the meter-plus iguana, its tail whipped violently over my shoulders and under my arms, smacking me on the back like a bull whip. With its front claws, it scratched both of my arms with several quick motions. I released my hold immediately and the iguana charged off across the ground, disappearing into the thorn forest. I had expected it to try and bite me, and I was ready for that—but it hadn't even opened its mouth! As I walked back to my field lab, arms covered with blood and long welts rising on my back, I realized how well prepared these putatively harmless herbivorous lizards were for encounters with large predators. If its tail or claws had been better aimed, I could easily have been blinded. *(VITT)*

Iguanas *(Iguana iguana)* and spiny-tailed iguanas *(Ctenosaura)* drop from trees overhanging rivers and large lakes and either swim across the water or descend to the bottom and walk across to the other shore—a behavior no doubt risky in waters infested with crocodilians or large predatory fish. Dropping from perches is also used by sleeping lizards awakened by predators. Because of the threat posed by nocturnal predators like snakes and marsupials, lizards select sleeping sites that either make it difficult for a predator to find them or that exclude predators. Most tropical *Anolis,* for example, sleep on twigs or thin tree branches, often near the end. When predators approach from within the tree, the branch moves, waking the anole, which then drops to the ground below and scurries off. Blunt-headed tree snakes *(Imantodes cenchoa)* get around the "safe" sleeping behavior of anoles by extending the thin anterior portion of their bodies from one limb or one shrub to another in search of sleeping prey. They effectively approach the anoles from the air, picking the sleeping lizards off the end of their nocturnal perches without disturbing the branch.

Lizard tails, which have diversified greatly and serve a wide variety of functions (including prehensility and counterbalance, which in some species further allows bipedality), are used in many ways for defense. Some lizards use their strong tails directly to protect themselves. For example, green iguanas whip their tails violently when grabbed. Spiny iguanas *(Ctenosaura)* also slap with their tails, which are large, muscular, and covered with large spiny scales that can inflict impressive wounds. Tails of *Uromastyx aegyptius* are similar and can also inflict painful wounds. To be whipped in the eyes by one of these lizards could ultimately prove fatal to a predator.

Some lizards use their tails more indirectly. The Australian desert gecko *Diplodactylus conspicillatus* has a nonglandular but very short and stubby bony tail. These nocturnal termite specialists hide, with heads pointed downward, in vertical shafts of abandoned trapdoor spider holes during the day, blocking off these tunnels with their tails. The climbing skink *Egernia depressa* accomplishes the same thing by wedging itself into tight crevices in mulga tree hollows (and rocks), and blocking off the entrance with its strong and spiny tail. Spinily armored tails are used by numerous other species of lizards in a similar fashion, including *Oplurus* from Madagascar, the Mexican iguanid *Ctenosaurus (Enyaliosaurus) clarki,* and the Saharan agamid *Uromastyx acanthinura.*

Many small lizard species have brightly colored tails like this Australian red-tailed skink, *Morethia storri*. (Dr. Hal Cogger)

Tails can also be used to distract predators. In many species of lizards (especially among juveniles), tails are brightly colored or otherwise very conspicuous, functioning as a lure to attract attention away from more vulnerable and less dispensable parts of the animal or to send a signal to a predator. At first glance, this behavior might seem paradoxical: a brightly colored tail would seem sure to attract a predator's attention. However, when viewed within the context of the evolutionary history of autotomy, or tail loss, advantages become obvious. Tail loss for escape from a predator encounter occurs in most groups of lizards, many salamanders, and even in a few snakes (Arnold 1988; Maiorana 1977; Greene 1973, 1997). In most species with brightly colored tails, consequently, tail loss as a predator defense mechanism likely occurred prior to development of the distractive color. Likelihood of escape by use of the tail was therefore probably high to begin with, and evolution of a brightly colored tail simply enhanced an already effective antipredator mechanism (Cooper and Vitt 1985, 1991).

In North America, many juvenile *Eumeces* have brilliant blue tails and perform behaviors nearly identical to those of some Australian skinks. Juvenile *Eumeces fasciatus* forage on and within leaf litter, often disappearing from sight as they pursue prey. As they enter leaf litter they lash their brilliant tails back and forth, presumably to distract the attention of any potential predator that may have already seen them. When a skink's head is within the leaf litter, it no longer can detect potential predators, but a bird or snake that has spotted it will attack the dispensable part of the lizard rather than its body or head

(Cooper 1998a). Blue tails have arisen independently multiple times among skinks and have also evolved among teiids, gymnophthalmids, gerrhosaurids, anguids, and lacertids.

Not all tail displays of lizards involve bright colors. *Amphisbaena alba* coils its snakelike body into a semicircle when attacked and raises both its tail and head, with mouth open, off the ground. Its tail is blunt, like its head, and it can inflict a painful bite. For an uninitiated predator, an attack on this lizard's tail could result in being bitten, whereas an attack to the head end would result in the tail being banged against it, giving the impression of being bitten. Either way, distraction offers an additional opportunity for escape. The behavior is well known among rural Brazilians, who call these lizards *cobras com duas cabeças,* or "two-headed snakes." Many small lizards, such as *Takydromus tachydromoides,* vibrate their tails rapidly, distracting a would-be predator's attention (Mori 1990).

Another escape tactic widespread in limbless lizards but understudied is saltation: some lizards literally spring off the ground, twisting their bodies about as they move through the air. Numerous observations of this behavior have been made on various species of flap-foots (Pygopodidae), those of Aaron Bauer being the most detailed. Using high-speed cinematography, he documented that saltation involves propagating a sine-wave motion through the body that ultimately lifts the lizard completely off the ground for just over a tenth of a second (Bauer 1986), with the entire sequence lasting just under a half-second. Forward displacement may or may not occur. The mo-

Vinegaroons (Uropygi) make no distinction between arthropod prey and this Mexican *Eumeces*. (Cecil Schwalbe)

ment of confusion a predator experiences provides an opportunity for the lizard's escape, and disorientation of the predator results in misdirected strikes. Rather than capturing the lizard, the predator may "capture" its tail, which is readily autotomized. An extended opportunity for escape is provided as the predator feeds on the tail. (Although pygopodids move slowly and are rather helpless without their muscular tails, regeneration occurs quickly.) Tail loss frequency is high in pygopodids that employ saltation. Saltation has also been observed in glass lizards *(Ophisaurus),* which are well known not only for their abilities to autotomize, but for having tails that subsequently break into several pieces. Saltation occurs in legless and nearly legless gymnophthalmids, but these species have been poorly studied. One Amazonian gymnophthalmid species, *Bachia flavescens,* can make jumps exceeding its total length by saltation (Beebe 1945; Avila-Pires 1995).

ESCAPING ONCE CAPTURED

The best known and most spectacular predator defense mechanism of lizards is tail loss, often referred to as tail autotomy (self-loss) or tail shedding. It is most effective after a predator has attacked and acquired a grip on a lizard. Tails of many, but by no means all lizards, break off easily. Indeed, some species can actually lose their tails voluntarily with minimal external force (Vitt et al. 1977; Arnold 1984b, 1988). The gecko *Coleonyx variegatus,* for example, raises its tail off the ground when first approached by a predatory snake. As the tail waves back and forth, the snake changes its orientation from the relatively large body of the lizard to its moving tail (Congdon et al. 1974). The snake strikes, biting the tail, and wraps coils of its body around the lizard to hold it. Surprisingly, *Coleonyx* rarely releases its tail immediately; instead it waits several seconds until the snake has committed to the tail. Then, by contracting muscles at the base of its tail, the gecko releases it, breaks away from the snake's grip, and runs to safety. Because tails are often used as fat storage organs, an autotomized tail is packed with energy. As nerves cause the tail to wriggle violently, the gecko's escape goes unnoticed by the snake. Rapid tail wriggling is supported metabolically by anaerobic respiration, since the tail is no longer connected to the lizard. Tails of some species can wriggle for as long as five minutes. The snake's effort wasn't in vain, because it has the tail; meanwhile, the lizard gains its life at the expense of losing its tail.

When lost, tails wriggle violently, moving considerable distances from the place where autotomy occurred. This movement certainly serves to distract predators (Dial and Fitzpatrick 1983). It can also benefit lizards. Laboratory experiments conducted by Benjamin Dial and Lloyd Fitzpatrick (1984) suggest that although some predators, such as snakes, rarely lose the tail once they have bitten it, predators such as birds that do not have the luxury of many needlelike teeth frequently lose both the lizard and its tail owing to this vigorous thrashing. In cases where the tail, too, is lost, some skinks, including many *Ctenotus* in Australia and *Scincella* and *Eumeces* in North America, return to the site of autotomy and swallow the remains, thereby regaining a portion of the energy lost when their tail was shed. Few, if any, other vertebrates display autoamputation and self-cannibalism. This phenomenon has been best studied in *Scincella lateralis,* a small-bodied terrestrial skink common in leaf litter. Although a number of herpetologists had reported skinks eating their own tails under captive conditions, Donald R. Clark (1971) conducted experiments aimed at understanding this phe-

Muscle, nerves, and blood vessels easily separate when lizard tails are autotomized at cleavage planes in vertebrae, as shown here for the Southeast Asian skink *Mabuya macularia.*

(Photo [composite of two] by Laurie Vitt)

nomenon. First, he confirmed that a high proportion of skinks *(S. lateralis* and four species of *Eumeces)* ate their own tails following removal. He then conducted a field experiment with *S. lateralis* in which skinks were collected from under surface objects, marked for individual identification, and then held by their tails until tails were released. Tails were left where the lizards were found. Some skinks returned and ate their own tails, indicating that it occurs in nature as well as in the laboratory.

Predators that capture a lizard tail gain energy by eating it, so, from their perspective, such a feeding event is in fact successful. Moreover, if a lizard is itself too large for many small-bodied predators, its tail might not be. The pygmy varanids *Varanus gilleni* and *V. caudolineatus* may actually "harvest" exceedingly fragile tails of geckos that are too large to subdue intact (Pianka 1969a). Likewise, small snakes, particularly juveniles of species like *Hypsiglena torquata,* may depend on tails of lizards such as *Coleonyx* or *Xantusia.*

Many lizards possess special adaptations that facilitate tail loss, including weak fracture planes within certain tail vertebrae (Etheridge 1967), muscular attachments that facilitate autotomy, and mechanisms for rapidly closing off blood vessels. Tail stubs heal readily and regenerate rapidly. Losing its tail has surprisingly little effect on a lizard's behavior, as individuals often resume basking and foraging within minutes, as if nothing had happened.

Primary costs of autotomy are (1) temporary absence of a tail for defense and (2) loss of an energy storage organ that might have social or reproductive consequences (Avery 1970; Bustard 1967b; Dial and Fitzpatrick 1981; Fox and Rostker 1982). A lizard that has lost its tail can be more susceptible to predation because it lacks that defensive mechanism, and sometimes it cannot run as fast as it could with a tail. Absence of a tail for defense is offset, however, by rapid tail regeneration. Although regrown tails can be almost indistinguishable from the original externally, their internal support structure is cartilaginous rather than bony. Contrary to popular belief, regenerated tails can be lost as well. Moreover, they need not sever within the region still containing vertebrae, though the precise mechanism by which the cartilaginous rod within a regenerated tail of at least some species separates is not yet known.

Because lizard tails contain energy—in some instances, a very substantial portion of an individual's total standing crop of energy and biomass—tail loss can be ener-

getically expensive. Energetic consequences of tail loss vary considerably among lizards and between juveniles and adults depending largely on the specific life history. In five-lined and broad-headed skinks *(Eumeces fasciatus* and *E. laticeps),* most energy consumed by juveniles is directed toward rapid growth. A juvenile that lost its tail might be expected to grow more slowly and either not reach sexual maturity at the same time as other individuals from the same cohort or, at best, reach sexual maturity at a smaller size. A series of controlled experiments on growth rates of juvenile *Eumeces* with broken tails, however, demonstrated that within the first five weeks of tail regeneration, tailless lizards regain the lost weight and "catch up" with tailed juveniles (Vitt and Cooper 1986). In one species, the tail regeneration group of juveniles even weighed more than those retaining their tails. The mechanism of this "catching up" remains unknown but could involve hormonal responses to injury affecting metabolic rates, with an associated increase in feeding rates. The point is, juveniles can compensate for tail loss and incur a negligible energy cost.

When skinks reach sexual maturity, tails rapidly change from the juvenile bright blue coloration to a dull brown, similar to the coloration of the body. The tail no longer attracts a predator's attention. But why would adults not reap the same benefit from a brightly colored tail? The answer lies in changes in energy priorities (Vitt and Cooper 1986). In adult skinks, energy is needed for reproduction: females use energy to produce eggs, while males use energy in social behavior related to finding and courting mates and interacting with other males. Females produce but a single clutch of eggs each year. Loss of the high-energy tail—which constitutes as much as half of skinks' total fat storage—during the breeding season could cause females to produce smaller clutches of eggs, or no eggs at all, and thus miss an entire year's contribution to future generations. For males, loss of the tail could result in lower social status, with negative consequences for reproductive success.

But what about lizards, such as numerous gymnophthalmids and some lacertids and skinks, that as adults have brightly colored tails—blue, green, orange, red, or yellow, depending on the species? This question cannot be answered definitively, but one possible explanation exists. Most gymnophthalmid females produce clutches of two eggs, and they appear to produce numerous clutches in rapid succession. Because they have small clutches, their investment in reproduction at any one time is low compared to lizards like *Eumeces.* Losing a tail, therefore, might only cost a female a single clutch in a season in which she will produce many. For males, similarly, because the reproductive season is long and females reproduce repeatedly, more opportunities to mate arise, making a temporary reduction in social status due to tail loss less critical. In this sense, adult gymnophthalmids are more similar to juveniles of lizards like *Eumeces,* with benefits of tail autotomy as a defense mechanism outweighing energetic costs of tail loss. Interestingly, most lizards with brightly colored tails are either juveniles of moderate-sized species, as in five-lined skinks, or adults of small species, as in gymnophthalmids.

Not all lizards that easily lose tails have brightly colored tails, either as juveniles or as adults. For example, the side-blotched lizard, *Uta stansburiana,* has a tail nearly identical in color to its body, yet it is easily autotomized. In natural populations, tail loss frequency is high, indicating that many lizards have escaped at least once by losing their tails. Laboratory experiments conducted by Stanley Fox and his colleagues at Oklahoma State University suggest that tail loss in juveniles incurs a social cost, with dominant juveniles becoming subdominant after losing approximately 66 percent of their tails. If, subsequently, the new dominant juvenile loses *its* tail, the demoted individual may regain its dominant status. Long-term effects in these lizards might include a social handicap that influences their ability to acquire good home ranges (Fox and Rostker 1982). Subadult female side-blotched lizards use their tail as a status-signaling badge, an indicator of their ability to defend resources. Tail loss for subadult females results in increased aggression from other females, presumably decreasing their ability to hold resources. When tails are artificially reattached to females that have lost their social status, previously subordinate females respond as though the tail had never been lost. In short, lizards regain their social status not as a result of behavioral changes, but because of the presence of their tail (Fox et al. 1990).

Throughout a lizard's life, social status changes for a wide variety of reasons. Losing a tail is a trade-off between temporary loss of social status and survival. Whether such loss of status affects an individual's lifetime reproductive success depends not only on subsequent events during its own life but also on a constellation of events affecting other lizards with which it interacts. Side-blotched

lizards that lose their tails as hatchlings survive better than those that do not lose their tails (Niewiarowski et al. 1997). Yet this seems paradoxical. If lizards suffer a reduction in social status as the result of tail loss, how can they possibly survive better? The answer lies in subsequent behavioral repertoires of individuals that survive predation. They might simply learn to avoid predators as a result of their experience, or they may stay closer to refuges during regeneration. Nearly every herpetologist who has conducted capture-recapture studies of lizards in natural situations knows that lizards become more difficult to recapture, often requiring new behaviors on the part of the investigator. Not only do lizards use alternative behaviors to escape predation while their tails regenerate, but, because of their drop in social status, they may increase the diversity of behaviors used in social interactions to regain status. Thus, by encouraging adaptive flexibility, experience gained from tail loss confers a survival advantage in future encounters with both predators and competitors.

Not all lizard tails are easily broken. Whereas most iguanids have fragile tails, agamids generally do not. Tails of varanids, helodermatids, and chameleons do not break easily either. Lizards with tough tails usually cannot regenerate a very complete tail if their original should happen to be lost. Evolutionary bases for these differences, sometimes between fairly closely related groups of lizards, are evasive. Easy tail loss is probably a primitive trait that has been lost for elusive, as yet unknown, reasons but regained repeatedly among various lizard lineages (Evans 1981; Arnold 1988).

Tail break frequency has been used to estimate predation intensity on lizard populations, although the procedure has serious problems and limitations (Schoener 1979; Schoener and Schoener 1990; Jaksić and Busack 1984; Medel et al. 1988). For one thing, efficient predators that leave no surviving prey obviously will not produce broken tails, but will likely continue to exert substantial predation pressures; broken and regenerated tails may therefore reflect lizard escape ability or predator inefficiency better than intensity of predation. Latitudinal gradients in tail break frequency and predator diversity shed some light on this issue. Predator densities increase from north to south in western North America (Pianka 1970a, 1986; Schall and Pianka 1980). Correlated with this latitudinal increase in predation, frequencies of broken and regenerated tails are higher at southern sites than at northern localities among four of five widely distributed lizard

Mexican horned lizard, *Phrynosoma asio,* after squirting blood from its eyes. (Wendy Hodges)

species. Frequency of broken tails in *Cnemidophorus tigris,* to cite one representative species, decreases with increasing latitude; moreover, diversity of predator escape behaviors used by *Cnemidophorus* also increases with frequency of broken and regenerated tails (Pianka 1970a; Schall and Pianka 1980). Therefore, a greater variety of escape tactics—a form of behavioral "aspect diversity" (Rand 1967)—rather than simple tail breakage, presumably reduces the ease with which predators can capture lizard prey. Presumably, the more lizards are at risk, the more likely they are to employ additional escape tactics.

CHEMICAL DEFENSE

Chemical defense, though poorly known in lizards, could be widespread, especially among species that specialize on insects containing noxious chemicals. Several species of horned lizards when disturbed squirt blood from sinuses surrounding their eyes. Explanations for this function,

ranging from none to defense against predators, have been controversial. Only recently has evidence suggesting its use in defense been convincing. Based on independent observations, a number of researchers have observed that blood squirting is elicited by attacks from canids, some of which are repulsed by the blood. However, observations of canids eating horned lizards suggest that at least some may not respond to the blood. In a series of clever experiments, George Middendorf and Wade Sherbrooke (1992) showed that *Phrynosoma cornutum* squirt blood from sinuses of their eyes in staged encounters with canids but discriminate between canid attacks and other attacks. Visual and tactile cues set off the blood-squirting response. *Phrynosoma cornutum* specialize on ants, many of which are known to produce a variety of toxic or noxious chemicals for defense. Among frogs that specialize on ants (poison frogs in the genera *Dendrobates* and *Phyllobates,* most bufonids, many microhylids, the ranoid genus *Mantella*), an association between ants and skin toxins appears clear (e.g., J. Caldwell 1996; Vences et al. 1997/98). When ants are removed from the diet of *Dendrobates,* the skin loses its toxic properties. *Phrynosoma* and other ant-eating lizards may take advantage of chemicals produced by ants as well. Several South American lizards in the genera *Plica* and *Uracentron* specialize on ants and produce feces that smell particularly pungent. These lizards (particularly *U. flaviceps*) could be using fecal smears of noxious chemicals to deter predators that might otherwise enter their arboreal retreats.

Some lizard species have developed effective means of chemical defense using their tails. Glandular tails of several members of the Australian gekkonid genus *Diplodactylus,* subgenus *Strophurus* (*D. ciliaris, D. elderi, D. strophurus,* and relatives), store and secrete a smelly noxious and sticky mucus. When disturbed, they squirt large amounts of this sticky odoriferous glop up to 50 cm by contracting longitudinal muscle blocks in caudal vertebrae, which push the contents of caudal glands dorsally (Rosenberg and Russell 1980; Rosenberg et al. 1984; Pianka 1986; Richardson and Hinchliffe 1983). Surprisingly, tails of these geckos are fragile and easily shed (but quickly regenerated). Regenerated tails also are capable of producing chemicals, so this chemical defense mechanism is only temporarily out of commission when a tail is lost.

SKIN ADAPTATIONS

Some lizards easily lose skin, escaping from a predator's grip, whereas others have skin that is so tough and slippery that predators have difficulty obtaining a good grip to start with. In some geckos, rather large pieces of skin can be lost with no apparent long-term effects. In Amazonia, *Gonatodes hasemani* frequently has large scars where skin has healed after being lost. The tiny gecko *Coleodactylus amazonicus* can lose 25 percent or more of its dorsal skin and not even seem bothered by the loss. Nearly all *Aristelliger, Geckolepis, Gehyra, Perochirus,* and *Teratoscincus* and some *Phelsuma* and *Pachydactylus* can lose rather large portions of skin with little apparent long-term effect (Bauer, Russell, and Shadwick 1989, 1990, 1993).

This regional integumentary loss occurs when lizards are grasped and then twist their bodies to escape, leaving the grasper with nothing but pieces of skin. Skin patches come off so easily that a distinct morphological mechanism must be at play. In a set of laboratory tests, Aaron

Bauer and his colleagues determined that geckos such as *Ailuronyx seychellensis* have exceptionally weak dorsal skin—indeed, only 10 percent the strength of that of another gecko, *Thecadactylus rapicauda*—especially along the long axis of the body. Skin comes off the dorsal surface of *Ailuronyx* because two easily separable layers of skin occur in the dermis, a thick outer layer and a thin inner layer. What appears to make the system work is high tensile strength of the inner layer, which remains intact even after the outer layer has been torn off. The outer layer regenerates, though scale structure is not identical to that of the original skin.

In many skinks, anguids, and gerrhosaurids, tough skin and relatively smooth scales make it difficult for predators to maintain a grip. By spinning rapidly, such a lizard can often escape once a predator has captured it. Although this behavior is effective against many predators, some colubrid snakes, including *Liophidium, Scaphiodontophis, Sibynophis,* and juveniles of *Xenopeltis,* have spatulate teeth that can move, maximizing contact with smooth-scaled skinks. Rather than the teeth being fused to the jaw as in most other snakes, a hinge made of connective tissue connects teeth to the jaw so that teeth can change orientation as skinks are manipulated (Savitsky 1981).

Spiny armor also protects some lizards from predators. Most horned lizards *(Phrynosoma)* respond to being grabbed by forcing their heads back, jabbing spines that line the back of their head into their captor. Although this probably does not directly harm most predators, it may startle them enough that they lose their hold on the liz-

This ***Gonotodes hasemani*** lost some of its hide, but lived another day. (Laurie Vitt)

The South African armadillo lizard, *Cordylus cataphractus,* bites its tail when confronted by a predator, making itself nearly impossible to swallow. (le Fras Mouton)

ard, allowing it to escape. More perilous for predators, however, is swallowing horned lizards, especially relatively large ones. Numerous observations have been made of snakes—some dead—with horns of *Phrynosoma* poking through the body wall. For example, a juvenile diamondback rattlesnake that ate a horned lizard later died (Vorhies 1948). One of the most spectacular observations on the effectiveness of spines involves a young roadrunner *(Geococcyx californianus)* found dead in Briscoe County, Texas, with a horned lizard *(P. cornutum)* stuck in its throat (Holte and Houck 2000). The roadrunner's neck was split open on both sides. Whether the bird died from trying to eat a prey item that was too large or died from being pierced by the horns remains unknown, but the slices in its throat suggest the latter.

Some anguids and cordylids bite their own tails when attacked by a predator, making themselves too large to swallow. The "armadillo lizard" *Cordylus cataphractus* is

one example (Rose 1962). *Elgaria* often take this a step further, incorporating a branch in the circle of their body closed by the bite on their own tail (Fitch 1935).

BALANCING PREDATOR ESCAPE BEHAVIOR WITH OTHER BEHAVIORS

Predation risks a lizard might take are influenced by social interactions with other lizards. Time and energy spent interacting is time not spent avoiding predation, and increased visibility during social interactions should put an individual at higher risk. Experiments on male *Tropidurus hispidus* show that males spend less time hiding following a predator attack if they have been engaged in aggressive interactions with another male of their species (Díaz-Uriarte 1999). This may increase their risk if a predator remains in the area while the lizard tries to maintain its hold on key resources. In broad-headed skinks *(Eu-*

meces laticeps), males involved in courtship or agonistic interactions with another male allow closer approach of potential predators than males not involved in these activities (Cooper 1999). Males take greater risks when a potential reproductive payoff exists.

SUBTLE AND INDIRECT PREDATOR ESCAPE MECHANISMS

Numerous questions remain on the evolutionary responses to predation and their impact on lizard species. For example, what is the role of predation in determining clutch or litter size, or even the size of individual offspring in lizards? How do predators influence where, when, or how often lizards deposit their eggs? What are the energetic and predation risks to females carrying eggs or offspring for long time periods? Why do most lizard eggs within a nest hatch nearly synchronously? Answers to these questions could be complex. Some are addressed, at least partially, in other chapters. Observations on individual lizard species provide at least initial insight into others.

A trade-off exists between number and size of offspring produced; that is, given a constant amount of energy available for reproduction, that energy can be divided into a few large offspring or many small offspring (Pianka 1976). Predation and how it occurs doubtless contributes to determining this trade-off, but surprisingly little is known about the particular dynamics of the process. If mortality of hatchlings and juveniles is relatively nonselective, or if all individual hatchlings in a population of lizards have the same probability of being eaten by a predator, then females that produce larger numbers of offspring relative to other females will leave more surviving offspring: a simple numbers game. The long-term result is that natural selection will favor females dividing their clutches into more smaller offspring. However, if larger offspring have a higher probability of surviving predators, then females that divide their clutches into a few large offspring could be favored by natural selection. Both are evolutionary responses in life history traits to predation, and both are long-term escape mechanisms. Of course, predation is not the only factor that determines the trade-off between offspring number and size. Availability of resources is another important factor. Competition among juveniles may be nonexistent if resources are abundant (favoring production of many small offspring) or it may be extreme if resources are limited (favoring larger, more competitive offspring). Responses similar to those based on predation can result. In the real world, both predation and competition operate simultaneously, rendering it difficult to tease apart cause and effect.

Aside from abiotic factors such as temperature and moisture that contribute to successful hatching, nest site selection can have a profound effect on survival of lizard eggs. Eggs are energy rich. A variety of snakes throughout the world are reptile egg specialists, including scarlet snakes *(Cemophora coccinae)* in the eastern United States, leaf-nosed snakes *(Phyllorhynchus)* in the arid southwestern United States, and the red Amazonian snake *Drepanoides anomalus.* Teiids, helodermatids, and varanids are all well-known egg eaters, with reptile eggs being the most common eggs eaten. Mammals, including coatis, foxes, raccoons, skunks, and a host of other species feed on reptile eggs. Even arthropods, particularly ants, are frequent predators on lizard eggs. Among nesting sites available, which appear to be limited, females make choices that determine whether any or all of their offspring sur-

An Australian *Ctenophorus cristatus* watches a raptor circling above. (David Pearson)

When Amazonian *Plica plica* hatch from their eggs, they immediately run off. (Laurie Vitt)

vive long enough to hatch. The impact of terrestrial predators on nesting success is best known in turtles, some of which suffer as much as 100 percent nest mortality in some years (Congdon et al. 1994). Predators can be denied access simply by placing nests off the ground, which is exactly what many lizards do. Female *Tupinambis teguixin* in the Amazon basin climb trunks of trees to as high as 3–4 m, digging cavities in nests of arboreal termites *(Nasutitermes)* and depositing their eggs in the cavity (Beebe 1945). Termites repair the nest, sealing the eggs in until they hatch. Many varanid lizards also do this. Not only are eggs protected from most predators, but they also develop rapidly because termite nests are warmer and more humid than surrounding habitat. *Uracentron flaviceps* deposits its eggs in cavities in trees 40 m above ground, and *Tropidurus hispidus, T. oreadicus,* and several other species deposit eggs under granitic cap rocks away from forest edges where predator abundance is high. Many geckos glue their single egg or eggs (up to two) to vertical or hor-

izontal surfaces above ground. Some, like tiny *Lygodactylus klugei,* even place their eggs in narrow crevices in tree branches, which restrict access to only the most streamlined predators.

One alternative to selecting a well-protected nest site is to place individual eggs in a variety of places ("don't put all your eggs in one basket"). This is exactly what most *Anolis* lizards do. Females produce clutches of a single egg but in rapid succession, with some species capable of depositing an egg every seven days. By depositing eggs in different places and at different times, females hedge their bets such that one or more clutches are likely to survive. Most gymnophthalmids appear to do this as well, although no detailed studies exist on their nesting behavior.

Other alternatives are to guard the nest and keep predators out or not deposit eggs at all but rather retain them until development is complete and give birth to live offspring. Egg-laying skinks in the genus *Eumeces* tend the nest until eggs hatch, thereby providing some protection from predation and disease, as well as enhancing hatching success by adding moisture to the nest through respiratory water loss (Somma and Fawcett 1989). Some other lizards brood eggs too; in most cases, however, the role of brooding as a means of protecting nests from predators is poorly understood. The large number of viviparous lizards, meanwhile, provide protection for their developing offspring by never making them available to predators in the first place. Even though carrying around developing embryos costs a female, behavioral adjustments help offset this cost.

Once lizards reach the appropriate stage of development, synchronous hatching also likely enhances juvenile survival. *Plica plica* deposits eggs in elevated piles of leaf litter and humus or inside of rotting palm logs. When a nest is disturbed, as long as development is nearly complete, all eggs hatch within seconds of each other and the young run from the nest. Although actual predation events have not been observed, imagine the confusion of a predator that starts on one egg only to see all eggs hatch and hatchlings scurry away, until finally it is left with a hollow shell in its mouth. Near synchronous hatching occurs

even in lizards like *Eumeces fasciatus,* in which females brood eggs until hatching. At that point, all hatchlings and the female abandon the nest. Because eggs contain nitrogenous wastes from development and remaining albumin, they likely produce strong chemical cues upon hatching that could attract predators to the nest.

We have only touched on the diversity of predator escape mechanisms in lizards. From the time that lizards enter the world as eggs or neonates they face invertebrate and vertebrate predators with a long evolutionary history of successful predation on lizards. The evolutionary arms race between lizards and their predators has produced a remarkable set of escape adaptations effective during each phase of a potential encounter with a predator. Because the set of potential predators changes as a lizard grows and gains experience, and energetic trade-offs dictate investing energy in reproduction (which has its own set of risks) when sexual maturity is attained, predator escape mechanisms themselves are labile; what was effective for a hatchling may not be effective for an adult. Many predator escape mechanisms used by lizards may never be known because they are intractable for study.

CHAPTER 5

SOCIAL BEHAVIOR

Varanus varius engage in wrestling matches during male combat. (Steve Wilson) Anyone who has visited Caribbean islands or southern Florida has witnessed the seemingly ridiculous antics of male *Anolis* lizards as they bob their heads and expand and contract their colorful and exaggerated dewlaps. These displays establish dominance among peers and attract attention of potential mates. From an evolutionary perspective, any behavior that results in more mating opportunities, and consequently production of more offspring, is at a selective advantage, and if a genetic basis to such a behavior exists, that behavior will be passed on at a greater rate than less successful behaviors. For an individual male, each mating opportunity can result in offspring carrying its genes. As a result, selection is strong on characteristics that give a particular male any sort of advantage, whether direct or indirect, in the mating game.

A bobbing lizard expanding a colorful dewlap, of course, is much more apt to attract a predator's attention than one that matches its background and never moves. A compromise must therefore be reached between potential gains in individual fitness associated with social behavior and potential costs incurred by it. If a male lizard displays and is captured by a visually oriented predator, clearly his potentially high fitness behavior is overshadowed by costs. A more subtle point is that if he *doesn't* get caught, he indeed has shown he has characteristics that provide him with a survival advantage, characteristics that can be observed and evaluated by potential mates. Because it is important for females to mate with highly fit males, they must have sensory capabilities to discriminate between successful and unsuccessful males. Basing their decision to mate solely on the bobbing and dewlap displays of males would be a mistake. Repeated displays, though, in a place with value for lizards (resource sites, refuges from predators, nesting sites), might indicate that a particular male has genes for accruing and defending resources that will increase the fitness of a female's offspring.

Because most lizards are diurnal, social behavior is easily observed and quantified. Of course, social behavior requires communication, and communication relies on signals and successful reception of those signals. The bobbing display of male *Anolis* constitutes one sort of signal, a visual one. However, not all lizards use visual signals as their primary system of social communication; the world of lizard behavior is much more complex and much more fascinating. Some lizards with deep blue dewlaps, for example, appear able to detect and use UV signals to communicate (Fleishman et al. 1993). Others live in a seemingly surreal chemical world where signals are transmitted at a molecular level, either on surfaces or through the air (Cooper and Burghardt 1990; Schwenk 1995). Tactile signals also transmit vital information that can make the difference between mating or not mating. All these modes are subject to natural selection.

A necessary requisite of successful social behavior is species identification: social behavior of lizards directed at different species usually has no genetic payoff and as a result should be selected against (put another way, correct species identification should be at a selective advan-

tage). Most social behavior is associated directly or indirectly with breeding. Underlying all social behaviors are hormonal regulatory systems that induce not only behaviors optimally timed for reproduction but also changes in coloration and morphology associated with breeding. At the most basic level, production of sperm by males and ovulation by females are synchronous (exceptions exist but require sperm storage), with timing varying partly as a function of geographic location and habitat. Evolutionary history constrains behavior and can be determined by examining the distribution of behavioral traits across a lizard phylogeny. For example, many high-latitude and montane lizards mate in early spring and deposit eggs in late spring or summer, or they are viviparous, producing young in summer. In these cases, reproduction is constrained to this time period by cold temperatures and relatively short growing seasons. Some other viviparous species, however, such as *Sceloporus jarrovi* and *S. torquatus,* ovulate and mate in fall, producing offspring as early as April (Guillette and Méndez-de la Cruz 1993). At some point, timing of hormonal events associated with reproductive behavior shifted in an ancestor of these *Sceloporus* and has been carried through to its descendant species.

Nearly all lizards use visual and chemical cues in social behavior, but importance of each varies considerably. The deepest split in the evolutionary history of lizards, iguanians versus scleroglossans, reflects a distinct contrast in communication and social behavior: whereas iguanians rely mostly on visual signals for social communication, scleroglossans rely to a much greater extent on chemical signals. Components of some visual signals of iguanian lizards are so dramatic and ritualized that they conjure up images of mechanical puppets performing identical skits over and over again. Many male lizards expand brilliantly colored dewlaps, flash patches of bright ventral coloration, and bob their heads and bodies in species-specific patterns to thwart other males or to attract attention of females. Chemical signals of most scleroglossan lizards, though much more subtle, are equally spectacular. Plumes of odiferous substances, usually produced in female cloacal glands, provide males with information necessary to identify species, sex, and reproductive condition of potential mates. Seemingly invisible chemical trails likewise allow males to track females through structurally complex environments or subterranean mazes where visibility is poor.

Acoustic signals, produced primarily by some gekkonids, usually in combination with behaviors, appear to work in the same manner as acoustic signals in frogs, though these have not yet been well studied. Tactile signals are also used, nearly always in conjunction with other signals. Signal and reception systems are constrained in part by geographic context and life history (e.g., brightly colored dewlaps would not work for nocturnal species), and as a result, the kinds of social behaviors seen in iguanians, gekkotans, and scleroglossans differ in important ways. Use of social signals is also hormonally controlled, with most social behaviors occurring during the breeding season when androgen levels are highest.

In what follows, we first consider signal and reception systems in lizards and provide examples demonstrating how these systems are used. We then examine social behavior and its consequences.

SOCIAL SIGNALS
AND RECEPTION SYSTEMS

VISUAL SIGNALS AND RECEPTION

Nearly forty years ago, Charles Carpenter of the University of Oklahoma began examining social displays in iguanian lizards using cinematography, thereby setting the foundations for modern lizard behavioral ecology. He found that each species produced stereotypic displays that identified individuals to species, established dominance, and attracted mates (Carpenter 1982). Visual signals are diverse. Size alone is one such signal. A large male, for example, is more likely to be sexually mature, plus he demonstrates at least some survival advantage simply by having lived long enough to get large. Brightly colored dewlaps, throat patches, dorsal and ventral coloration, protuberances such as the enormous "horns" of male Jackson's chameleons (Chamaeleo jacksoni), and a long list of behaviors including body bobbing (push-ups), neck and head bobbing, erection and contraction of dewlaps, throat expansion, full-body shuddering, inflation of the body, open mouth threats, rapid color change, and even outright fighting—all these signals transmit information to males or females depending on context.

By definition, the reception system for visual displays is the optic system. Not only do most lizards appear to detect and discriminate colors, but some also detect UV light, adding a dimension beyond what we ourselves perceive. Establishing the neurophysiology of color vision in lizards has been elusive at best (Peterson 1992; Schwenk 2000). Retinas in lizard eyes have high cone density—

suggesting high visual acuity—but lack rods. Although some lizards have rodlike structures, evidence is lacking that they are homologous to rods in other tetrapods. In addition, diurnal lizards (but not nocturnal ones) have oil droplets in the cones of their retinas; these are pigmented and most likely act as filters contributing to color vision.

Behavioral observations on lizards and certain ecological traits of lizards strongly suggest that color vision exists and is used in lizards. Many diurnal lizards are brilliantly colored, particularly during the breeding season, and appear to assess sex, relative age, and even breeding condition on the basis of colors (Cooper and Greenberg 1992). Most nocturnal geckos, in contrast, are drab in coloration; brightly colored nocturnal geckos like Calodactylodes and Ptenopus may conduct their social behavior late in the afternoon while light is still available, or perhaps at night their colors appear as various shades of gray, rendering them highly cryptic. Rapid color change in social contexts in otherwise cryptic lizards also suggests that colors are detected and used as conspecific signals. Difficulties in assessing how color vision works center on attributes of color itself. For example, because color varies in intensity depending on available light, the immediate context will influence just how it is perceived. Such variables are usually not controlled in experiments with lizard color vision (Fleishman et al. 1998).

Visual signaling is ancestral in squamates; reduction of the importance of visual signaling in most scleroglossans and loss of visual displays in snakes are derived conditions associated with the increased influence of chemical communication (see below). This history of the relative importance of different sensory systems is indicated by what occurs in the sister taxon to the Squamata. Male Sphenodon punctatus (one of two living rhynchocephalians) use body inflation and erection of the dorsal crest of enlarged scales to increase their apparent size and combine these with head shaking, face-offs, and mouth-gaping exchanges during territorial defense. Many iguanians incorporate similar behaviors into their visual displays as do some geckos, suggesting their historic origins (Gillingham et al. 1995).

In most iguanians, males have bright coloration or highly modified morphology compared with females, characteristics that assume importance in social communication. Dewlaps of varying sizes and colors occur in a diversity of iguanian lizards ranging from Asian flying

As in many lizards, the crest of dorsal scales on the neck of **Calotes calotes** is larger in males (shown) than in females.

(Christopher Austin)

lizards (*Draco,* family Agamidae) to New World anoles (*Anolis,* Iguanidae). An anole dewlap carries several kinds of information, including size, color, and pattern. In addition, because these lizards can control movement of their dewlap, head, and body, the frequency and intensity of dewlap expansion, head bobs, and even body movements convey additional signals. Dewlap size alone is sex dependent: males have large dewlaps and females have a small dewlap or none at all.

Exact information conveyed by a dewlap display varies among anole species and is influenced by many variables, including the number of species occurring together. When two or more species coexist, dewlap color is used in species recognition. Two nearly identical anoles, *Anolis marcanoi* and *A. cybotes,* occur together in the southwestern Dominican Republic. *Anolis marcanoi* have red dewlaps and *A. cybotes* have white or yellow dewlaps. Intraspecific aggressive encounters are common under laboratory conditions, whereas interspecific encounters are much less common. By painting male dewlaps, investigators can essentially transfer the dewlap color code of one species to another and so bring about expected behavioral changes, suggesting that dewlap color alone is sufficient as a species recognition signal in these two anoles (Losos 1985). In *A. carolinensis,* by contrast, which until recently occurred in isolation from other anole species in the southern United States, dewlap color plays no role in species recognition— since for the most part there have been no other species to recognize (Greenberg and Noble 1944)!

Throughout most of the Neotropics, more than two

Anolis species occur in close proximity and their dewlap colors are much more complex, with a secondary background color occupying part of the dewlap as a continuous blotch or broken into spots. Additional differences in stereotyped head-bobbing patterns occur as well. Taken together, these differences allow species recognition in multispecies systems (Echelle et al. 1971; Jenssen 1983).

Six species of flying lizards occur together in some areas of Sarawak, and the male of each has its own distinctive dewlap (Inger 1983): short, triangular orange or yellow (*Draco volans*); elongate black (*D. melanopogon*); elongate triangular, red basally but changing to yellow green distally (*D. haematopogon*); clavate grayish brown (*D. obscurus*); elongate pale brown or gray (*D. maximus*); and elongate yellow (*D. quinquefasciatus*). In most species, dewlap colors extend onto a male's neck. In some species, females have small dewlaps, much smaller than those of males. Other differences among species include body size and dorsal and ventral patterns of wing coloration, but the roles of these features, if any, in social behavior are unclear. Like anoles, *Draco* have distinctive display-action patterns of dewlap presentation. *Draco volans sumatranus,* for example, display their dewlap in a series of nine events that include a medium- and a long-duration full dewlap extension followed by a short partial extension (Mori and Hikida 1994), augmented by push-up displays.

As one might expect, potential exists for confusion in species-specific displays among closely related species, especially in regions where their geographic ranges come into contact. In many areas of Colorado and Utah in the western United States, the two phrynosomatines *Sceloporus undulatus* and *S. graciosus* occur alone, but in some

The widespread South American teiid **Cnemidophorus lemniscatus** is highly sexually dimorphic in color. (Laurie Vitt)

Both male and female *Anolis nitens tandai* in the western Amazon have bright blue dewlaps, but the dewlap is larger in males, as shown here. (Laurie Vitt)

areas they occur together. Where they occur alone, stereotypic push-up displays used in territorial interactions between males are strikingly similar, consisting of a sequence of short and long push-ups (Ferguson 1973). However, in Mesa County, Colorado, both species occur together, and their push-up displays are quite different: whereas *S. graciosus* uses a series of rapid push-ups, *S. undulatus* uses a series of interspersed short-long push-ups. At the Mesa County site, moreover, push-up displays differ from those of the same species in areas where they occur alone, suggesting that display behavior has shifted in sympatry. Such a shift is referred to as character displacement and can occur in ecological, morphological, or behavioral traits.

Many geckos, particularly diurnal species, use a variety of visual signals (Marcellini 1977). Male *Phelsuma* have bright red or blue dorsal coloration, which they display to other individuals by tilting their bodies. When not in the presence of other males, male *Gonatodes humeralis*

are drab colored, nearly solid gray on the dorsal surface of the body and head. Within seconds of encountering another male, however, a brilliant complex pattern of red, yellow, and purple appears, and males initiate body and neck arching and side-to-side tail wagging. If this level of interaction fails to elicit retreat of one male, aggressive attacks and chasing follow. When males first encounter females they also turn on bright coloration, but females do not. Bright coloration alone may be the social signal used to discriminate sex. Males often approach a female in a jerky animated walk, and if the female is receptive, he grasps her neck in his jaws and copulates, with the female appearing to remain relatively passive. Because male *G. humeralis* are territorial, females within an individual male's territory are probably familiar with the territory holder.

Adult males are not the only lizards that use visual social communication. Females, and in some instances juveniles, also use visual signals for some types of communication. Female *Anolis aeneus,* for example, which can be territorial or nonterritorial, use head bobbing during encounters with other females: territorial females perform a "jerkbob," a series of up-pause-down-pause jerks of the head, whereas nonterritorial females perform a "multibob" display consisting of a rapid series of head bobs (Stamps 1973). Juvenile *A. aeneus* also perform aggressive head bobs in a manner similar to adults (Stamps 1978). Because juvenile home ranges overlap, dominance hierarchies develop based on size. Visual displays change with age in these and likely other lizards. In females, frequency of aggressive displays is greatest when they are young, decreasing as they age and nearly disappearing when they reach sexual maturity.

Gravid females of a variety of iguanid lizards including *Gambelia, Crotaphytus,* and *Holbrookia* develop red or yellow spots just before ovulation or just after mating (Cooper et al. 1983). In the keeled earless lizard, *Holbrookia propinqua,* bright female coloration functions in sex recognition, as it does in other studied iguanids (Cooper 1984, 1986). Early suggestions that female colors did not function in sex recognition were inconclusive (e.g.

Clarke 1965). Nevertheless, males court females independent of bright female coloration, which implies that other stimuli are also involved. In *Gambelia,* females with red coloration aggressively reject courtship attempts by males. Color change in females is controlled by reproductive hormones (Cooper and Ferguson 1973; Medica et al. 1973).

CHEMICAL SIGNALS AND RECEPTION

The understanding of chemical signals, or pheromones, in lizard social behavior is in its infancy. Both male and female lizards produce chemical signals. When produced by males, signals appear to be directed at other males, but when produced by females, signals are directed toward males.

The overriding importance of chemical cues in the behavior of squamates is suggested by the fact that they have three well-developed systems for detecting chemicals (Schwenk 1995). A well-developed nasal olfactory system (sense of smell), consisting of sensory epithelial cells in the nasal cavities, detects airborne chemicals (fig. 5.1). The gustatory system (taste) is composed of taste buds, reach-

Following fertilization, some lizards, such as this North American leopard lizard *(Gambelia wislizeni),* develop bright coloration and aggressively display at males that try to court them. (Cecil Schwalbe)

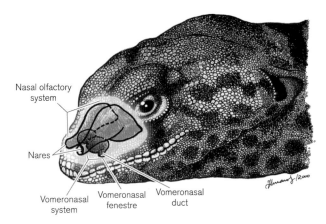

Figure 5.1 Lizard chemosensory systems. Two of the three chemosensory systems in lizards, the nasal olfactory system and the vomeronasal system, are located at the anterior end of the head. The third, the gustatory system (not shown), consists of taste buds located on the tongue-tip. (Illustration by José Pedro Sousa do Amaral)

ing their highest density on tongue-tip papillae (see chapter 14). Complementing these two systems is the vomeronasal system, consisting of pocketlike structures in the roof (palate) of the mouth, the paired vomeronasal organs (also referred to as Jacobson's organs). Each contains a dorsal sensory epithelium. Nerve fibers transmit information via accessory olfactory nerves to accessory olfactory bulbs in the brain. Vomeronasal organs open inside the oral cavity through vomeronasal fenestrae (openings) located in front of the internal openings to the nasal olfactory system (internal nares). The vomeronasal system is not unique to lizards and snakes—it occurs in lizard ancestors and amphibians as well—but ducts open into the mouth only in squamates.

Lizard tongues pick up heavy (nonairborne) chemicals, bring them into the mouth, and pass them over openings to the nasal olfactory system and the vomeronasal system after exposing them to the taste buds. This is accomplished by tongue flicking in which the tongue is extruded, exposing it to air- and surface-borne chemicals. Most people, including many "naturalists," associate tongue flicking with snakes, not lizards, yet many lizards tongue-flick frequently. Tongue morphology, of course, varies greatly among lizards (see chapter 14), as does ability to detect and respond to chemical signals. Some lizards, such as collared lizards *(Crotaphytus),* have short, thick tongues that protrude very little during tongue flicking, whereas others, such as monitor lizards *(Varanus),* have long, highly flex-

ible forked tongues that can be extended and swept from side to side, very much like snake tongues (fig. 5.2).

Detection and interpretation of chemical signals is likely hierarchical, with volatile chemicals detected by the nasal olfactory system stimulating a lizard to sample further with its vomeronasal system. The vomeronasal system allows greater discrimination of less volatile compounds and, through sampling of surfaces, allows discrimination of other compounds as well. This well-developed system in lizards and snakes allows discrimination of species, sex, social status, sexual receptivity, and in some instances, individual identity. Squamates even use their forked tongues as edge detectors, comparing signals from both sides to follow scent trails (Schwenk 1994b).

Like other aspects of lizard biology, chemosensory systems of lizards are tied to phylogeny (see chapter 14). True chameleons, for example, have relatively nonfunctional vomeronasal systems and instead rely nearly completely on visual signals, not just in social behavior but for prey discrimination as well. Because their ancestors had well-developed chemosensory systems, this is a derived condition. Iguanians, for their part, have not developed the vomeronasal system to nearly the degree that scleroglossans have, probably as a result of constraints on tongue development associated with feeding (Schwenk 1995). Iguanian tongues, supported by tongue and throat musculature, are used to capture and pick up prey (a function taken to the extreme in chameleons). Scleroglossan tongues are not used in feeding but are free-floating, and the tongue is extended by hydrostatic elongation rather than

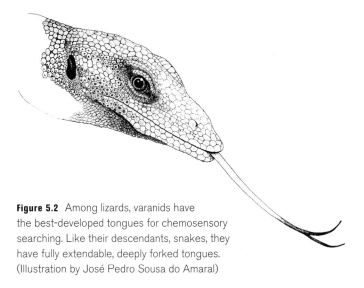

Figure 5.2 Among lizards, varanids have the best-developed tongues for chemosensory searching. Like their descendants, snakes, they have fully extendable, deeply forked tongues. (Illustration by José Pedro Sousa do Amaral)

requiring movement of the entire tongue for extrusion. Its primary specialization appears associated with chemoreception (Schwenk 1993), a function no doubt occasioned by the freeing of the tongue from its role in prey capture.

Chemical cues are produced in enlarged femoral pores, skin, or the cloacal region. Femoral pores on the posterior edge of the underside of hind legs in many male lizards produce waxy substances during the breeding season that contain nonvolatile chemicals. These are smeared on surfaces within or at the edge of male territories, presumably to mark specific sites. Desert iguanas, *Dipsosaurus dorsalis,* produce waxy territory markers from femoral pores that absorb ultraviolet (UV) wavelengths (Alberts 1989). This adds another dimension to chemical communication: production of substances visible only to species capable of detecting UV light. If individual variation in femoral-pore chemicals is as great in most species as in the green iguana, species, sex, and individual identity might be determined on the basis of chemicals alone (Alberts 1992; Alberts et al. 1993). Individual discrimination of femoral-pore chemicals likely facilitates formation of dominance hierarchies in some lizards (Pough et al. 1998). Exactly how femoral-pore secretions of nonterritorial lizards like teiids and gymnophthalmids function remains unknown. Male leopard geckos *(Eublepharis)* determine sex of conspecifics on the basis of chemicals picked up from the skin by tongue flicking (Mason 1992; Cooper 1997a, 1998b).

Production of signaling chemicals by females of chemically oriented lizards occurs in urodaeal glands in walls of the cloaca (Trauth et al. 1987). These chemicals achieve central importance in species, sex, receptivity, and individual discrimination. Their importance has caught the attention of numerous naturalists. In a classic study published nearly seventy years ago, Gladwyn Kingsley Noble and E. R. Mason (1933) observed use of chemical signals during social behavior in five-lined skinks *(Eumeces fasciatus)* in the laboratory, and nearly fifty years ago Henry Fitch (1954, 1955) observed males tracking females in the field using chemical cues. These and other studies led to an intensive investigation of chemical signal use in the closely related North American skinks *Eumeces fasciatus, E. laticeps,* and *E. inexpectatus,* largely by William E. Cooper Jr. and colleagues. During the breeding season males develop bright orange or red head coloration and their chemosensory system becomes enhanced (Cooper et al. 1987), both due to increases in testosterone levels. Males follow females through complex habitats by tongue-flicking the pheromone trail left by a foraging female (Cooper and Vitt 1986a). Males discriminate species and sex and can determine reproductive state of females (receptivity) by chemical cues (Cooper and Vitt 1986b, 1986c, 1987a). They also discriminate odors of familiar and unfamiliar conspecifics (Cooper 1996). Other scleroglossans, including gekkonids, cordylids, gerrhosaurids, and amphisbaenians, have equally acute discriminatory chemical sensing systems (Cooper 1997a,b; Cooper and Burghardt 1990; Cooper and Trauth 1992; Cooper et al. 1996; Graves and Halpern 1991; Mason 1992).

From a human perspective, the notion that individual lizards may recognize each other by smell might seem farfetched. Yet although we tend to recognize individuals based largely on facial and other obvious morphological characteristics, we seem best at recognizing our own kin—and we do so partly through nonvisual cues. Mothers, for example, can recognize their babies from odors, and humans can usually identify their mates based on odors. Kin recognition has some obvious evolutionary advantages, especially in species with extended parental care. If a female is to invest energy raising offspring, she should invest that energy raising her *own* offspring. Very few lizards have parental care that extends beyond egg brooding (e.g., *Ophisaurus* and *Eumeces*) or aiding offspring in breaking out of placental membranes (e.g., South American *Mabuya, Elgaria coerulea*). However, in the best-studied viviparous lizard, *Tiliqua rugosa,* and some others (e.g., *Egernia stokesii*), newborn lizards discriminate between their mothers and other females by means of chemical cues (Main and Bull 1996). Similarly, females discriminate their own offspring from offspring of other females even when their offspring have been housed separately with foster mothers. Likewise foster mothers, having spent time with other mothers' offspring, can recognize their own when given a choice.

ACOUSTIC SIGNALS AND RECEPTION

Acoustic communication is limited primarily to Gekkota, although lacertids in the genus *Gallotia* on the Canary Islands produce sounds that may be used in courtship, and varanids can produce hissing sounds during aggressive interactions. In at least twenty gekkotan genera, sounds are produced by vocal cords and vary from nearly inaudible squeaks (e.g., *Hemidactylus*) to loud barking sounds (e.g., *Gekko*) (Böhme et al. 1985; Bauer et al. 1992). Ears comprise the reception system and are well developed in most lizards, though those without external ear openings (e.g., the greater earless lizard, *Cophosaurus texanus*) still appear to detect sound.

Some geckos—e.g., *Phelsuma, Lygodactylus,* and *Hemidactylus* (Marcellini 1974, 1977)—commonly produce sounds during intraspecific social interactions, but surprisingly little is known about these. Male calls are known for many species, and female calls are known for some. Male, female, and juvenile Mediterranean house geckos, *Hemidactylus turcicus,* produce sounds (Frankenberg 1982). When directing a call to another male, males produce multiple regular and irregular clicks. When a call is directed to a female, however, it is more regular, with reduced frequency of irregular clicks. Juveniles and females produce squeaks when approached by other adult geckos. Females also produce squeaks when called to by a male. Male calls function to establish territories, maintain dominance hierarchies, and attract females. Experiments by Dale Marcellini show that while female *Hemidactylus frenatus* do not respond to male calls, males respond neg-atively to calls of other males (Marcellini 1977). In *H. frenatus,* calls serve to maintain spacing of individuals by advertising territorial ownership. Calls are often produced when geckos emerge from retreats, and in some instances may be made in response to another gecko's call. South African *Ptenopus* call in large numbers just after dark, producing choruses similar to those of frogs (Haacke 1969; Frankenberg 1982).

In addition to producing calls in a strictly intraspecific social context, some lizards produce distress calls when approached or subdued (Frankenberg 1974, 1975, 1982). These include nearly all vocalizing geckos, leopard lizards *(Gambelia),* some anoles *(Anolis),* the pygopodid *Delma,* and even some *Cnemidophorus.* Distress calls usually consist of squeaks, which would hardly seem to cause a predator to release its grip. Sounds are also produced by the cloaca in some species as they void in response to being grabbed.

Other acoustic signals might include seismic signals generated when large lizards interact aggressively; this possibility is virtually unstudied except within snakes, which appear capable of detecting low-frequency waves generated in substrates.

HITTING BELOW THE BELT

The large arboreal gecko *Thecadactylus rapicauda* frequently moves into human structures, particularly at the edge of lowland tropical rain forest. During fieldwork in the Brazilian state of Pará in 1995, my colleagues and I lived in a small building, a former grain storage facility, adjacent to an active pig sty. The building was of typical Amazonian styling, with a set of rafters open on the inside, and in these rafters lived a number of very large *Thecadactylus.* They would move about over our heads at night as we hung in our hammocks, and on many occasions one or more could be heard barking loudly—indeed, they often woke us up.

One night two large individuals were chasing each other across the rafters, frequently vocalizing in an apparent at-tempt to enhance a postural display that involved raising the body off the rafter and opening the mouth. We did not observe actual biting, but anyone who has experienced the bite of these geckos (I have, many times!) can attest to their ferocity. After watching these two geckos' antics for some time, I finally retired to my hammock, continuing to hear their mutual pursuit. Early the following morning, everything was quiet—and next to my hammock on the floor lay the entire tail of a large *Thecadactylus.*

Although it is impossible to know for certain, my impression was that while fighting one gecko managed to bite the other's tail, resulting in its release. Regenerated tails—which are slightly larger than original tails (Vitt and Zani 1997)—are common in *Thecadactylus,* often, perhaps, as a result of male-male aggressive encounters. Losing one's tail in such an encounter likely decreases the gecko's ability to successfully interact with other males. Not only would his total weight be decreased by as much as 25 percent, thereby reducing his overall impression of size and increasing the likelihood that other males would not retreat in an encounter, but the loser would also need to invest substantial energy in tail regeneration instead of in reproductive-related activities. This phenomenon may well be quite common in geckos but unreported because it is difficult to observe nocturnal geckos. *(VITT)*

Tactile communication occurs when one individual rubs, presses, or hits a body part against another individual. Tactile communication is common in turtles and snakes (e.g., ritualized combat in viperids) but also occurs commonly in amphibians and many lizards. After visual, acoustic, or chemical contact has been established between two individual lizards, rubbing, pressing, or hitting a body part may follow. During male-male interactions, individuals often butt up against each other in attempts to establish dominance, and when two males are similar in size, tactile interactions can escalate into potentially life-threatening fights (see chapter 13). In male-female interactions, tactile communication often occurs late in courtship when males attempt to copulate. Tactile signals may consist simply of chin rubs or body rubs against the female, but they may include the male biting the female's neck as he positions himself to insert a hemipenis into her cloaca. Because all lizards have internal fertilization, intromission itself is an intimate form of tactile communication.

SOCIAL INTERACTIONS AND THEIR CONSEQUENCES

TERRITORIALITY

Anyone who has observed lizards over extended periods in the field recognizes that individuals use the same areas day after day, and they most certainly know where they are within areas they use. Taken together, all places that a lizard might go in the course of its daily activities — engaging in social interactions, feeding, sleeping — define an area known as the home range. Home range size varies among and within species in accordance with numerous ecological factors, including body size, resource distribution and abundance, and time available for activity (Ruby and Dunham 1987).

The highly territorial iguanians usually defend some or all of their home range, whereas most scleroglossans, which are relatively nonterritorial, defend only their specific location, wherever that happens to be at the moment. Territoriality is the defense of a specified area — one that tends to remain stable through time. Territories can include the entire home range (resulting in home range defense), specific resources (resource defense), or areas within the home range (specific site defense) (Stamps

1977). Territoriality is usually associated with the sit-and-wait foraging mode, as exemplified by the Iguania. Evolutionarily, however, territoriality is ancestral in lizards, so territoriality did not evolve within Iguania; rather, it was lost within Autarchoglossa (some skinks, cordylids, and a few other autarchoglossans reverted back to territoriality, converging on iguanians) (Martins 1994). Loss of territoriality is one of many consequences of enhanced development of the vomeronasal chemoreception system and the evolution of an active foraging mode.

Among territorial lizards, large males establish and maintain territories, which they generally defend from other males. Territories are discrete areas established around resources important in attracting females. These resources vary considerably, but include prey items, good feeding sites, retreats from predators, nest sites, and even good basking sites. The advantage to individual males in acquiring and holding high-quality territories is that they have greater opportunities to mate than males holding poorer territories or those unable to hold territories at all. The disadvantage is that territorial defense requires activities that consume time and energy and make territorial males conspicuous to potential predators. Females, meanwhile, gain an advantage from male territoriality in being given an opportunity to assess male quality based on the quality of his territory and his ability to defend it.

Many visual displays discussed above allow male iguanians to delineate territorial boundaries. Assertion displays, nearly ubiquitous in iguanians, advertise these boundaries and let other males know that the territory has been claimed (Stamps 1977). Whereas in iguanids visual displays do not typically involve color change, within agamids and chamaeleonids dramatic, sudden color changes occur. Male rainbow lizards, *Agama agama,* can change nearly instantaneously from gray to bright blue when encountering another male during the breeding season (V. Harris 1964). Bright coloration signals territorial ownership, with head bobbing and fighting often following. In *Agama planiceps,* the bright orange head and blue-black body coloration of dominant males fades rapidly when an individual is captured.

Although most anoles are territorial, exceptions exist. The Jamaican species *Anolis valencienni* is cryptically colored and moves about slowly searching for prey rather than using a strict sit-and-wait foraging mode (Hicks and Trivers 1983). Not only do females and males overlap considerably in home ranges, but two or more males often use nearly the same space—though the males will engage in aggressive interactions. Males and females copulate frequently, and their lack of territoriality may reflect benefits of foraging behavior.

Although most phrynosomatines are highly territorial during the breeding season, in most species only males defend territories (Stamps 1977, 1983). In *Sceloporus jarrovi,* however, females are also territorial, excluding other females. This species, moreover, differs from many temperate-zone lizards in that courtship and mating behavior occur in fall. Consequently, these lizards are active for several months before breeding takes place. At first, males defend the entire home range, making it their territory. A male's territory usually overlaps with those of several females. Yet as the season progresses and the breeding season ensues, males shift their positions such that overlap with female territories is high and aggression, activity, and home range size increase (Ruby 1978).

As has been mentioned, most autarchoglossans are not territorial, but again, exceptions exist. *Ctenotus fallens,* an Australian skink, has recently been shown to be territorial (Jennings and Thompson 1999), as are emerald tree skinks, *Lamprolepis smaragdina,* in the Solomon Islands (McCoy 2000). Perhaps many more skinks will be found to be territorial in future studies. In addition, an interesting convergence on iguanians has taken place in cordylids, most of which are territorial. Male Augrabies flat lizards, *Platysaurus broadleyi,* for example, aggressively exclude other males from their territories with a combination of status-signaling and chases (Whiting 1999).

Rock lizards *(Lacerta)* of the Caucasus occur at high densities on surfaces of rock outcrops. Once considered to be a single species, *L. saxicolous* (Darevsky 1967), this group now includes at least twenty-five species. Unlike many terrestrial lacertids, these perform behaviors that might be interpreted to indicate territoriality, though the few general studies that exist on these lacertids don't allow discrimination between territorial defense and typical scleroglossan defense of personal space. Although rock lizards occupy common refuges in deep crevices with no indication of aggressive interactions, when on rock surfaces, males defend well-defined areas from other males. Similar behavior has been observed in the European wall lizard, *L. muralis* (Steward 1965). Apparently, high density, restriction to open surfaces, and limited refuges have all played a role in the evolution of social behavior in these lizards.

Juvenile lizards engage in social interactions, some of which significantly influence their reproductive success as adults. Many anoles, for example, display within minutes of hatching. On Barro Colorado Island in Panama, green iguanas *(Iguana iguana)* hatch from eggs nearly synchronously at communal nesting sites (Burghardt et al. 1977). They then move from nest sites in various-sized groups, remaining together for extended periods. Tactile communication, including rubbing of various body parts, scratching, forelimb waving, and even nipping of the body, occurs. Because most green iguana predators are solitary, social behavior reduces the probability that any single iguana will be killed. A group of juveniles may also intimidate some predators, whereas a single individual might not. The large number of eyes available for detecting predators in a social group of young iguanas may also serve as an early warning system. In addition, remaining in social groups probably facilitates transfer—through ingestion of feces—of gut symbionts necessary for digestion of cellulose in the plants the lizards eat.

Interactions among juvenile lizards can determine which ones will be successful territory holders as adults, and may even determine survival. The side-blotched lizard, *Uta stansburiana,* occurs in open, relatively arid habitats of the southwestern United States. Adult males are territorial, and females usually deposit eggs within the territory of their mates. In low-resource years, home range quality spells the difference between survival and death (Fox 1978, 1982). When observed in laboratory enclosures, juveniles with higher-quality home ranges were dominant over juveniles with lower-quality home ranges. Home range characteristics of dominant juveniles, moreover, were similar to those of surviving juveniles from the previous year, indicating that social dominance and occupancy of high-quality home ranges has a substantial payoff. In this case, behavioral differences among juveniles are directly related to differences in home range quality and, ultimately, survival. For example, high proportions of juvenile side-blotched lizards lose their tails, presumably during unsuccessful predation attempts. Not only does tail loss reduce overall body mass, but it also means this effective predator escape mechanism is not available to the lizard, at least for a time. Tails regenerate rapidly. In addition, juveniles with high social status (i.e., dominant over other juveniles) lose that status following tail loss, though they can regain it when their tails have grown back

(Fox and Rostker 1982). Consequently, losing a tail as a juvenile can have a cascading effect on a lizard's future success. Juveniles that maintain high social status—and avoid losing it along with their tails—are more likely to enter the adult population than those that cannot achieve high social rank.

COURTSHIP AND MATING BEHAVIOR

Unlike rhynchocephalians, which mate by cloacal apposition (no intromittent organ), mating in squamates requires insertion of a male's hemipenis—one of a pair, and a key characteristic of squamates—into a female's cloaca. Hemipenes are bizarre structures, often covered by recurved spines; when not in use, they are inverted and situated inside the tail base. Why squamates have two hemipenes remains something of a mystery, considering that only one is used at any one time. It may simply be a developmental consequence of bilateral symmetry.

Sex in lizards doesn't involve just mating; indeed courtship, the set of behaviors leading up to mating, is often complex and can last several days. The immediate function of courtship is to stimulate a sexually receptive female into mating with a particular male. The long-term function is to ensure that females mate with the best males—those carrying genes for desirable traits.

In social systems where a male's quality is directly related to the territory he holds, prolonged courtship may not occur, particularly if a female lives in his territory. This is why, for example, male anoles approach familiar females, grasp them by the neck, and copulate with them all in one rapid act. In this instance, possession of a good territory is all that is required for females to accept a given male as a mate. In the case of unfamiliar females, in contrast, substantial display behavior may be necessary.

Within nonterritorial lizards such as most autarchoglossans, other characteristics of males assume importance. With no territory to defend, the mere presence of a male does not indicate that he is the highest-quality male around. Although his size and morphology may suggest his worth as a mate, without other males to compare him to, a female cannot make a choice. By prolonging courtship, however, she gains an advantage, in essence giving higher-quality males the opportunity to displace the male pursuing her. Prolonged courtship is thus especially common in species that cover large areas while foraging and do not defend specific resources. Presumably, this is why

male broad-headed skinks *(Eumeces laticeps)* and western whiptails *(Cnemidophorus tigris)* follow females for as much as three days prior to successfully mating (Anderson and Vitt 1990; Cooper and Vitt 1987b). Once mating has occurred, males often guard their mates against mating attempts by other males to ensure paternity.

Mating behavior is also stereotypic in green iguanas. A male approaches other males and females similarly, performs some head bobs, and on the basis of the reaction to his head bobs either initiates an aggressive interaction (against males) or courtship behavior (toward females). In the case of courtship, both male and female perform some vibratory head bobs, after which the male approaches the female, moves his head across her neck in a searching fashion, and then bites her neck skin. While holding on to the skin, the male backs up, straddles the female, and attempts to lift her tail with his own. The female eventually raises her tail, whereupon the male inserts one of his hemipenes. If a female is not receptive, however, she will reject the male with a head-swinging display during early stages of the encounter.

The Bengal monitor, *Varanus bengalensis,* has a complex set of courtship behaviors involving visual, chemical, and tactile communication (Auffenberg 1994). A male will approach a female from behind while the female is resting on the ground. When he reaches her shoulder region, the male flicks her neck, shoulder, and posterior head region (the tympanum) with his tongue. The female then turns her head away, and the male puts his head over the female's neck and head and pushes down with his chin while moving his head from side to side. The female now often moves away, but the male follows, climbs on top of her, holds her forelimbs down with his, and lifts his tail forcefully; this apparently stimulates the female to lift her tail, at which point mating occurs. In some instances, the male bites the female's neck. Occasionally, a female will hiss when approached, often exhibiting gular extension—a threat posture—as well. The entire sequence is short, lasting from about twenty seconds to two minutes. Visual cues help males locate females, chemical cues allow identification of species and sex, and tactile cues stimulate the female. Courtship in other monitors is similar (Auffenberg 1981, 1988; King and Green 1999).

For most lizard species studied, morphological traits appear closely linked with behavioral traits, largely because males often have structures that females don't have. *Anolis* dewlaps, chameleon horns, and large heads of male

Tupinambis and *Dracaena* are examples of morphological traits that play a central role in social interactions. Because of their apparent functional linkage, these traits and behaviors have been assumed to be evolutionarily linked as well. Phrynosomatines present a rich opportunity for testing this idea, both because their evolutionary relationships are well established (Wiens and Reeder 1997) and because they employ a wide variety of display behaviors and vary considerably in morphological traits (particularly ventral coloration) used in such displays. Dividing phrynosomatine display behaviors into fifteen categories, John Wiens (2000) examined the relationship between display behaviors and presence or absence of belly patches, an indicator of display coloration. Analysis revealed that evolutionary changes in display behaviors were not linked with evolutionary changes in display morphology. Species of *Sceloporus,* for example, that lose their display coloration nevertheless retain all the basic components of display behavior. The assumption that display behavior and color are linked simply doesn't hold.

Of course this conclusion may not apply to all lizards, and may not apply to less labile morphological traits. Nevertheless, it raises interesting questions about complexity of lizard social behavior. Perhaps display behavior and morphology are not linked because each provides a different signal. Alternatively, perhaps display behavior and morphology provide the same signal and are thus redundant, such that one or the other is adequate to communicate the intended message.

AGGRESSIVE INTERACTIONS

Aggressive interactions, usually between sexually mature males, occur in most lizards. The associated behaviors include visual displays, posturing, chasing, grappling, and in some instances fighting, with potential for serious injury. Aggressive interactions in territorial species—most Iguania—are usually directed toward establishing and maintaining territorial boundaries. At the heart of aggressive interactions between individuals is competition for resources in short supply, including females. Aggressive interactions in nonterritorial species—most Scleroglossa—also occur over scarce resources, or rather, one particular resource: individual females (Anderson and Vitt 1990). In both cases, relative body size is among the most important variables determining success

in aggressive encounters between males, even though hormone levels and established residency play roles as well (Ruby and Baird 1993; Cooper and Vitt 1987c).

Head bobs and push-ups are used by various iguanines in aggressive displays. In general, large terrestrial species *(Cyclura)* use low-amplitude head bobs, large arboreal species *(Iguana)* use exaggerated head bobs, and smaller terrestrial species *(Dipsosaurus* and *Sauromalus)* use push-ups. Interactions among males of some large-bodied iguanines can be quite impressive. Male green iguanas *(Iguana iguana)* initially display at each other with a series of rapid head bobs (Distel and Veazey 1982). Once sex is established based on response to this signal, a male struts in a circle around the challenging male with his large dewlap fully erected and body expanded to give the impression of maximum size. The males may hiss, lash their tails, and continue to bob their heads. As males circle, the diameter of the circle of interaction becomes smaller and smaller and the interaction intensifies proportionally, with pushing and leaning now being interspersed with head bobs and possibly tail twitching. If the interaction continues to escalate, which becomes more likely the more similar in size the two males are, one male will attempt to mount the other and bite it on top of its neck or back. If this male succeeds in inflicting a bite, it pulls back, causing the opponent to rapidly flee. The defeated male usually darkens in color. If the mounting male is thwarted before biting, the escalating interaction is repeated until one male wins.

On the Galápagos Islands, male land iguanas, *Conolophus subcristatus,* have abundant food available and no apparent natural predators as adults. Male territories, which are established early in the breeding season, are based on ecological factors, especially location of burrows. Females choose males primarily on burrow quality, which they can compare because burrows are clustered. The result is effectively a lek-type breeding system. Interestingly, on one island, Isla Fernandina, females nest in a crater many kilometers away from burrows, which are used strictly for overnight retreats. An individual female may travel 10 km, ascending 1500 m in elevation, to reach a good nesting site (Werner 1983). Thus nesting sites are not key resources attracting females to males (Werner 1982), and their scarcity results in aggressive interactions among females as they vie for a superior nest (Pough et al. 1998).

Many phrynosomatines engage in shudder displays—three or more very rapid up-and-down nods of the head—

when confronting conspecifics, whether stationary or on an approach. For a long time, these were considered strictly courtship displays performed by males, but a combination of observations and experiments conducted by Douglas Ruby (1977) on *Sceloporus jarrovi* proved this interpretation false. For one thing, rather than being associated mainly with mating rituals, shudder displays in this species reflect a high level of excitement in a context in which attack by another individual is imminent. For another, they are performed by both males and females. Although displays are nearly identical, differences exist in the frequency of their performance by each sex as well as the context. Females, for example, never shudder during courtship or late in the breeding season.

Male *Chamaeleo gracilis* of Sierra Leone engage in impressive aggressive displays and often fight, inflicting potentially lethal injuries. On first interaction, the body changes to a lighter color, with blotches disappearing and spots intensifying; the gular region is expanded and the back arched, giving a general impression of increased size as males approach each other; the tail and body are lifted off the perch and the tail is moved around (Bustard 1967a). As long as both chameleons assume the expanded display posture, the interaction escalates, with lunges being made at the throat and bites inflicted on the gular region until one of the lizards loses, possibly with serious injuries. If a male assumes a submissive posture, however, stretching his body out longitudinally, the winning chameleon ceases to attack. Although some other chameleons (e.g. *C. chamaeleon*) engage in similar fighting, not all do. *Chamaeleo bitaeniatus* and *Bradypodion pumilum,* for example, display but do not fight; *Chamaeleo jacksoni* interlock their three horns and push on each other during aggressive interactions; and *C. hohnelii* engage in a circling behavior with mock biting and extreme color change but no actual biting.

Without doubt, the most impressive aggressive interactions are between male Komodo dragons *(Varanus komodoensis)* simply because of their large size. Such interactions occur most frequently during feeding, when several or many individuals, aided by their chemosensory systems, converge on dead prey (Auffenberg 1981). Although not territorial in the typical sense (no real estate is defended), *V. komodoensis* does respond to carrion as a temporary resource in short supply, with a temporally restricted social hierarchy consequently arising at the site of the resource. The threat display in itself is breathtaking.

The head is held at a 45° angle to the ground, body expanded and arched upward, and the tail held straight out. The throat may be enlarged as well, giving the impression of a laterally compressed inflated balloon. Depending on the type of threat, additional behaviors may be added as well. In most instances, the threat display by a large male is enough to cause smaller individuals to present conciliation displays, which establishes a hierarchy. If interactions escalate, the tail may be lashed out, and in extreme cases biting with infliction of severe and sometimes lethal wounds follows.

Other large varanid males engage in what appear to be wrestling matches, with the anterior parts of both lizards locked together as they stand on their hind legs (King and Green 1999). These contests are invariably over access to females. Australian pygmy monitors have a somewhat different type of ritualized male combat, in which two grappling males wrap both their front and hind legs around each other and roll along the ground entwined (King and Green 1999; Murphy and Mitchell 1974; Thompson et al. 1992).

Establishment of territorial boundaries lessens the likelihood of fights—which, because they require energy and are always risky, are engaged in only when there is a reasonable chance of winning—especially when potential opponents are familiar with each other: a phenomenon known as "dear enemy recognition." In collared lizards *(Crotaphytus collaris)* and green anoles *(Anolis carolinensis),* for example, males holding adjacent territories seldom interact aggressively, unless an intruder is a stranger (Fox and Baird 1992; Qualls and Jaeger 1991). Similarly, territorial male *Platysaurus broadleyi* in South Africa allow familiar neighbor males to approach more closely than unfamiliar males, and more readily attack the latter (Whiting 1999). Members of the opposite sex are not "enemies," thus response to an unfamiliar female is quite different. In *Holbrookia propinqua,* males court unfamiliar females more intensely than resident females, possibly to increase the chance of successful mating or to induce the female into becoming a resident of the male's territory (Cooper 1985). In the end the dear enemy phenomenon, though possibly common in territorial lizards, remains poorly studied.

Occurrence and intensity of aggressive interactions are determined in part by lizards' relative ages. In uniform rocky habitats, established adult male collared lizards defend well-defined nonoverlapping territories (Baird et al. 1996). Yearling males, which are not yet territorial, inhabit home ranges within territories of larger males. They avoid aggressive interactions with territorial males by maintaining low rates of movement and display and by retreating from territorial males. Whereas adult males interact frequently with females having home ranges that overlap with the male's territory, yearling males in the same area interact little with females. Females, for their part, generally reject yearling males and mate with older males. Yearling males that remain in the area survive by keeping a low profile, gaining a familiarity with the area, and growing large; only in this way will they gain the potential to inherit a territory and females associated with it.

MATING SYSTEMS

Polygyny is the most common mating system in lizards; in most studied cases, males have multiple mates. The obvious result is that some males breed much more than others, or, to put it in behavioral terms, males exhibit higher variance in reproductive success than females. Strictly speaking, this assumes that nearly all females produce similar numbers of offspring during each potential reproductive episode (although reproductive studies of lizards suggest that that is not always the case). Polygyny can result from males holding large high-quality territories that attract multiple females; lekking, which offers females the opportunity to compare individual males; and mate searching, in which nonterritorial males seek females, often in complex habitats, mate with them, and then seek other females (sequential polygyny). Polygynous mating systems predominate in short-lived species and those that breed multiple times within a season. Monogamous or nearly monogamous mating systems are also known but remain poorly documented. Possibly much more common than currently recognized, they appear to predominate in long-lived species that produce few large eggs or offspring. Some long-lived species reproduce once per season and are effectively monogamous for that season, but mate with a different mate each subsequent season. Mating systems directly impact sexual selection.

Iguanine mating systems in general are a form of resource-defense polygyny: in other words, by defending resources that females use, males gain access to those fe-

males (Dugan and Wiewandt 1982). Consider, for example, the spiny-tailed iguana, *Ctenosaura similis,* of the dry forest of Costa Rica (Gier 1997). Females are nonrandomly distributed, with their burrow systems and basking sites comprising one area of activity and feeding sites another. Male territories are organized around these burrow and basking sites such that dominant males have access to several females. The result is a two-tiered dominance hierarchy, with larger and older males maintaining high-quality territories and younger, smaller males opting for an alternative mating strategy and not maintaining territories. Because females move between burrow sites and feeding sites, however, and subordinate males remain at the periphery of dominant male territories and also spend time at female feeding sites, opportunities for mating by subordinate males do exist.

In desert iguanas *(Dipsosaurus dorsalis),* which live in extremely arid environments, the resource-defense polygyny mating system centers on relatively cool microhabitat patches. Female home ranges are organized around desert shrubs—*Larrea* and *Chilopsis*—offering the coolest midday temperatures (Gier 1997). Female home range overlap is therefore highest under these shady shrubs, even though food is not necessarily most abundant there. By establishing territories around these plants, dominant males gain access to more females than subordinate males.

In *Anolis carolinensis,* reproductive success is much more variable for males, which are promiscuous, than for females, which mate only once (Ruby 1981). During the extended breeding season, females consecutively deposit single-egg clutches; meanwhile, males shift their territories in response to female location, in an effort to maximize their mating opportunities. Male anoles successful at mating are larger, older, and more active than those that fail, as in most other lizards (Ruby 1981, 1984).

Mating success of *Uta palmeri* on the small island of San Pedro Mártir in the Gulf of California is determined by location of male territories with respect to food resources (Hews 1990, 1993). Food is most abundant near nest sites of the blue-footed booby *(Sula nebouxii)* and brown booby *(S. leucogaster),* seabirds that nest on beaches. The lizards eat insects and other arthropods associated with booby nests, as well as fish scraps found in and around the nests. Males vigorously defend territories, and most matings involve territorial males. Females respond to increased food availability by shifting their home ranges from low- to high-resource areas. Males occupying areas with more abundant food—in other words, areas in which boobies have elected to nest—therefore experience higher mating opportunities due to the influx of females. In this mating system, males capable of holding on to high-resource territories gain a selective advantage.

One of the most spectacular instances of monogamy and overall complex social behavior known in lizards occurs in the large (280–330 mm SVL) skink *Tiliqua rugosa* of Australia, commonly known as the sleepy lizard. *T. rugosa* are viviparous, with broods of from one to four extremely large offspring (Bull 1988, 1994). Most adult females remain with a single male for extended time periods, not just for the breeding season (September through November) but also, apparently, during subsequent years. Using DNA fingerprinting, Michael Bull and his colleagues have determined parentage of offspring produced by putative monogamous pairs. In most instances, males observed with females for extended periods sired all offspring produced by that female (Bull et al.

1998). Nevertheless, some females produce offspring not fathered by their observed partner, indicating that extra-pair matings do occur. Some individual males also attend more than one female. Thus, even though monogamy predominates among *T. rugosa,* polygyny occurs at low levels.

For an individual male *Tiliqua rugosa,* of course, attending a female partner continuously increases his chance of siring all of her offspring. Long-term pair bonding has other advantages as well. When males and females are together, males feed less than they do when alone, and females—heeding signals from the vigilant male, who watches for both predators and other males while attending the female—respond to the presence of an intruder more quickly than when they are alone (Bull and Pamula 1998). Heightened vigilance has its roots in the historical need to escape predators, including aboriginal people, dingos *(Canis familiaris dingo),* and carpet pythons *(Morelia spilota),* even though today adult *T. rugosa* are virtually immune from predation by most if not all preda-

Considering that colors of chameleons, in this case *Chamaeleo oustaleti*, often reflect their "mood," one has to wonder why mating elicits such a different color response in males and females. (Bill Love)

tors they might come in contact with. Once females give birth to their large offspring, the offspring remain within her home range, adding another level of complexity to this lizard's social organization (Bull and Baghurst 1998).

SEXUAL SELECTION AND DIMORPHISM

The relationship between sexual selection and sexual dimorphism in lizards has caught the attention of naturalists for nearly seventy years, beginning with insightful experiments conducted at the American Museum of Natural History in the first third of the twentieth century (Noble and Bradley 1933). Many lizard species are sexually dimorphic in one way or another. Male chameleons often have spectacular ornamentation in the form of horns or other protrusions on their heads, while male basilisks and iguanas have impressive ornamentation on their heads, backs, or tails. Heads of many male lizards are larger than those of females; this is the case with *Tiliqua rugosa,* in which males with the largest heads also have the highest reproductive success (Bull and Pamula 1996). In many species, too, males exhibit bright coloration and can change color, whereas females cannot, and males are also often larger in overall body size.

Contrary to early suggestions (Stamps 1983), sexual dimorphism is not restricted to territorial species. Examples of sexual dimorphism abound throughout our taxonomic accounts and include nonterritorial lizards. Most such cases are attributable to sexual selection, a special case of natural selection in which differences among individual males result in differential reproductive success. Sexual selection can occur in two ways. If interactions between individuals of the same sex determine which one breeds and which one does not, it is called intrasexual selection. An example would be a species in which males fight over territories and only territory holders get mates. Males that are larger and have larger heads would be expected to win such aggressive interactions. When a male breeds because a female has chosen him, it is called intersexual selection. Say she makes her choice based on overall size. This characteristic is usually related to age in lizards, and regardless of what an old male lizard might look like, the simple fact that he survived so long is an indicator of relatively high fitness. Both types of selection lead to the evolution of enhanced characteristics in males, in this case attainment of larger body and head size. The payoff is relatively more offspring for these males than for other individu-

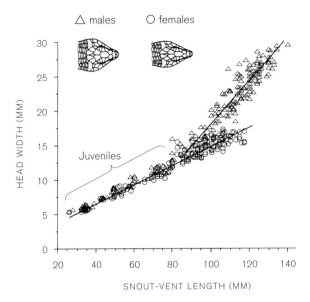

Figure 5.3 Sexual dimorphism in broad-headed skinks. Both head size and shape diverge between male and female *Eumeces laticeps* just after they reach sexual maturity.

als. Regardless of how long a lizard might live or how spectacular it might look, if it doesn't breed, it doesn't pass on genes for its "good" characteristics.

Among the best-known examples of sexual dimorphism in nonterritorial lizards is the broad-headed skink, *Eumeces laticeps,* of the southern United States. Prior to reaching sexual maturity, body and relative head size of males and females are nearly identical. Following sexual maturity, heads of males continue to increase in size, whereas female heads grow very little (fig. 5.3). Males also reach larger body size than females, and in the breeding season their heads transform from brown to brilliant orange or red, and even muscle mass in the head increases. The last two characteristics, and maybe the others as well, are hormonally induced (as a result of natural selection).

Male broad-headed skinks court females and remain with them for extended periods. Although these lizards are nonterritorial, when another male is encountered a series of interactions follows. If the two males differ much in size, the smaller one retreats, partly because he risks great danger by interacting with a larger male and partly because of his social status as a youngster—though that will change as he ages and grows larger himself (Cooper and Vitt 1987c). If males are nearly equal in size, however, an aggressive encounter follows in which males may attack each other, biting at each other's head and body.

When two male *Ctenosaura similis* of similar size interact, the encounter can escalate into a fight. Note how one male appears to be assessing the size of the other, using its own gape as a caliper. (Paul J. Gier)

(Although the jaws on the enlarged head are extremely powerful, skull ossification is increased as well.) Such aggression can produce impressive scars and, in some instances, major injuries, including large open wounds on the neck, torn-off limbs, and torn-off tails. Even death can occur. The winner continues to court the female.

The most obvious sexual selection in *E. laticeps* is intrasexual: males compete with other males to achieve reproductive success. However, more subtle intersexual selection also occurs. Females pursued by relatively small males will often reject them entirely, thus using a very basic criterion to choose a mate: size (Cooper and Vitt 1993). Prolonged courtship offers females an opportunity to see what other males are out there, another manifestation of intersexual selection. If a larger male with better fitness characteristics comes along and displaces a courting male, a female has effectively made a choice.

After mating, male broad-headed skinks remain with the female, guarding her from other males. A large male's ability to exclude rivals prior to fertilization can be taken as yet another measure of male quality (Cooper and Vitt

1997). If another male should successfully mate with a female, competition among sperm of the two males might determine parentage, but this has yet to be explored in any lizard. Multiple paternity is known in some lizards, suggesting that sperm competition might well occur.

Males of one Australian skink species, *Ctenotus fallens,* defend territories and engage in combat similar to that of *Eumeces* (Jennings and Thompson 1999). In a fight males square off head against tail and lunge to bite at the base of the opponent's tail. These lizards have no inguinal fat bodies and most of their fat is stored in their tails. Males that lose their tails presumably lose not only the fight but substantial fitness as well because of the need to regenerate their tails to again be competitive.

Nearly all teiid lizards that have been studied are sexually dimorphic, even though, with one possible exception (*Cnemidophorus lemniscatus;* Stamps 1983), none maintain territories. In most *Cnemidophorus, Ameiva, Kentropyx,* and *Tupinambis,* males are larger in body size and have larger heads than females (e.g., Anderson and Vitt 1990; Censky 1996). In *A. ameiva, Dracaena guianensis,* and some

During an expedition to the Rio Xingu in the central Amazon of Brazil in 1987 I observed a particularly enlightening interaction among several large teiid lizards *(Ameiva ameiva)* that combined courtship behavior, mating systems, and chemical communication, such that major insights into the social behavior of this spectacular lizard could be gained in just a few minutes. I was hidden in the shade under a low tree in a patch of undisturbed rain forest. Open patches along the river were receiving direct sun; as a result, the air and ground were very warm—ideal conditions for teiid activity. I was engaged in a biological survey of the area, one of my primary goals being to collect voucher samples of all species of reptiles and amphibians I could find. The site was destined to be flooded for a hydroelectric facility, which of course had environmentalists and indigenous peoples all in a rage. Our samples would likely be the only ones ever to be collected there. As I was pondering the socioeconomic consequences of a hydroelectric system in the Amazon, I heard a large lizard moving through the understory vegetation. As I watched, a large female *Ameiva* emerged from the vegetation and moved across a portion of the clearing. It was too far away for me to shoot it with a powerful BB gun meant to humanely kill lizards for research. Just as the female disappeared into another patch of vegetation, the largest male *Ameiva* I had ever seen emerged just where the female had first appeared. Even though it could easily have seen her across the clearing, it tongue-flicked the ground and followed in the female's footprints, disappearing into the vegetation after her. They were moving toward me in a roundabout fashion. Another male then emerged from the same area where the other two had come from, but I didn't think much about it. He was also an adult but smaller than the first one.

After a few minutes, I heard the original lizard moving toward me. She emerged from the vegetation with the large male alongside her, trying to mount. The male would move up alongside the female and attempt to bite her neck, attempting at the same time to wrap his tail under hers. At that point, I shot the female, which died immediately. The male ran off. As I walked toward the female to pick it up, I heard movement again and dropped back into the shade. The large male had returned. It tongue-flicked the area, located the dead female, and began trying to mate with her. I shot him too. Once again I stepped out, only to hear the third lizard approaching. He came out of the vegetation tongue-flicking the trail left by the female. He too found the dead female, flanked by a dead male, and attempted to mate with the female. I shot him as well.

After bagging all three specimens and collecting basic ecological data (body and air temperatures, time of activity, microhabitats), I thought about what I had seen. First, because *Ameiva* are active foragers and nonterritorial, males must find females when they are receptive. In this case, the environment was structurally complex: it would be nearly impossible to locate and keep tabs on a female based on vision alone. Both males had therefore pursued the female using a chemical trail. The first male was not only huge (measuring 190 mm SVL, the largest I've ever measured), but his head was also disproportionately large. He was much larger than the female, even though she was as large as females of this species get. While this huge male was tracking the female, another smaller male was maintaining a discreet distance from the large male but tracking the female as well. When males interact, larger males chase off smaller males, and if they are similar in size, serious fights

can occur. The difference in size between the two males was apparently great enough that the smaller male never moved close enough to elicit an aggressive attack. When the large male finally caught up with the female, he did his best to mate with her, only to be frightened off by my intervention. In short order, his hormones got the best of him and he was back after the female, taking risks that resulted in him becoming another fine specimen of the Museum of Zoology in São Paulo. Then the other male showed up. For males, large size alone gives them an advantage over smaller males, and pursuing females directly proves a profitable mating strategy. Small males adopt a different strategy: by remaining on the periphery, then can wait until an opportunity arises to sneak in for a mating, but in a way that minimizes risk.

In this case, hormones were getting the best of both males. When they tracked down the female, their nervous systems apparently never converted the observation that dead lizards were present into the notion that continuing the pursuit might be risky. Moreover, both males' hormonal systems had completely overridden any genetic propensity for assessing risk and payoff. There simply *was* no payoff in mating with a dead female! The plume of chemical cues originating in the sexually receptive (but dead) female thus set in motion a hormonally mediated behavior pattern that was unstoppable.

Observing this series of events and examining the extreme size differences in these lizards initiated my interest in sexual dimorphism and sexual selection in lizards and formed the basis for several later studies, including one with *Ameiva* and other teiids (Anderson and Vitt 1990). *(VITT)*

Some teiids, like this Amazonian caiman lizard, *Dracaena guianensis*, exhibit extreme sexual dimorphism, with males having heads that seem almost too large for their bodies. (William W. Lamar)

species of *Tupinambis* (particularly *T. merianae* and *T. duseni*), not only are male heads larger, but their jaw musculature has hypertrophied to such an extent that males look as though they are bogged down by their oversized heads. Courting males remain with females for extended periods of time as females move through their home ranges foraging, defending the female from advances by other males. Mate defense escalates to aggressive encounters involving posturing and biting when males are nearly equal in size.

Although in general male lizards are larger and more brightly colored (at least during the breeding season) than females, exceptions are not uncommon. Numerous examples of reverse sexual dimorphism in size and coloration exist, and one particularly interesting example of reverse sexual dimorphism in relative head size has been reported. In *Draco melanopogon,* a smaller species of flying lizard, males are smaller in body size than females, and the underside of female "wings" is more intensely colored (Mori and Hikida 1994). Moreover, heads of females are relatively larger than those of males (Shine et al. 1998). Be-

cause social displays take place on the lizards' perches, the underside of wings does not play a role; hence, males do not need noticeable coloration underneath. When gliding, the underside of wings is visible and bright colors identify females as to sex. In *D. melanopogon,* adult females have wider heads than adult males, larger wings, and relatively longer tails (Shine et al. 1998). Large size reduces the extent of wing loading while a female is burdened with eggs, and relatively larger wings enhance gliding with the added mass of the clutch. Finally, the larger head stabilizes flight by counterbalancing clutch mass, which is centered in the abdomen. In larger species of *Draco,* males increase in size disproportionately—most likely because of intrasexual selection and the relationship between size and male reproductive success, though studies have not yet been conducted to test this idea.

Females are larger than males in many geckos (e.g. *Gonatodes hasemani,* Vitt, Souza et al. 2000; *Thecadactylus rapicauda,* Vitt, Zani, and Avila-Pires 1997; *Nephrurus,* Pianka, pers. obs.), gymnophthalmids (e.g. *Prionodactylus eigenmanni,* Vitt, Sartorius et al. 1998), some iguanids (e.g. *Enyalius leechii,* Vitt, Avila-Pires, and Zani 1996; *Polychrus acutirostris,* Vitt and Lacher 1981), some agamids (e.g. *Moloch horridus,* Pianka et al. 1998), and numerous species in other families. In most of these cases, reasons for reversed sexual dimorphism remain unknown. For geckos and other lizards with small, invariable clutch size, larger female body size may evolve independently of male body size based on selection for increased egg size. However, if intrasexual selection favors large body size in males, males could still grow larger than females, as occurs in *Anolis* lizards. For some extremely low density lizards, such as *E. leechii* and *P. acutirostris,* the explanation is quite different. In these species, females produce clutches of many eggs, and clutch size generally is associated with female size. Consequently, selection has favored large size in females because of high reproductive payoff (more offspring). However, because of the very low population density, males don't appear to interact, at least on a continual basis, and there is no indication of male territoriality. When a male locates a female, he courts her, mates, and then literally rides on her back for several days. Thus no advantage to large size exists. Although it hasn't been examined, riding females likely ensures paternity: in the event that another male attempts to court the female, the riding male can defend his investment.

Even though evolution of sexual dimorphism by sexual selection appears straightforward, an interesting paradox exists. If a characteristic, such as relative head size, is favored by sexual selection, why aren't lizard heads even larger than what we observe? The answer is simple: overlarge heads would interfere with a lizard's ability to escape predation; a balance must therefore exist between benefits of sexual selection on a given character and costs of further enhancing that character. Furthermore, why don't all males have giant heads if this trait is so important? In fact, close examination of figure 5.3 reveals that at any given adult male body size, considerable variation in head size exists. Clearly, mating systems, sexual selection, and sexual dimorphism in lizards are much more complex than indicated above. Some males without socalled good traits (large heads) appear generation after generation, which suggests a certain degree of mating success, even if these individuals seldom win aggressive encounters with other males. On the other hand, diversity of male morphologies in some natural lizard populations might point instead to the use of alternative mating strategies. Male tree lizards *(Urosaurus ornatus)* in one Arizona

population, for example, exhibit two very different, developmentally fixed morphologies: while some males have orange dewlaps, are large in body size, and are nonterritorial, others have a well-defined blue spot in the center of their orange dewlaps, are smaller in size, and are territorial (Hews et al. 1994). In other populations, additional male phenotypes exist, rendering the mating system even more complex (Hews et al. 1997). A combination of hormonally mediated developmental events and genetic background determine both dewlap color and size of male tree lizards. Exactly how these differences translate into reproductive success in natural populations remains intriguing.

Throughout this discussion, we have emphasized sexual dimorphism as resulting from male-male competition

SOCIALITY AMONG ROCK LIZARDS

On rock surfaces in northeastern Brazil, the lizard *Tropidurus semitaeniatus* occurs at such high densities that a dozen or more individuals can easily be observed spread across a single boulder. Each social group contains a large dominant male with enlarged testes, several subordinate males with small testes, several females, and a number of juveniles of various ages. Subordinate males behave like females and appear not to be attacked by dominant males. A dominance hierarchy maintains social order on the rock surface, and each lizard uses a particular portion of the rock, with individuals maintaining a minimum distance from one another. When an intruder approaches, the lizards move together in a wavelike manner across the rock surface and enter crevices. Many of these extremely flat lizards enter a single crevice—often packing themselves in like sardines—and any social interactions that might have existed on the boulder break down while in the crevice. After ten to fifteen minutes they come back out on the rock surface and quickly reestablish the dominance hierarchy.

In this particular social system, all lizards scan the habitat for predators. When one lizard responds to a threat, all lizards follow that lizard's lead and enter crevices. Thus the eyes of many have a payoff for each individual lizard. Lizards living alone on a rock surface are probably much more likely to suffer predation than lizards in such a social group. Also, even though social structure is certainly important for enhancing individual fitness, escaping predators is far more important. Consequently escape behavior overrides social behavior until any threat is gone. *(VITT)* (intrasexual selection) or female choice of desirable male characteristics (intersexual selection). However, males choose from among available females as well. The most obvious basis for such choice is female size, for two reasons: (1) because size is related to age, a larger female has demonstrated her ability to survive; and (2) in species with variable clutch size, large females produce more eggs than small females. Even in lizards with invariant clutch size (anoles, geckos, gymnophthalmids), offspring size may be related to female size, providing a fitness reward for selecting large females. When given a choice, for example, *Platysaurus broadleyi* and *Eumeces laticeps* males pair more frequently with larger, older females than with smaller, younger ones, in the first case because these larger females likely produce larger eggs, in the second case because they produce a greater number of eggs (Whiting and Bateman 1999; Cooper and Vitt 1997).

Social behavior of lizards is extremely complex, and we have only touched on it here. Because most of the information available is on New World iguanians, chameleons, and the spectacular large-bodied species of varanids, our view of lizard social behavior is highly biased by our limited knowledge. Relative importance of visual, chemical, tactile, and acoustic sensory systems varies among evolutionary groups of lizards, with associated influences on how their social systems have developed. Iguanians, for example, are primarily visually oriented and territorial, whereas scleroglossans (particularly Autarchoglossa) are primarily chemically oriented and nonterritorial. Polygyny is by far the most common mating system in Iguania. Increased acuity of the chemosensory system and associated behavioral, physiological, and ecological traits provided novel opportunities for social behavior, freeing lizards from constraints imposed by territoriality. Polygyny remains the most common mating system in relatively short-lived scleroglossans that produce many relatively small offspring (leading to a low parental investment). Long lives, high adult and juvenile survivorship, production of large offspring (i.e., high parental investment), and chemically based individual recognition resulted in evolution of monogamy and long-term pair bonds in other scleroglossans. As more species are studied, particularly in poorly represented families, we will no doubt discover patterns of social organization and behaviors that we can't even imagine with our current limited knowledge. Dominance hierarchies in African flat lizards *(Platysaurus)* and South American flat lizards *(Tropidurus semitaeniatus* and *T. helenae)* occurring at high densities on rock surfaces remain unstudied, for instance, as do nighttime antics of most nocturnal geckos. What really goes on under the cover of Joshua tree rubble where tiny-bodied long-lived *Xantusia* interact year after year, or within the soil where amphisbaenians respond to chemical signals in the world they will never see? Only the imagination and creativity of the investigator limit research opportunities when it comes to lizard social behavior.

CHAPTER 6

REPRODUCTION AND LIFE HISTORY

In the North American skink *Eumeces laticeps*, males are larger than females, have relatively larger heads, and remain with females for extended periods while courting and mate guarding. (Laurie Vitt)

People often ask questions like "Why am I here?" and "What is the purpose of my life?" Biologists have a simple answer: all life forms exist to perpetuate their genes in future gene pools. The value of an organism lies solely in its contribution to future generations. Reproduction has primacy. Although we might wish that natural selection favor beauty, brains, or brawn, it need not. If ugly, dumb, weak individuals leave behind more babies, in time their genes will dominate the gene pool. Of course, "reproduction" in an evolutionary sense is much more complex than two individuals combining their genes to produce another generation. Social behavior (chapter 5) leading up to reproduction is often complex and certainly as important as fertilization, as long as fertilization does occur. Reproduction and events leading to it involve taking risks and paying costs. Many of these costs affect probability of survival in the face of predators (chapter 4), but some involve energy trade-offs that balance a parent's own life against production of young. The diversity of reproductive "strategies" in lizards, ranging from early-maturing, short-lived, high-fecundity, egg-laying species with low parental investment (small offspring) to long-lived, late-maturing, low-fecundity, live-bearing species with high parental investment (large offspring), has provided evolutionary biologists some of the most challenging problems in biology.

Several very basic facts set the stage for a discussion of lizard reproduction. Unlike amphibians, in which most fertilization is external, squamates rely on internal fertilization. Unlike tuataras (members of the sister taxon of squamates), which like birds mate by cloacal apposition, male squamates have a pair of hemipenes, one of which injects his sperm deep into the female's cloaca. Because gametes are always in the reproductive tracts of one parent or the other, males and females have much more control over what happens to their reproductive investment. As in all tetrapod vertebrates, females invest much more in individual offspring than males. Whereas males produce millions of tiny sperm during each reproductive bout, females produce only from one to about sixty large eggs, which in most cases are fully supplied with nutrients necessary for development prior to fertilization. Because of this high investment, females should do more than males to ensure that fertilized embryos survive to parturition. This also means that females should be much more choosy about mates than males because they have much more to lose in any given breeding bout than males.

Energy allocation to individual offspring occurs in a variety of ways in lizards: most species ovulate eggs containing all materials and energy necessary for development, but some spread their energetic investment over a longer time period—occasionally as long as an entire year—by passing nutrients to developing offspring across varying degrees of placental membranes. Such prolonged gestation has its own set of costs.

A Galápagos tropidurine *(Microlophus)* assesses its territory from atop a rocky perch. (Eric Pianka)

SEASONAL CYCLES IN REPRODUCTION

Exactly when reproduction takes place during the year varies considerably among lizard species. Although intuition suggests that species in seasonal environments (e.g., temperate zones, tropical areas with distinct wet-dry seasonality) should reproduce during the season when offspring survival is likely to be greatest, whereas species in aseasonal environments (e.g., wet tropics) might be expected to reproduce continually, this is not strictly the case. Nevertheless, cold winters in temperate zones do tend to constrain reproduction in most lizards to spring and summer, extremely dry conditions in wet-dry tropics are not conducive to lizard reproduction, and aseasonal wet tropics allow those species capable of continual reproduction to do so. High-elevation species reproduce seasonally as well. Generally speaking, however, evolutionary history has influenced lizard reproduction in major ways such that one lizard clade does things very differently than other clades even in the same environments.

Examples of highly seasonal reproduction in temperate zones abound and cut across taxonomic groups and continents. In most Temperate Zone desert lizards, mating occurs in early spring and eggs are laid during summer. The leaf-tailed gecko, *Phyllurus platurus,* of southeastern Australia, for example, undergoes vitellogenesis (yolk formation) and ovulates in spring, with egg laying occurring during summer (Doughty and Shine 1995). In female *Cordylus polyzonus* of the southwestern Cape Province of South Africa, vitellogenesis occurs during winter, ovulation occurs in spring, and live young are born in late summer (Fleming and van Wyk 1992). Other South African cordylids, such as *Pseudocordylus melananotus,* have similar seasonal cycles (Fleming 1993). Along the Pacific coast of the United States and British Columbia, *Elgaria coerulea* ovulates in spring, carries its young during summer, and gives birth at the end of summer (Vitt 1973). At Antwerp, Belgium, *Lacerta vivipara* ovulate in spring, with parturition occurring in late July and early August; however, young females reproduce slightly later than older females, even though gestation periods are the same (Bauwens and Verheyen 1985). In this species, the entire reproductive cycle is slightly offset in younger females as a result of greater investment in growth prior to reproduction compared to older females, which grow very little prior to ovulation. In all these and many other examples, potential phylogenetic variation in seasonality of reproduction is obscured because the seasons themselves are short.

A much more illuminating story, involving a diversity of seasonal reproductive patterns, emerges in the case of tropical lizards, such as those of the Alligator Rivers Region of the Northern Territory of Australia and the semiarid caatinga of northeastern Brazil. Among tropical Australian skinks, *Cryptoblepharus* breeds year round, *Carlia* and *Sphenomorphus* breed in the wet season, and *Lerista, Morethia,* and most *Ctenotus* breed in the dry season. Among agamids, two, *Diporiphora* and *Gemmatophora,* breed in the wet season, whereas *Chelosania* breeds in the dry season (James and Shine 1985, 1988). Tropical caatinga teiids show similar variability: whereas *Ameiva* and *Cnemidophorus* reproduce continually with some annual variation related to wet-dry seasonality, *Tupinambis* breeds during the late dry season, with eggs hatching in February. Among iguanids, *Polychrus acutirostris* is highly seasonal, depositing eggs at the beginning of the wet season, with eggs hatching in May and June (Vitt and Lacher 1981), while *Tropidurus hispidus* and *T. semitaeniatus* deposit their eggs over an extended period during the dry season (Vitt 1992a). The skink *Mabuya heathi* ovulates between October and January and carries embryos for nearly a year before giving birth between September and November (Vitt and Blackburn 1983).

In both Australian and Brazilian lizards, seasonality of reproduction for each species is more like that of closely related species living in different habitats than distantly related species living in the same habitat. For clades capable of extended reproduction, such as *Cnemidophorus,* increasing season length increases the reproductive period. A particularly nice demonstration of the effect of increased season length on seasonality of lizard reproduction occurs in the tropical Mexican lizard *Sceloporus variabilis* (Benabib 1994). At high elevation (1000 m above sea level), the reproductive season is about eight months, with the first females containing oviductal eggs in February. At a lower-elevation site (45 m above sea level), oviductal eggs appear in January and the reproductive season is a month longer (nine months).

A number of high-elevation *Sceloporus* in central and northern Mexico and southern Arizona have much more complex reproductive cycles. Mating occurs in spring or early summer in some (*S. aeneus* and *S. bicanthalis*), midwinter in others (*S. torquatus*), and late summer in still others (*S. mucronatus*) (Méndez-de la Cruz et al. 1998). Hatching in oviparous species (e.g., *S. aeneus*) occurs in late summer or early fall, but in viviparous species, parturition takes place in summer (*S. bicanthalis*) or late spring (*S. torquatus* and *S. mucronatus*).

Anolis species and most scincids, gekkonids, and gymnophthalmids have largely tropical distributions and reproduce nearly continuously as long as the season is favorable. Extreme dry seasons curtail reproduction in many of these. Aside from climatic factors, small clutch sizes (1–2 eggs) allow females to produce many clutches in rapid succession and may allow them to reproduce continuously.

In general, male testicular cycles are synchronized with vitellogenesis and ovulation in females. In most Temperate Zone lizards, testes of males enlarge during spring at the same time that ova yolk up in females. Among tropical lizards with prolonged breeding seasons, males produce sperm for extended periods and testes remain enlarged. However, in some species, testes enlarge and mating occurs when female gonads are quiescent, such as at an overwintering site. In such species, female lizards store

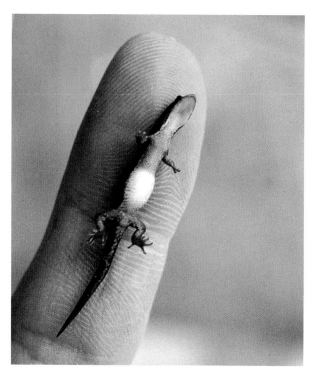

The egg of the tiny Amazonian gecko _Coleodactylus amazonicus_ can be seen through the ventral body surface. (William E. Magnusson)

a season. In southern Nevada, for example, female _Uta stansburiana_ that survive more than a year can deposit up to seven clutches in years with high resource availability (Turner et al. 1973). Many tropical lizards with larger clutch sizes also produce multiple clutches, including _Ameiva ameiva, Cnemidophorus ocellifer,_ and _Tropidurus hispidus_ (Vitt 1992a).

FECUNDITY AND EGG/OFFSPRING SIZE

Female investment in reproduction reduces energy available for growth, maintenance, and fat storage. In most lizards, gravid females are wary and cryptic; living off stored lipids, they forgo feeding and hide until after their eggs have been deposited.

Although clutch volume or litter weight (a measure of instantaneous reproductive investment) varies greatly among lizard species—ranging from as little as 4–5 percent of body weight to as much as 40–50 percent (Vitt and Price 1982)—within single species it is more or less constant. A widely foraging _Cnemidophorus_ might have a low clutch volume, whereas a sit-and-wait foraging _Phrynosoma_ may have a large clutch volume, but the ratio of clutch mass to body mass (called relative clutch mass) stays the same within each species. Ratios of clutch or litter weight to female body weight are strongly correlated with various energetic measures (Ballinger and Clark 1973; Tinkle and Hadley 1975; Vitt 1978) and have often been used as crude indices of a female's instantaneous investment in current reproduction. Relative clutch mass is sometimes equated with the elusive notion of "reproductive effort" (Vitt and Congdon 1978; Vitt and Price 1982).

Surprisingly few reproductive studies on lizards provide data necessary to determine exactly what causes the trade-off between egg number and offspring size. Estimating reproductive effort requires examination of energy invested in reproduction within the context of a lizard's complete energy budget. A relatively constant clutch mass can be divided into many small offspring or a few large offspring (Pianka 1970a, 1976, 1992; Pianka and Parker 1975a) (fig. 6.1). Because for most lizards studied offspring size is constant over time, we can conclude that evolution has favored some "optimal offspring size" for each species or population. When resources are abundant, moreover, natural selection will favor production of many offspring, even if individual offspring are smaller as a consequence. Under these conditions, competition for food

sperm for extended periods of time. Sperm storage allows a female to reproduce without locating a male for mating. It might also facilitate founding of new populations in rare species (Cuellar 1966a). In one such gecko, _Christinus marmoratus,_ mating actually follows egg laying (Greer 1989; King 1977; King and Hayman 1978). Sperm storage also sets the stage for sperm competition, a phenomenon not yet demonstrated in lizards but one that is certainly possible.

Number of clutches produced during a reproductive season varies considerably among species and is tied to a great extent to season length. In geographic areas where the reproductive season is limited to just a couple of months, most lizard species produce a single clutch per year and in some cases one every other year (although reasons for biennial reproduction may differ). Where seasons are longer, some species produce multiple clutches. As indicated above, _Anolis_ lizards, gekkonids, gymnophthalmids, and many small skinks produce small clutches in rapid succession. Nevertheless, even some of these are highly seasonal, with egg production often associated with fluctuations in rainfall and insect abundance. Some species with larger clutch sizes also produce several clutches within among offspring is low; all can feed at maximal rates, grow, and reach sexual maturity rapidly. This is often referred to as *r*-selection, which maximizes the intrinsic rate of population increase *(r)*. High rates of predation on juveniles effectively increase resource availability to survivors, so actual resource levels need not be high for food supplies to be virtually unlimited. High juvenile mortality can be a particularly powerful selective agent on offspring number if it is nonselective with respect to size. When resources are limited, in contrast, offspring compete with each other for food, and larger, more competitive offspring have an advantage. In this case, natural selection favors increased offspring size (competitive advantage), which reduces number of offspring that can be produced. This is referred to as *K*-selection because natural selection favors individuals best at competing for limited resources. *K* is the carrying capacity based on resources. The trade-off between number and size of offspring is not always simply a matter of resource availability, of course.

Other evolutionary scenarios, such as bet hedging, make the same predictions as those based on *r*- and *K*-selection (Stearns 1976, 1977). Bet hedging occurs when

juvenile survivorship is low and unpredictable (Murphy 1968); in such situations, the optimal reproductive tactic may be to spread reproductive investment out in time and space, thereby increasing the chance that at least some offspring survive.

Average clutch or litter size varies from one to over fifty among different species of lizards (Fitch 1970). The largest clutches, up to sixty eggs, are produced by *Varanus niloticus* (Cowles 1930; FitzSimons 1943). The smallest are produced by lizards that appear to have genetically fixed clutch size. All *Anolis,* for example, produce clutches of one egg, and clutches are produced in rapid succession (as little as seven days apart), partly as a result of allochronic ovulation, a process in which ovaries alternate in their production of eggs (Andrews 1985; Andrews and Rand 1974; Jones et al. 1979). Apparent "fixed" clutch size is not so clear in other groups. Gekkonines usually produce clutches of two eggs, one from each ovary, but *Gehyra variegata, Ptenopus garrulus,* and *Thecadactylus rapicauda* lay only a single egg (Pianka 1986; Vitt and Zani 1997). The central Brazilian gecko *Gymnodactylus geckoides darwinii* deposits clutches of two eggs even though *G. geckoides geck-*

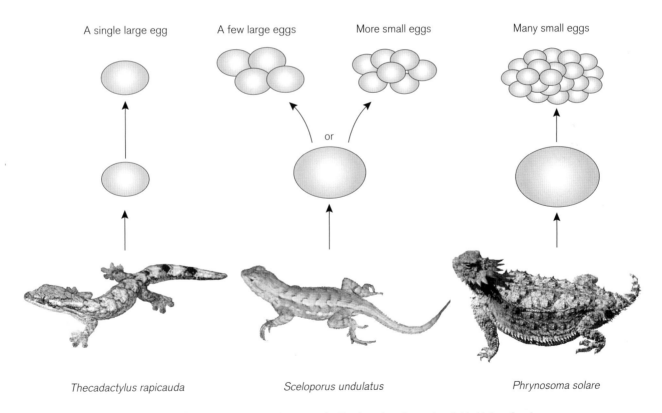

Figure 6.1 Egg size versus number. Any given investment in reproduction by a female can be divided into a few large eggs or many small eggs. This important trade-off reflects a long evolutionary history of natural selection acting on offspring size.

oides from northeastern Brazil produces single-egg clutches (Vitt 1986; Colli, pers. comm.). Most gymnophthalmids lay clutches of two eggs, but *Bachia* and some other fossorial genera produce single-egg clutches. Fossorial lizards in several other families (Scincidae, Anguidae, Amphisbaenidae, Dibamidae) also have very reduced clutch or litter sizes. The Kalahari skink *Typhlosaurus gariepensis,* for example, produces only a single large offspring. Reduced clutch size in fossorial lizards is not simply a consequence of elongation, because similar elongation in terrestrial snakelike lizards (e.g., *Ophisaurus*) has not resulted in decreased clutch size. For some fossorial species, a long history of competition in resource-limited environments may have selected for large offspring size.

Considering all lizards, the most common clutch or litter size is from one to four, but this figure is heavily biased by the large number of species producing clutches of one egg (*Anolis* and sphaerodactylines) or two (most gymnophthalmids, pygopodids, diplodactylids, and gekkonines), a pattern likely prompted by single evolutionary events. Across species, modal clutch size among lizards is two, whereas it is about six to eight among snakes (Fitch 1970).

Within some species, however, clutch size can be quite large. In the Kalahari agamid *Agama hispida,* clutch size averages 13, and in certain horned lizards it is still larger, averaging 24.3 in the Texas horned lizard, *Phrynosoma cornutum,* for instance. One of the most fecund iguanians is *Ctenosaura pectinata,* one female of which had 49 eggs in her oviducts (Fitch 1970). Among lizards with variable clutch sizes, number of eggs produced is correlated with female body size across species (Fitch 1970; Tinkle et al. 1970). Thus small species usually produce a few small eggs and large species produce many eggs or offspring. *Scincella lateralis, Cnemidophorus sexlineatus,* and *Goniocephalus liogaster* average three offspring; *Uta stansburiana, Scelotes mira,* and *Takydromus tachydromoides* average four; *Gambelia wislizeni, Cordylus warreni,* and *Xenosaurus grandis* average five; and so on, with *Iguana iguana, Ctenosaura pectinata, Tupinambis teguixin,* and *Varanus niloticus* averaging more than thirty. Within lizard species, clutch size usually increases with female size unless clutch size is small (e.g., 3–4 eggs). Age- or size-related variation in clutch size among females within a lizard population provides the basis for male choice of mates based on a potentially high fecundity payoff (see chapter 5).

The Australian live-bearing skink *Cyclodomorphus branchialis* produces two very large offspring. (Dr. Hal Cogger)

Australian *Varanus tristis* are alert lizards even as they first behold their surroundings. (Eric Pianka)

Reflecting the trade-off between clutch size (number) and offspring size, lizard hatchlings often vary among populations within species, even though relative clutch mass remains constant. Hatchling *Sceloporus occidentalis* are larger in the southern part of their range than in the north. Moreover, southern neonates exhibit higher locomotor performance (both speed and stamina) than their smaller northern cousins. An interesting study of this phenomenon was conducted by Barry Sinervo and Ray Huey (1990), who made hatchlings in southern populations smaller by extracting yolk from developing ovarian eggs. The performance of these miniaturized offspring was compared with that of normal northern hatchlings. Although they no longer ran faster than northern lizards, they still maintained higher stamina. In these lizards, therefore, offspring size alone determines running speed, but stamina is independent of size.

Number and relative size of eggs are not determined by obvious environmental factors, at least across species. In deserts, for example, lizard mean clutch size varies widely, from 1 egg in several species to 13.4 eggs in *Agama hispida* (see appendix F in Pianka 1986). Expenditure per offspring (egg mass/female mass) likewise varies, from only 0.015 in *Varanus tristis* to 0.173 in *Typhlosaurus gariepensis* (representing an inverse measure of the extent to which a juvenile lizard must grow to reach adulthood). Relative clutch mass (clutch mass/total body mass) ranges from 0.053 in *Gehyra variegata* to 0.307 in *Agama hispida*.

Despite this variability, local environments do have some impact on lizard reproduction, both temporally and spatially, particularly if we look at individual species. Fe-

cundity can be adjusted to fluctuations in food supplies and local conditions from year to year or spot to spot: for example, in the North American whiptail, *Cnemidophorus tigris* (Teiidae), females tend to lay larger clutches in years with above-average precipitation and presumably ample food supplies (Pianka 1970a). Similar phenomena have been documented in the side-blotched lizard, *Uta stansburiana,* and doubtlessly occur in many other lizard species as well (Hoddenbach and Turner 1968; Pianka and Parker 1975a; Dunham 1980). In an *Uta* study conducted by Dr. Fred Turner and colleagues (1973), mean clutch frequency for all age groups of females was positively related to total rainfall, particularly during November–February, with fewer clutches being produced in dry years than in wet years. Many female *Sceloporus undulatus* in Kansas produce two clutches per season. Although eggs in the first clutch (spring) are relatively small, averaging 0.228 g, clutch size is relatively large, averaging 8.7 eggs; in the second clutch (summer) these variables are reversed: the eggs are larger, averaging 0.289 g, but clutch size is smaller, averaging 7.0 eggs. Total investment in eggs remains constant, at about 2 g (Ferguson et al. 1980). Seasonal differences in the relationship between egg number and size result from differences in resources available to young at different times of year (Ferguson and Bohlen 1972). Production of smaller offspring early in the season may be possible because they will have a greater opportunity to feed and grow, while progeny produced later may require a "head start" to reach an appropriate size before winter.

In many lizard species, like this *Varanus dumerilii* from Southeast Asia, hatchlings are brightly colored compared to adults. (R.D. Bartlett)

REPRODUCTION AND LIFE HISTORY

Like other lizards, hatchling Australian *Hypsilurus spinipes* resemble miniature adults when they emerge from their eggs. Notice the egg tooth in the one just breaking through. (Dr. Hal Cogger)

The Australian agamid *Ctenophorus isolepis* also lays two clutches (Pianka 1971b): the first (August–December) averages 3.0 eggs, whereas the second (January–February) averages 3.9 eggs. The explanation is straightforward: females increase in size during the season, and, as in many lizard species, larger females tend to lay larger clutches. Individual females invest relatively more on their second clutch: first-clutch volumes average only 11.2 percent of female weight, but second clutches average 15.1 percent (Pianka and Parker 1975a). Offspring from second clutches are slightly larger than those from first clutches.

Even if the genetic component is ignored, not all offspring are equivalent. Progeny produced late in a growing season often have a lower probability of reaching adulthood than those produced earlier, hence they contribute less to enhancing parental fitness. Likewise, larger offspring cost more to produce, but they are also "worth more." How much should a parent devote to any single progeny? For a fixed amount of reproductive effort, av-erage fitness of individual offspring varies inversely with the total number produced. One extreme would be to invest everything in a single very large but extremely fit progeny—a tactic adopted by the skinks *Typhlosaurus gariepensis* and *Tiliqua rugosa*. The opposite extreme would be to maximize the total number of offspring produced by devoting a minimal amount of maternal energy to each, as seen in exceedingly fecund lizard species like *Phryno-soma* and *Ctenosaura*. More often, parental fitness is max-imized by producing an intermediate number of off-spring of intermediate fitness: the best reproductive tactic is thus a compromise between quantity and quality. The exact shape of the trade-off curve relating progeny fitness to parental expenditure in any organism is influenced by a plethora of environmental and life history variables, in-cluding spatial and temporal patterns of resource avail-ability, length of life, body size, survivorship of adults and juveniles, and population density. The competitive envi-ronment of immatures is likely to be of particular importance because larger, better-endowed offspring should usually enjoy higher survivorship and generally be better competitors than smaller ones.

Any two parties to this triad—clutch size, female reproductive investment, and expenditure per progeny—uniquely determine the third; however, forces of natural selection molding each differ substantially. Clutch or litter weight presumably reflects an adult female's best current investment tactic in a given environment at a particular time, whereas expenditure on any given individual progeny is more closely attuned to the average environment encountered by a juvenile. Put slightly differently, the energy a female invests in her clutch depends on how much energy is available to her plus risks associated with producing and carrying around that clutch. From a female's perspective, it makes little difference whether her clutch is divided into many small eggs or a few large eggs; it is the total mass that matters, for in the end it cannot interfere with her ability to live long enough to deposit the eggs (or, in the case of live-bearers, give birth). Natural selection acting on females determines how much energy, mass, or volume they can afford to invest in reproduction. An optimal parental tactic is to invest as much as possible in reproduction while balancing risks, so that lifetime reproductive success is maximized. How that reproductive investment is divided up is a different matter, and size of individual propagules (eggs or embryos) determines offspring size, at least in a proximal sense. Ultimately, size of individual offspring is determined by natural selection acting directly on them and is relatively independent of selection based on a female's total investment in reproduction. In other words, the ability of a given-sized offspring to compete with siblings of larger or smaller size will determine whether that offspring survives to pass on its genes. In a sense, then, clutch (or litter) size is the direct result of the interaction between an optimal parental reproductive tactic (which determines clutch volume) and an optimal juvenile size (which determines how that volume is divided up).

EGG-LAYING SPECIES

Most lizards deposit eggs in carefully constructed nest chambers and abandon them. Developing embryos receive all their nutrients from yolk, a process known as lecithotrophy. However, not all eggs are the same. Eggs of many geckos, for instance, have calcified shells resistant to water loss and are often deposited in relatively open dry places. *Gonatodes humeralis* deposit their eggs under bark and loose pieces of wood on tree trunks, and *Phyllopezus pollicaris* deposit their eggs under caprocks on granitic boulders, in places without moist soil (Vitt 1986; Vitt, Zani, and Barros 1997). *Hemidactylus mabouia* often "glues" its eggs to the surface of boards in buildings shared with humans. Eggs of most other lizards have leathery shells designed to allow considerable transport of water into developing embryos. Eggs of *Eumeces laticeps,* for example, gain more than 60 percent of their initial mass during development by taking up water (Vitt and Cooper 1985a). Most oviparous lizards with leathery-shelled eggs deposit them in moist nests either in the ground or in rotting vegetation. The large Amazonian lizard *Plica plica* lays eggs under palm debris on boulders or inside rotted palm trunks. *Ameiva ameiva* and *Kentropyx calcarata* dig shallow burrows in sand exposed to sun and lay their eggs. *Tropidurus hispidus* and *T. semitaeniatus* lay their eggs under caprocks on granitic boulders but always in places within crevices where moist soil exists. *Tupinambis* in South America and *Varanus* in Australia and Africa often deposit their eggs inside termite nests.

Advantages of egg laying include reduction of the time during which reduced performance places gravid females at risk and increased opportunities to produce additional clutches within a season. Disadvantages center

Many lizards, like this Madagascan *Oplurus cuvieri*, deposit their eggs in shallow nests that they dig. (Bill Love)

on inability of females, in most cases, to protect their eggs from either predators or potentially lethal fluctuations in microclimate.

Parental care of any kind is uncommon in lizards. In a few species, females defend their nest sites or attend their eggs. All known oviparous *Eumeces* and some anguids brood their eggs, providing protection from some potential predators and regulating microclimate within the nest. Brooding female prairie skinks, *Eumeces septentrionalis,* for example, facilitate embryonic development by maintaining high humidity within the nest through respiratory water loss (Somma and Fawcett 1989). *Varanus varius* return to their nesting sites in termitaria when their

Female North American *Eumeces fasciatus* attend their eggs until they hatch. (Laurie Vitt)

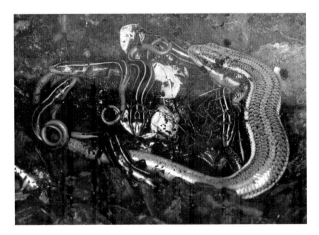

Just after hatching, neonatal *Eumeces fasciatus* abandon the nest, as does their mother. (Laurie Vitt)

offspring hatch (about 290 days after eggs are laid!) and tunnel into the termite mound to assist in their neonates' escape (Carter 1999).

VIVIPAROUS SPECIES

The bearing of live young, or viviparity, has arisen repeatedly among squamates, even multiple times within a single genus. Viviparity probably arises via egg retention, because a female that "holds" her eggs can bask, thereby warming them and enhancing development (Huey 1977). In the Australian skink *Lerista bougainvillii,* populations vary from oviparous to viviparous. An increase in the time eggs are retained in the oviduct results in a decrease in shell thickness, providing a mechanistic basis for the evolution of viviparity from oviparity (Qualls 1996). Live bearing and egg retention are more prevalent in cooler regions and at higher elevations and high latitudes (Tinkle and Gibbons 1977; Shine and Bull 1979).

Nearly 20 percent of all lizard species are viviparous, producing their young as live miniatures of adult lizards (Blackburn 1982, 1985). This reproductive strategy has arisen in at least eleven different families—Agamidae, Anguidae, Chamaeleonidae, Cordylidae, Diplodactylidae, Iguanidae, Lacertidae, Scincidae (many times), Trogonophidae, Xantusiidae, and Xenosauridae—but it does not exist at all in Helodermatidae, Varanidae, *Lanthanotus,* Gekkonidae, Eublepharidae, Pygopodidae, Amphisbaenidae, Rhineuridae, Bipedidae, Gerrhosauridae, Teiidae, Gymnophthalmidae, and Dibamidae. Although a total lack of viviparity in small families might simply be an artifact of small sample size (e.g., Helodermatidae comprises only two species), absence of viviparity in larger families suggests a developmental constraint. Given that evolution of viviparity requires egg retention, the most obvious constraint is in development while eggs are in the oviducts. In some lizards, such as *Calotes versicolor* and *Urosaurus ornatus,* development of embryos within oviductal eggs ceases when they are about 70 percent developed, and completes after eggs are deposited (Radder et al. 1998; Mathies and Andrews 1999). In *U. ornatus,* the mechanism causing arrest of development is likely lack of oxygen, the partial consequence, perhaps, of having a shell. Of course, developmental arrest does not occur in all lizard species, or viviparity would be unlikely to develop. Indeed, a propensity for evolution of viviparity is suggested in lizard clades in which development is not ar-

All studied *Mabuya* in the New World, like this *M. carvalhoi,* have a complex placenta providing all nutrients to developing offspring.

(Laurie Vitt)

rested. When *Sceloporus scalaris* females are forced to retain eggs, for example, development continues, and this species is in a group in which viviparity has arisen several times (Andrews and Mathies 2000). The mechanism causing reduced shell thickness with increased time in the oviduct also appears to be a necessary requisite for evolution of viviparity. Once shell deposition disappears altogether, the potential for maternal-offspring nourishment across embryonic membranes exists.

About 14 percent of iguanids, 17 percent of chamaeleonids, 43 percent of scincids, and more than 70 percent of anguids are viviparous. All cordylids are live-bearers, except *Platysaurus.* All xantusiids and xenosaurids are viviparous. More interesting, however, is the number of origins of viviparity within lizards. Shine and Bull (1979) conservatively estimate at least twenty-eight independent origins of viviparity among squamates, including snakes. Blackburn (1982) suggests that viviparity has arisen at least forty-five times just among lizards, about half of those in skinks alone. More recent data indicate that viviparity has arisen at least fifty-six times in lizards (Stewart and Blackburn 1988). Lower estimates of the number of origins

failed to take into account differences in the structure of placentae in some groups, which would indicate a higher number.

The number of viviparous species, of course, is much greater than the number of origins. When several closely related species are viviparous and their most recent ancestor is as well, a single evolutionary event accounts for all viviparous descendent species in a given clade.

Viviparity has arisen in different ways even though selective pressures favoring its evolution were probably similar. Females of many viviparous species ovulate eggs that contain all nutrients necessary for complete embryonic development in the form of yolk. Development takes place within a female's oviducts. Such lecithotrophic embryos are really no different from embryos that develop in eggs deposited in the external environment, except that females can provide heat, water, and some minerals to developing embryos. Live-bearing lizard females provide nutrients to embryos across a variety of placental types, distinguishable by which portions of embryonic membranes come into contact with maternal membranes facilitating nutrient transfer (Stewart and Blackburn 1988).

Nutrient provisioning by the mother, regardless of how it occurs, is called "matrotrophy."

The most extreme case of matrotrophy in reptiles occurs in skinks in the genus *Mabuya*. All Neotropical *Mabuya* have a complex placenta that involves transfer of nearly all nutrients across the chorioallantois (Blackburn and Vitt 1992). *Mabuya heathi,* a skink of the semiarid caatinga of northeastern Brazil, ovulates the smallest known eggs in reptiles, measuring only about 1 mm in diameter (Blackburn et al. 1984). More than 99 percent of a full-term embryo's dry mass comes from nutrients transferred directly from its mother—a degree of matrotrophy similar to that of eutherian mammals!

Evolution of viviparity in most lizards appears tied to adaptation to cold climates, either at high latitudes or high elevations (Shine 1983, 1985; Shine and Bull 1979). The basic logic is that the likelihood of eggs hatching will rise if females in cold places retain them within their bodies longer. Basking in the sun enhances development, whereas eggs laid in the ground can't move to patches of sunlight to gain heat. Moreover, because basking females can maintain higher body temperatures than eggs in nest sites, internally incubated offspring are produced earlier in the season than hatchlings from eggs. Eventually eggs are retained until they "hatch" while still within the female's body, thus maximizing reproductive success. As time goes on, shell glands disappear and the species becomes locked into viviparity.

Viviparous species in warm climates are usually descendants of species that evolved in cold climates. New World *Mabuya* might be exceptions; additional data are necessary to determine exactly what drove the evolution of viviparity in these skinks. Because all New World species have similar forms of viviparity and no present-day Old World species have anything even resembling this advanced placental type, New World species may have descended from a single, as yet unidentified ancestor that likely reached South America by rafting.

A particularly interesting distribution of viviparous species occurs in *Sceloporus* lizards. Viviparity has arisen independently at least four times in this genus, twice in the *S. scalaris* species group alone (Benabib et al. 1997). Within this group, the oviparous *S. aeneus* lives at relatively low montane elevations in central Mexico. At higher elevations, *S. aeneus* is replaced by the viviparous *S. bicanthalis.* In *S. aeneus,* neonates appear in fall, while in *S. bicanthalis* hatchlings appear through summer and early fall, when food is abundant. The fact that *S. aeneus* lacks characteristics suggesting the propensity for prolonged egg retention means that no evolutionary intermediates exist between these two species. In other words, viviparity in *S. bicanthalis* probably did not evolve in direct response to cool temperatures via egg retention (Méndez-de la Cruz et al. 1998). Rather, the driving force may well have been a shift to earlier vitellogenesis coupled with longer egg retention such that offspring could be produced early enough in the fall to take advantage of wet-season resources. This explanation is key in understanding the reproduction strategies of both *S. bicanthalis* and certain other *Sceloporus* species.

Advantages of viviparity center on female control over development and predator avoidance. Parental females can move among microhabitats, spending time in microhabitats most appropriate for egg development. Because females of viviparous species are present when young are born (by definition), the possibility exists for parental care. Although few squamates in fact exhibit such behavior, mothers of some, if not many, viviparous species do assist their young in breaking out of fetal membranes (Reboucas-Spieker and Vanzolini 1978). In terms of predation, viviparity has the obvious advantage that embryos carried within a female can escape along with her, whereas eggs deposited in the external environment are defenseless and subject to easy predation as long as a predator can find them. Not only that, but a female lizard is often larger than many predators that could eat an egg. Following birth, juvenile Australian *Egernia* skinks sometimes stay in their mother's burrows or hollows for a period of time, where they may gain some additional protection from predators. Disadvantages of viviparity center on reduction of female mobility during often prolonged gestation and loss of the ability to produce multiple clutches within a season.

COSTS OF REPRODUCTION

Activities related to reproduction require energy that could be used for growth and maintenance. These activities also can be risky, though often the risks involved are subtle. Evolutionary biologists refer to these risks as "costs of reproduction." In simplest terms, an evolutionary balance exists between potential payoffs from reproductive-related activities and their costs (Shine 1980; Shine and Schwarzkopf 1992; Schwarzkopf 1994). Defining exactly what those activities are and identifying direct costs is fascinating, and has built the careers of many well-known herpetologists. Risky behavior used by a territorial male to attract females or maintain his status is one cost of reproduction, just as diverting energy into eggs is for females. Females that cannot perform as well (e.g., run as fast) while burdened with eggs must pay a cost in diminished ability to respond effectively to predators. When extended parental care is involved, as is the case with species that brood eggs, a cost is the reduction in energy intake resulting from reduced foraging opportunities. Another is exposure to predators that search for lizard nest sites.

Although reduced performance associated with increased mass of a female carrying eggs is usually considered a result of clutch or litter weight, this may not always be the case. Gravid females of the Tasmanian skink *Niveoscincus microlepidotum,* for example, have lower running speeds than nongravid females, suggesting a reproductive cost in performance (Olsson et al. 2000). However, female performance remains low just after giving birth, so the cost is not attributable simply to the physical burden of the litter. Moreover, running speeds decrease early in pregnancy and remain reduced; if the change in performance were strictly physical, performance should continue to decrease as the burden increases rather than remaining constant. In addition, females burdened with large fat bodies do not exhibit reduced performance, again suggesting that excess weight is not the key factor affecting performance. Instead physiological changes, primarily hormonal, associated with reproduction, particularly in viviparous species with extremely long gestation times like *N. microlepidotum,* may reduce performance independent of any effect of clutch mass. In this case, the "cost" appears to be physiological rather than simply physical.

One of the most interesting comparisons is between sit-and-wait foragers and wide-foraging species. Sit-and-wait foragers move very little while scanning the landscape for mobile prey and rely on crypsis to escape predation. Presumably, increased clutch volume would have minimal effect on their ability to avoid predators. Thus, extreme sit-and-wait foragers like *Phrynosoma* can "afford" to have robust morphology and high clutch volumes. Widely foraging lizards, in contrast, rely on speed and agility for escape from visually oriented predators and as a result have streamlined morphology and low clutch volumes.

SEX DETERMINATION

Sex is not always determined genetically in squamates as it is in birds and mammals. Squamates exhibit both male heterogamety (XY and XXY) and female heterogamety (ZW). However, some squamate species appear to lack sex chromosomes entirely, which presents the opportunity for sex to be determined by other means, and intriguing possibilities arise.

In many reptiles including some squamates (among lizards, the Agamidae, Eublepharidae, Iguanidae, and Lacertidae), sex is determined by the thermal environment during development (J. Bull 1980; Crews et al. 1994; Janzen and Paukstis 1991). When eggs are incubated at low temperatures, they hatch out as females; most eggs incubated at intermediate temperatures become males; and at still higher incubation temperatures, females again become preponderant. Thus, a mother's choice of location for a nesting burrow could influence the sex of her progeny, effectively giving her control of sex ratio (which is impossible under genetic determination of sex), though no data exist showing this to be the case.

PARTHENOGENESIS

The vast majority of lizards reproduce sexually. However, the lizard genera *Leiolepis* (Agamidae); *Brookesia* (Chamaeleonidae); *Hemidactylus, Hemiphyllodactylus, Heteronotia, Lepidodactylus,* and *Nactus* (Gekkonidae); *Lacerta* (Lacertidae); *Cnemidophorus* and *Kentropyx* (Teiidae); *Gymnophthalmus* and *Leposoma* (Gymnophthalmidae); and *Lepidophyma* (Xantusiidae) contain about thirty intriguing all-female species, or parthenoforms, that reproduce asexually without males. Just before meiosis, a premeiotic doubling of chromosomes occurs, temporarily forming 4N germ cells. These then undergo standard meiosis, the reduction division that normally produces haploid gametes, but in this case produce diploid ova with 2N (Cuellar 1974). These ova develop into offspring genetically identical to their mother and sisters, forming true clones. Skin from individuals of the parthenogenetic whiptail lizard *Cnemidophorus uniparens,* for example, can be removed and successfully grafted onto the bodies of *C. uniparens* from hundreds of miles away, whereas similar skin grafts between individuals within a single population of *C. tigris* are rejected (Cuellar 1976). This lack of compatibility at the cellular level (histocompatibility) has been the greatest challenge in human organ transplant

This Amazonian gymnophthalmid, *Gymnophthalmus underwoodi,* is parthenogenetic: no males are known. (Janalee Caldwell)

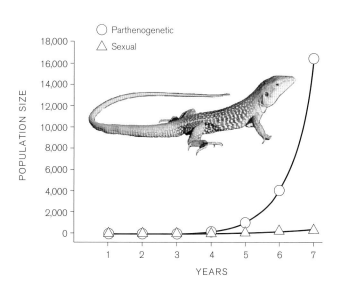

Figure 6.2 Population growth in parthenogenetically reproducing lizards versus sexually reproducing lizards. Starting with a single female (fertilized in the sexual species), setting clutch size equal to four eggs in each, and assuming that each individual lives long enough to reproduce only once and is then eliminated from the population, the difference is dramatic: the parthenogenetic species reaches 16,384 individuals after just seven years, whereas the sexual species reaches only 256 individuals.

therapy. Histocompatibility has been demonstrated with skin grafts within parthenogenetic geckos as well, indicating that the clone originated from a single individual (Cuellar 1984).

Many if not all parthenogenetic species arose by hybridization of two sexual parental species. Some are diploid, but others are polyploid: they have more sets of chromosomes than either parental sexual species, including chromosome sets from both parental species (Lowe and Wright 1966; Cole 1984; Cole and Townsend 1990; Cole et al. 1990). Polyploids usually result from backcrosses between parental species and existing parthenoforms. Because such parthenoforms possess copies of both parental genomes, they are highly heterozygous and may therefore be able to cope with ecotonal edge habitats found between the habitats of their parental species (Wright and Lowe 1968). Clonal reproduction perpetuates and maintains heterozygosity perfectly, even better than sexual reproduction would! Many unisexual species are thought to be of relatively recent origin (only hundreds of years old), suggesting that unisexual lizards might be short lived. In parthenogenetic lineages, a single gravid female can found a new population; moreover, populations of all-female species increase much more rapidly than populations of standard sexual species because all individuals produce eggs and young, all of which are females (fig. 6.2).

Sex ratios in sexual species usually approach 1:1, with half of the breeding population (males) not producing eggs. If the environment faced by offspring of parthenogenetic females is similar to that in which the parental female survived to reproduce, her daughters, being genetically identical to the mother, would be expected to do well also. However, environments are extremely complex and typically change through time; even though the physical landscape itself might remain constant, the biotic environment including predators, parasites, and competitors is in constant flux. For this reason, parthenogenetic individuals are at a selective disadvantage relative to sexually reproducing individuals because they have no genetic variability and thus cannot adapt to changing environments via natural selection. Unisexual clones simply cannot evolve. Some parthenogenetic species may persist either because they live in relatively stable environments (such as human structures) or because they can track habitat patches that remain stable through time even though their spatial distribution changes. Parthenogenetic species might be expected to do particularly well in disturbed habitats (e.g., ecotones, riparian habitats) where resources undergo periodic blooms. The so-called weed hypothesis asserts that unisexuals exploit newly opened disturbed habitats, rapidly breeding, but then moving on to new patches of recently disturbed habitat (Wright and Lowe 1968).

LIFE HISTORY STRATEGIES

"Life history strategies" are not strategies in the sense that lizard species plan out how their lives will proceed. Rather, suites of reproductive characteristics and age-specific survivorship schedules result in populations persisting through time in the face of either predictable or unpredictable mortality. Patterns of age-specific mortality are linked to reproductive variables such as clutch size, egg size, and clutch frequency. W. Frank Blair (1960) examined this aspect of lizard biology in central Texas during the 1950s in a pioneering study of survivorship in the arboreal lizard *Sceloporus olivaceus*. More than three thousand lizards were captured, toe-clipped, released, and then followed and observed over five years. Survivorship in this multiple-clutched lizard was low, with only about 2.5 to 3 percent hatchling survival to sexual maturity and

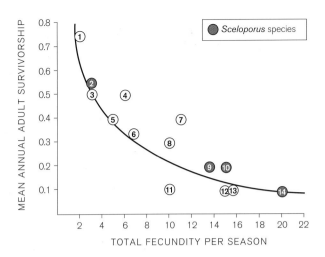

Figure 6.3 Survivorship and fecundity trade-off. As shown here, lizards that produce many offspring in a season have low annual survivorship, whereas those that produce few offspring usually have high adult survivorship. Species 1 = *Xantusia vigilis*, 2 = *Sceloporus graciosus*, 3 = *Cnemidophorus tigris*, 4 = *Cnemidophorus sexlineatus*, 5 = *Scincella lateralis*, 6 = northern *Uta stansburiana*, 7 = *Tachydromus tachydromoides*, 8 = *Emoia atrocostata*, 9 = *Sceloporus occidentalis*, 10 = *Sceloporus undulatus*, 11 = *Anolis carolinensis*, 12 = southern *Uta stansburiana*, 13 = *Crotaphytus collaris*, and 14 = *Sceloporus olivaceus*. The curve fitted to data points represents stable populations in which lizards replace themselves during their lifetimes. (Note: In graphs like this, various data points may not be independent, owing to phylogenetic relatedness. For example, four species of *Sceloporus* are shown, which are much more closely related to one another than to other species represented. Nevertheless, *Sceloporus* exhibit almost the entire range of survivorships and fecundities, demonstrating considerable evolutionary lability of life history tactics.) (Redrawn from Tinkle 1969 by permission of University of Chicago Press)

an adult annual survivorship of only about 10 percent. In contrast, a similar mark-recapture study on a population of desert night lizards, *Xantusia vigilis*, in the lower Mojave Desert (Zweifel and Lowe 1966) showed that these secretive little viviparous lizards enjoy quite high survivorship, with an annual survival rate of nearly 90 percent and some individuals living to be seven to eight years of age. In a similar study of the New Zealand gecko *Hoplodactylus duvaucelii*, most individuals marked were still alive thirty years later (Barwick 1982). In general, therefore, lizard species with high survivorship exhibit low fecundities, whereas those with high fecundities have lower survivorship (fig. 6.3; Tinkle 1969).

Many, perhaps most, aspects of reproductive strategies

have a phylogenetic basis. Donald W. Tinkle of the University of Michigan recognized that lizard species can be clustered into several rather well defined groups on the basis of life history characteristics, with some species (such as xantusiids, anguids, helodermatids, and varanids) classifiable as long lived and late maturing, reproducing year after year, whereas others (such as iguanids, agamids, and gekkonids) are short lived, mature early, and confine their reproduction to a single season (Tinkle et al. 1970; Dunham et al. 1988). A large number of variables contribute to variation in life history traits within and among lizard species (Ballinger 1983; Niewiarowski 1994; Bauwens and Díaz-Uriarte 1997). In a study of reproductive tactics of ninety-one species of lizards, Arthur E. Dunham and Donald B. Miles (1985) undertook a discriminant function analysis, which reduces variation in many correlated variables into less variation within a smaller number of uncorrelated variables, to describe an axis with actively foraging species with large body sizes and small clutches at one end and sit-and-wait foragers with small body sizes and large clutches at the other end. This axis could be used as a dimension for construction of a "periodic table of lizard niches" (see also chapter 7).

In this chapter we've seen that many lizards deposit eggs, others are viviparous, and some even provide a modest degree of parental care. Number of offspring produced during each reproductive episode varies considerably among and within lizard species but is tied to body size. Sex is determined by developmental temperature for some lizards but not for others. Although a vast majority of lizards reproduce sexually, some reproduce by parthenogenesis, sending out generation after generation of exact replicas of themselves. Small offspring thrive in environments with unlimited resources, but highly competitive environments require larger progeny. Taken together, body size, age at sexual maturity, and number of offspring produced define distinct life history strategies in lizards, each of which reflects generations of natural selection responding to immediate changes in biotic and abiotic environments. Interestingly, longevity, viviparity, and parental care are much more common in autarchoglossans than in gekkotans and iguanians.

CHAPTER 7

REFLECTIONS OF THE REAL WORLD

Alert to the presence of an intruder, an adult male *Agama aculeata* peeks out from behind a bush in the Kalahari. (Eric Pianka)

"Ecology" and "environment" are terms often misused and abused by politicians and other anthropocentric advertisers, who tend to reduce them to "clean air, clean water, no beer cans, and no pollution." Ecology extends far beyond the human sphere, however: it is nothing less than the study of all interactions between organisms and their environments. "Environment" includes everything that impinges on a particular organism, both abiotic (e.g., sunshine, wind, rain) and biotic (prey, competitors, parasites, predators). Any organism's environment is exceedingly complex and multidimensional.

Highlands of Sri Lanka are home to some of the strangest lizards, like this *Ceratophora tennentii*. Little is known about its ecology.

(Christopher Austin)

Indeed, entire lifetimes could be spent studying the ecology of a single species!

Ecologists are not very interested in captive animals. Their subjects are wild organisms in natural settings—the environments in which these creatures have evolved and to which they have become adapted. Philosopher of science Holmes Rolston (1985) made a useful analogy: he likened life on earth to a book written in a language that humans can barely read. Each page in this book of life represents a species, describing not only its phylogenetic relationships, but also its interactions with its physical and biotic environments, while each chapter represents a biome. Biologists are just now acquiring the skills necessary to read and decipher this book, which is, sadly, tattered and torn, with pages missing (extinct species such as passenger pigeons) and entire chapters ripped out (e.g., tall-grass prairies of midwestern North America). We must save as much of this vanishing book of life as possible, and study it well before it is gone forever.

"Natural history" is almost synonymous with "ecology," though it tends to be less rigorous and not as quantitative. Natural history includes knowing where and when to look for individuals of a particular species, realizing why they are there and not someplace else, and understanding such processes as foraging, thermoregulation, reproduction, and predator escape. When we were boys, both of us roamed wild places and collected anything that moved, as well as some things that didn't (rocks, for example). In that natural laboratory, we discovered the powerful biological logic described in our introduction. And studying the natural history of lizards drew both of us into more rigorous areas of science.

Every lizard species has its own fascinating story, only part of which can be told by means of data collected on existing populations. The evolutionary history of each species holds much more of the story than we will ever be able to recover. Present-day interactions between lizard species coexisting in the same environments are only the momentary result of an ongoing game of one-upmanship that makes even the best chess game seem simpleminded by comparison. In the long-term "game" of lizard interactions, moves alternate with countermoves in a constantly escalating manner, with each evolutionary episode featuring the survivors from the previous episode as the central players. Losers (nonsurvivors) never get to play the game twice.

Because birds are conspicuous and readily observed, they have been studied so intensely that for many decades "bird stories" dominated ecology books and classes. Now, lizards are finally coming into their own, and saurian examples are gradually supplementing or even replacing avian examples in the study of ecology. During the past few decades three symposia resulting in books have been devoted exclusively to lizard ecology (Milstead 1967; Huey et al. 1983; Vitt and Pianka 1994), and lizards have become model organisms for the study of a wide variety of phenomena, including energetics, reproductive tactics, thermoregulation, foraging, predation, parasitism, competition, spatial ecology, territoriality, life history, social behavior, community structure, and biogeography. Even psychology is playing a role: a recent symposium on snake ecology alluded to "lizard envy."

Lizards have proven almost ideal organisms for ecological studies, largely because of their basic traits and remarkable diversity. For one thing, lizards do not fly away, plus they are amenable to field manipulations, thus facilitating important ecological experiments. Because lizards are ectotherms, moreover, they are often abundant, making them relatively easy to locate, observe, and capture. Ectothermy may well give lizards an advantage over birds in areas of low and unpredictable productivity such as desert regions. Food a small bird consumes in one day supports a similarly sized lizard for a full month. Not only that, but these low-energy animals usually swallow their prey items whole, which allows stomach content analysis of diet.

Lizards also exhibit a wide range of morphological, behavioral, physiological, and ecological variation. Some species are annuals, living only one year, whereas others live for decades. Reproductive tactics are quite varied and readily studied. Some species, such as geckos, pygopodids, the iguanid *Anolis,* gymnophthalmids, and certain skinks have small clutch or litter sizes, while others are exceedingly prolific, laying large clutches of several dozen eggs year after year. Some species reproduce only once each year, while others have multiple clutches. Viviparity has arisen repeatedly among different lineages, as has leglessness and fringed toe lamellae. Arboreal and terrestrial lizards occur among both nocturnal and diurnal species. Some species, such as the Galápagos marine iguana, *Amblyrhynchus cristatus,* are highly aquatic as well as living comfortably on land. Many are fossorial, living a subterranean existence. Lizards inhabit a broad range of habitats, including deserts, grasslands, chaparral, rock outcrops, deciduous forest, and rain forest.

Lizards have varied diets. Although most species are generalist predators on arthropods, some are dietary specialists, eating only one prey type, including ants, termites, scorpions, other lizards, birds, mammals, and even some plant foods. Some lizards are ambush hunters, catching prey by sitting and waiting for it to move past. Others are more active, searching out prey by foraging widely. Although active foraging is energetically expensive and somewhat hazardous, in that it more readily attracts the attention of predators, it increases contacts with potential prey, especially relatively sedentary prey.

An important concept in ecology is the ecological niche, which describes how an organism conforms to and uses its environment. A lizard species' niche can be defined with respect to a single axis, such as temperature. Body temperatures of *Cnemidophorus lineatissimus* in thorn forest habitats of Jalisco, Mexico, for example, vary from about 33.5°C to 40°C, and average 36.5°C (Casas-Andreu and Gurrola-Hildago 1993). This lizard's thermal niche

Acanthosaura armata, **one of the Southeast Asian tree lizards (Agamidae), is a sit-and-wait foraging predator.** (Louis Porras)

is therefore defined within those limits, and it is not active at temperatures below about 33.5°C. Temperature is only one of many possible niche axes. The dietary niche of *C. lineatissimus* is defined by what it eats, its spatial niche by where it lives, and so on. Lizard niches are multidimensional, as the following list of variables makes clear:

1. *Spatial Niche*

   Habitat (deserts, shrubby habitats, forests, grasslands, etc.; sandridges; sandplains; rocky outcrops)

   Microhabitat (arboreal → terrestrial; open versus vegetation; fossorial; aquatic; diurnal versus nocturnal retreats)

   Anatomical correlates: Convergent evolution (fringed toe lamellae, shovel snouts, prehensile tails)

2. *Temporal Niche*

   Time of activity

   Nocturnal versus diurnal species

   Thermoregulatory tactics (thermoconformers → thermoregulators)

3. *Trophic Niche*

   Dietary niche breadth (generalists → specialists: ants, termites, scorpions, other lizards, birds, mammals, some plant foods)

   Mode of foraging (ambush hunters or sit-and-wait predators versus widely foraging; search versus pursuit; energetic costs and profits, etc.)

   Anatomical correlates: head length versus prey size; hinged teeth

4. *Reproductive Tactics*

   Egg-layers versus live-bearers (clutch size, reproductive effort, expenditure per progeny), bet hedging

5. *Predator Escape Tactics*

   Speed, agility, mimicry, camouflage, spines, tail length, body shape, tail autotomy

Frequently, two or more lizard species are similar with respect to one or more niche axes but different on others. The food niches of the Australian lizards *Egernia depressa* and *Diplodactylus conspicillatus,* for example, are both dominated by termites (85.5 and 99.5 percent, respectively; Pianka 1986), but whereas *E. depressa* is diurnal, *D. conspicillatus* is nocturnal; thus, their "time" niches differ. In the field of ecology, autecological studies tend to describe several niche axes for a particular species, whereas community ecology studies compare niche axes for all species in a particular place. Each provides a different perspec-

tive on lizards in the real world. As more niche axes are examined, differences among coexisting lizard species become more evident, and it may eventually be possible to construct a "periodic table of lizard niches" (Pianka 1993).

Examining niche axes independently is always instructive. For example, lizards exhibit a wide range of thermoregulatory tactics, ranging from totally passive thermoconformity (poikilothermy) to active temperature regulation. In an analysis of costs and benefits of lizard thermoregulatory strategies, Ray Huey and Montgomery Slatkin (1976) compared body temperature with ambient environmental temperature to investigate the degree of passiveness in thermoregulation. They found that among active diurnal heliothermic lizards (even very small ones), there was often almost no correlation between air and body temperature, suggesting efficient thermoregulation. Among nocturnal species the correlation was typically quite high, suggesting true poikilothermy. Some species—especially Australian ones, and including nocturnal and diurnal as well as crepuscular—were intermediate, filling in this continuum of thermoregulatory tactics. Further statistical (regression) analysis of body/air temperature data for 82 species of desert lizards from ten families and 54 species of Neotropical lizards from six families (Pianka 1986; Vitt, unpubl. data) rendered a very intriguing intersection at 44°C (fig. 7.1), defining the critical thermal maximum faced by all lizards—a limit that may well represent an innate design constraint imposed by lizard physiology and metabolism (Pianka 1985, 1986, 1993).

Ecology of individual lizard species is not independent of evolutionary history, and as a result, closely related species are often ecologically quite similar. All chameleons, for example, are much more like each other in the way they capture prey, move, and look than they are like more distantly related lizards such as teiids. Exactly how similar species are ecologically, and why, is not always obvious because of long and sometimes obscure evolutionary histories. Phylogenies (evolutionary trees) based on morphological or molecular traits offer an exciting new framework for elucidating evolutionary pathways, leading to ecological traits observed today.

PROXIMATE AND ULTIMATE FACTORS

Unlike some other sciences, such as chemistry and physics, biology is hierarchical. Biologists study events and phenomena over a vast range of spatial and temporal scales, from molecules to communities to the biosphere itself. Many biologists adopt a reductionistic approach, focusing on a relatively narrow range of phenomena. Thus we have molecular biologists, cell biologists, physiologists, behavioral biologists, population biologists, and community ecologists. To truly understand any biological phenomenon, however, an interdisciplinary attitude is necessary.

The terms "proximate cause" and "ultimate cause" (or effect), though frequently misused and misunderstood, are critical to understanding biological phenomena, for it is the interplay between proximate and ultimate factors that determines the how and why of observed events. Consider your automobile: the apparent reason it starts is that you turn the key. In fact, however, a series of proximate events are involved in that action. Turning the key connects energy stored in the battery to the starter, which spins, throwing a gear to connect with the flywheel on the engine, which in turn causes the engine to turn over. The same battery energy supplies high voltage to the spark plugs, and an ingenious distributor times production of sparks with injection of fuel into piston chambers where fuel ignites, forcing the pistons to move and turn the crankshaft, thus generating continuing power to run the engine. Turn the key off, and it all comes to an immediate stop because no spark is delivered to the plugs. This is a mechanistic explanation of how your car works based on a sequence of proximate events. From a driver's perspective, it is not necessary to even realize that more than one event (turning the key) enables use of the car. As for ultimate causes, one might mistakenly consider the oxidation process itself, whereby the spark ignites the fuel to power the engine, to determine whether the car will run simply because it is so far removed (in a mechanistic sense) from turning the key. However, oxidation, a physical process, is in fact still a proximate effect, part of the mechanistic train of events. If anything at all about gasoline engines can be considered an "ultimate" cause, it might be the long history of thought and experimentation that allowed humans to harness oxidative reactions—including the invention of engines and the engineering expertise that goes into designing them. Given the large number of people who burn themselves up by lighting matches near gasoline and other explosive compounds each year, natural selection has not done well in shaping our abilities to effectively harness the oxidative process at a population level.

Let us now apply these notions to lizards. Most Tem-

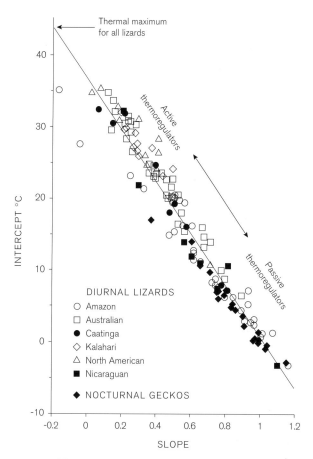

Figure 7.1 Variance in lizard thermoregulation tactics. Each data point in this graph represents the least-squares linear regression of body temperatures against air temperatures for a given species of lizard. Active thermoregulators, such as most North American diurnal lizards, have a low slope (i.e., their body temperatures are relatively constant regardless of air temperatures), whereas passive thermoregulators, such as geckos, have a high slope (their body temperatures are always close to air temperatures). A direct negative relationship exists between the slopes and intercepts of regressions of lizard body temperatures and air temperatures for lizards. These data indicate a spectrum of differing thermal strategies among different groups of lizards. Positions along this continuum can be used as an index reflecting complex use of spatial and temporal niche dimensions.

perate Zone lizards become inactive during winter, retreating underground to safe sites that will not freeze. (Because they are ectotherms, this is not called hibernation but brumation.) Consider factors evoking such behavior. Lizards have a vestigial third eye (the parietal eye), located beneath a scale on the top of their heads between their paired eyes used in vision. A lizard perceives shortening day lengths by comparing photoperiod to its internal biological clock. Decreasing day length and cooler temperatures elicit a pituitary response and hormonal changes that trigger the urge to become inactive. For a physiologist, this constitutes an effective answer to the question "Why do lizards brumate?" This mechanistic explanation focuses on the immediate environmental cue, or proximal cause, for the behavioral response.

Evolutionary biologists, however, would explain lizard brumation somewhat differently. In their view, winter constitutes a predictable long-term selective force. Individual lizards that do not evade winter's icy clutches die, leaving fewer offspring (or none) in the population gene pool. Individuals that avoid winter mortality, in contrast, survive to breed the next spring and pass on their genes to future generations. Any genetic predisposition to become inactive would thus be favored and would become incorporated into the gene pool. Over eons, natural selection fine-tuned lizards so that they behaved appropriately and became adapted to their natural environments. In this case, long-term predictable winter mortality is considered the ultimate cause for lizard brumation.

A physiologist's answer is mechanistic, whereas an evolutionist's is strategic. Ernst Mayr (1961) termed these the "how?" and "why?" approaches to biological questions. At a certain level, each answer seems complete and adequate. But to really understand lizard brumation, one needs to know both answers. Day length is a short-term, mechanistic, proximate cue eliciting brumation, whereas winter mortality is the long-term, strategic, ultimate factor that led to the evolution of the proximate mechanisms controlling the behavior. Any biological phenomenon can be "explained" across a range of levels of approach. Modern ecology requires integration of proximate and ultimate factors to understand the complex interactions that allow life to develop meaningful and effective strategies for its continued existence.

LIZARD COMMUNITIES

An ecological "community," consisting of all the organisms that live together at any given spot, can be visualized as a complex network of interacting species. This set of species, coupled with the surrounding physical, abiotic environment, constitutes an "ecosystem." Ecosystems are classified into different "biomes," such as tundra, desert, grassland, savanna, chaparral, thorn scrub, deciduous forest, coniferous forest (taiga), and rain forest. Lizards are conspicuous components of most natural communities, especially those in warmer places, though lizards occur in some very cold places as well, such as high in the Andes.

An important point to keep in mind is that, unlike individual organisms, communities are not designed directly by natural selection. Individuals exhibit differential reproductive success, but communities do not. It is tempting, but misleading, to view ecosystems as "superorganisms" that have been "designed" for efficient and orderly function. In fact, antagonistic and asymmetric interactions at the level of individuals and populations (such as competition, predation, parasitism, and even mutualism) frequently impair certain aspects of ecosystem performance while enhancing other properties.

Community structure involves all the various ways that members of communities relate to and interact with one another, as well as any community-level properties that emerge from those interactions. Finding ways of simplifying communities for more thorough understanding remains a considerable challenge in the emerging field of community ecology. Defining and examining niche axes of component species is a good start because it allows us to make direct comparisons among species on various dimensions.

One way to begin to understand something as complex as a community is to adopt the reductionistic "take it apart, then try to put all the pieces back together again" approach. This has led to the idea of compartments, or taxonomic subsets of the entire community, one type of which is trophic level. According to this scheme, plants, which use energy directly from the sun and nutrients in their environments to produce plant tissue, are primary producers or autotrophs ("self-feeding"), whereas herbivores and primary carnivores are primary and secondary consumers, respectively. Because consumers do not feed themselves, they are called heterotrophs ("other feeding"). Another, vital, heterotrophic trophic level consists of decomposers: bacteria and fungi that recycle dead organic materials back into nutrient pools where these resources can once again be taken up by primary producers. This compartmental approach emphasizes one niche axis, food, but examines it not from the perspective of a single species but rather on a communitywide basis.

Communities can be compartmentalized in other ways as well. One useful concept is the guild, a group of functionally similar species that exploit the environment (use

A thin, streamlined body and an extremely long tail allow this Australian *Diporiphora superba* to move about in tall grass. (G. Shea)

specific niche axes) in similar ways. Diurnal terrestrial insectivorous lizards constitute a guild distinct from diurnal arboreal insectivorous lizards, for instance, and both of these are different from guilds represented by nocturnal or fossorial or herbivorous lizards. Within any given guild, species have the potential for stronger interactions than they do with members of different guilds simply because they use at least one resource in a similar way. Thus guilds provide the potential for strong interspecific competition, with species-specific solutions. Because lizards in a termite-eating guild, for example, must compete for termites, they might gradually adapt to searching for food at different times of the day. Because real communities contain many species, ecologists often study subsets of species, such as members of a given guild.

Another type of taxonomic subset is the assemblage, which in the case of lizards would be all species of lizards in the community. Still another useful natural compartment is a "sink subweb," constituting everything eaten by a particular top predator plus everything eaten by its prey and by their own prey, etc. (see fig. 7.5 for a partial example of such a subweb).

"Community structure" is another important term used by community ecologists, one that we all likely understand intuitively, though we may never have called it that or thought about it as such. Consider a stroll at about 11 A.M. into the Sonoran Desert after having just examined the many fine exhibits in the Arizona-Sonora Desert Museum west of Tucson. In the museum, you saw many desert plants and animals and learned a lot about each, but in an entirely artificial environment: a leopard lizard (*Gambelia wislizeni*) or a western whiptail (*Cnemidophorus tigris*) in a cage tells you virtually nothing about where each species really lives or how it survives. On your stroll, you observe a large adult leopard lizard sitting on the ground in the shade of a desert shrub. At the same time, you watch an adult whiptail wander around, stop at the base of a creosote bush, and dig into a termite mud tube casing and eat several termites. The leopard lizard, seeing the smaller whiptail lizard digging, rushes out and grabs the whiptail in its powerful jaws, ultimately killing and swallowing it.

This seemingly trivial set of observations provides the basis for recognizing community structure. Each lizard did something very different: one actively searched for and found small prey; the other sat in one place and when a large prey item came within range ambushed it and ate it. If you remained in the area long enough, you would see that whiptails and leopard lizards typically and repeatedly do what you observed; they do different things, and they do them purposefully. Leopard lizards could dig out termites, but they do not; nor do whiptails ambush and eat leopard lizards. In simplest terms, this is structure. If all species used all resources randomly, there would be no structure. By observing all species in this

habitat, you would find that structure in the Sonoran Desert lizard assemblage has many dimensions. Lizard species eat different things, live in different microhabitats, and are active at different body temperatures and times. Examination of simple lizard assemblages often allows an investigator to determine underlying causes for structure with a minimum number of extraneous variables, whereas examination of more complex systems provides a means to assess the importance of often more obscure variables.

THE POWER OF SIMPLE SYSTEMS

One of the more interesting questions in community ecology is whether independent communities that appear to be structured similarly have similar histories; that is, is community development predictable? This seemingly simple question has in fact been among the most challenging in the field of ecology. Because each species has its own independent evolutionary history, the set of potential variables is different for each species. To make

SERENDIPITOUS ACQUISITION OF A FIELD ASSISTANT

In graduate school at the University of Washington in Seattle, I became interested in the problem of species diversity: why are there more species in some places than in others? A prominent geographical pattern, repeated in many different groups, became the focus of my research: latitudinal gradients in species diversity. Deciding to study ecology and diversity of flatland desert lizards in western North America, I selected a series of ten study sites along a 1,000-km latitudinal transect from southern Idaho through southern Arizona. I took field notes and collected samples of lizards so that I could later determine such things as clutch size and stomach contents.

When I began my fieldwork in the early 1960s, the U.S. desert southwest was still pretty much wide open and unfenced. One could pull off the road virtually anywhere and find relatively pristine desert. This gradually changed with encroachment of urbanization, grazing, agriculture, and land speculation. By the next decade, unfenced desert had become very hard to find. Many of my study areas no longer support any lizards, and the collections I made, now deposited in major museums, represent a sort of modern-day fossil history of what was once present at these sites before human encroachment on the habitat.

For three field seasons from 1962 through 1964 I was a desert rat. Living out of my blue Volkswagen van, Elizabeth (Betsy for short), I drove up and down this transect collecting data for my Ph.D. To escape the desert heat during midday, I would sometimes drive into a nearby town, where I could also usually find access to a typewriter.

One day I went to the air-conditioned library in the small town of Mojave, California, and asked the librarian if I could use her typewriter to type a few pages. She quizzed me about what I was doing, and when I told her I was a graduate student studying lizards, she became excited and told me that her oldest son was extremely interested in lizards. She invited me to come over for dinner to meet him. Bill Shaneyfelt was a high school student, eager to assist. I put all his siblings to work stringing tags (each lizard specimen had its own unique number).

It was summertime, so Bill was out of school, and he volunteered to accompany me on a three-week circuit of several study sites in southern California and southern Arizona. He was a dead-eye with a BB gun and helped me collect hundreds of lizards. I recall him matters worse, the degree of difference varies depending on the species' level of evolutionary relatedness. Consequently, searching for repeated historical patterns has always been confounded by use of evolutionarily distant species reflecting different historical constraints.

During the latter half of the twentieth century, Harvard University herpetologist Ernest E. Williams and his students initiated a series of studies on *Anolis* lizards in the Caribbean islands that avoided many earlier problems in comparative studies. The studies had two advantages in particular: First, lizards under investigation are relatively closely related, so differences in evolutionary histories were minimized. Second, because Caribbean islands have different histories and different anole faunas but similarly structured physical environments, the potential existed for determining whether community structures were similar on different islands.

Williams (1969, 1983) identified what he called "ecomorphs," anoles with specific morphologies associated with like microhabitats on different islands (fig. 7.2). Trunk-ground anoles, for example, which live at the in-

High in the canopy and on trunks of Amazonian trees, *Anolis punctatus* (Iguanidae) represents a dominant portion of the canopy lizard fauna. This one is molting. (Laurie Vitt)

popping off a fast, wary zebra-tailed lizard at a long range of 15 m or more (Bill had to aim high to allow for the BB's falling trajectory). Another time, Bill brought in a peculiar lizard that he couldn't identify. For a brief moment we thought it might be a young leopard lizard; but it turned out to be a gigantic male *Uta* that had somehow managed to survive longer than a year. Lizards have relatively indeterminate growth and continue to grow most of their lives. This was by far the largest male *Uta* I ever observed.

We discovered a way to catch fast and cautious whiptail lizards. If pursued, they would often dart down a kangaroo rat's burrow, pausing just inside the opening to look back and see if the pursuit would be continued. Noting the burrow's direction, we crept up from a side the lizard could not see, leapt up,

and jumped, smashing the surface in on top of an unsuspecting lizard. Then quickly down on our knees, all hands poised and ready to grab, we'd paw through the sand. Once, just a second before Bill committed to his lunge, a flickering forked tongue warned him to hesitate as a sidewinder's head emerged from the loose sand!

A few years later I finished my Ph.D. I found that the species of lizards living together in flatland desert habitats varied in number from four in the cold shrubby deserts in the north to as many as ten in the warm Sonoran Desert of southern Arizona. The number of species of lizards occurring together was correlated with the structural complexity and spatial heterogeneity of the vegetation in desert habitats. Northern deserts are structurally simple, supporting a homogeneous vegetation consisting of small

chenopodeaceous shrubs, such as sagebrush (*Artemisia*) and saltbush (*Atriplex*). Southern deserts, in contrast, support a much more complex vegetation that includes a variety of small semishrubs, larger woody shrubs including creosote bush (*Larrea*), scattered trees such as Joshua trees, palo verde, and ironwood, plus various sorts of cactus. Such spatial heterogeneity offers a greater variety of microhabitats, thus supporting more lizard species (Pianka 1967). Several species of climbing lizards require large shrubs and trees for perches. In addition, southern sites supported nocturnal lizards. With Bill Shaneyfelt's help, I was able to determine why more lizard species occur in some places than in others. *(PIANKA)*

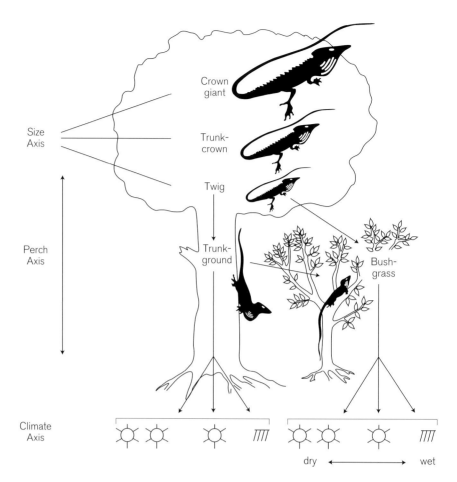

Figure 7.2 Evolutionary radiation in *Anolis*. Depending on where anole species live, their morphology varies, producing characteristic body forms, called ecomorphs or ecotypes. Ecological factors contributing to formation of ecomorphs are shown here (Williams 1983, 239). (Redrawn from Huey et al. 1983 by permission of Harvard University Press, © 1983 by the President and Fellows of Harvard College)

terface between tree trunks and the ground, are relatively small and chunky bodied. Crown giants are very large bodied, long tailed, and live in the canopy. Williams's observations that ecomorphs were similar on different islands, but that the actual species of anole varied from island to island, suggested that similar community structure had arisen independently a number of times.

In the 1990s an energetic ecologist from Washington University in St. Louis, Jonathan Losos, used modern molecular techniques to examine evolutionary relationships among anoles on islands of the Caribbean to determine the sequence of events that led to the similar structure of these anole assemblages. Figure 7.3, for example, shows probable sequences of evolution of various *Anolis* ecomorphs for Jamaica and Puerto Rico—sequences that exhibit striking similarity (Losos 1992). Except for the additional presence of the grass-bush ecotype on Puerto Rico, extant *Anolis* assemblages on these islands are very similar. This example suggests that lizard assemblages

build up in a predictable and repeatable fashion and that certain ecotypes emerge during evolution to fill various "empty" niches.

THE POWER OF COMPLEX SYSTEMS

Lizards can easily be categorized as living in specific microhabitats, being active at particular times of day, and even eating certain kinds of prey. An observant naturalist could determine, for instance, that the desert horned lizard, *Phrynosoma platyrhinos,* is active in morning and late afternoon, eats nearly exclusively ants, and lives in open spaces on the desert floor. Slightly more sophisticated observations form the basis for studies in lizard community ecology. For example, ants found in lizard stomachs can be identified to species and counted, providing a quantitative measure of prey use. Microhabitat and activity data can be collected and quantified in much the same manner. By collecting similar data for all species

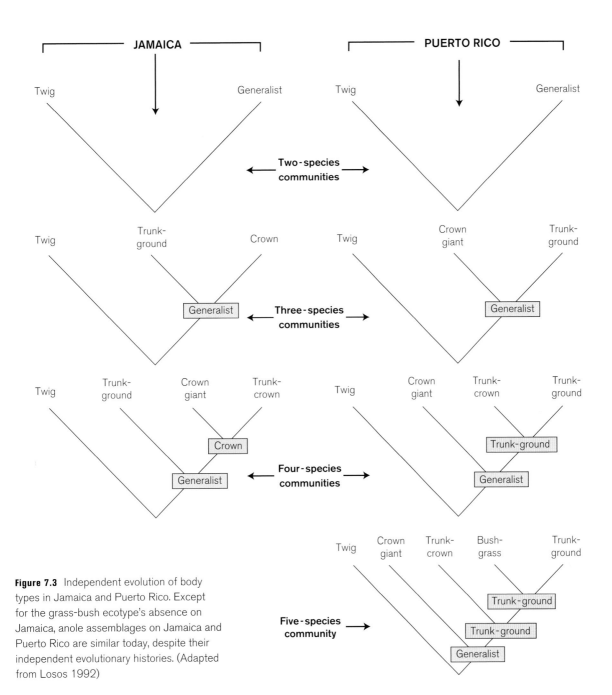

Figure 7.3 Independent evolution of body types in Jamaica and Puerto Rico. Except for the grass-bush ecotype's absence on Jamaica, anole assemblages on Jamaica and Puerto Rico are similar today, despite their independent evolutionary histories. (Adapted from Losos 1992)

within a defined area, a naturalist can make the transition from single-species observation to community ecology and begin to ask questions about observed patterns of resource use, such as: How many species do particular assemblages contain? Do the different species in an assemblage use available resources in different ways—or as a community ecologist would put it, do they engage in "resource partitioning"? If so, what are the proximate and ultimate causes of that partitioning behavior?

A useful way to depict an assemblage of consumer species (e.g., lizards) is to construct a consumer-resource matrix, a table quantifying relative use (utilization coefficients) of various resources (table 7.1). "Resource states" can be prey types or microhabitats. Say ants constitute 20 percent of the diet of one lizard species but 90 percent of the diet of another species. The resource state in this example is ants, and the utilization coefficients are the proportions of ants used by each lizard species (0.2 and 0.9,

TABLE 7.1

A resource matrix for four species of North American desert lizards

| | PREY TYPE* | | | | |
SPECIES	ANTS	GRASSHOPPERS	BEETLES	TERMITES	LARVAE
Cnemidophorus tigris	0.0	0.2	0.2	0.5	0.1
Uta stansburiana	0.2	0.4	0.3	0.0	0.1
Gambelia wislizeni	0.0	0.3	0.1	0.0	0.0
Phyrnosoma platyrhinos	0.9	0.0	0.1	0.0	0.0

*Entries represent proportion of diet for each species.

respectively); the consumer-resource matrix simply tabulates utilization coefficients for all resources (including ants) used by all lizard species in the assemblage.

Such a matrix allows easy calculation of niche breadths and overlaps among species. A niche breadth is a quantitative measure of the variety of resources used by members of a particular lizard species. For example, in a desert lizard community, a species like *Phrynosoma platyrhinos,* which eats 90 percent ants, has a narrow dietary niche breadth because it relies largely on a single prey category, even though many other prey are available. The side-blotched lizard, *Uta stansburiana,* in contrast, has a broad dietary niche breadth because it eats many different kinds of prey. Niche breadth (B) for each species can be calculated using a simple formula,

$$B = 1/\Sigma p_i^2$$

where p_i is the proportional use of resource i. (Although the p_i values for individual species are called utilization coefficients [u_i] in the consumer-resource matrix, the values are the same.) If a lizard is a specialist like *P. platyrhinos,* feeding primarily on a single resource, Σp_i^2 approaches 1, so B is also near 1, indicating a narrow dietary niche. However, if a lizard species uses each of ten resource categories evenly (each $p_i = 0.1$), then Σp_i^2 is 0.1 and B is 10, indicating a broad dietary niche breadth. The first is a specialist and the second is a generalist.

Dietary overlaps indicate how similar each species is to all others in its use of particular resources. Species with identical diets (ones that eat only caterpillars, for example) have overlaps of 1.0 (unity); species that share no resources have zero overlap; and species with some shared resources have overlaps of less than 1 but greater than 0. Species with high overlap, moreover, belong to the same guild and are potential competitors.

The question arises whether species partition resources. In other words, in a real-world community, where interactions among species do occur, is observed overlap lower than would be expected if there were no interactions? Put yet another way, are patterns of resource utilization in a community random, or are they "coadjusted" to mesh in a meaningful way?

Answering these questions requires the use of randomized replicates (pseudocommunities) based on observed prototype communities, together with some creative statistics. First we return to our original consumer-resource matrix and randomize the order of utilization coefficients within each species' row. In table 7.1, for example, a randomization for *Phrynosoma platyrhinos* would result in the lizard eating 90 percent beetles and 10 percent ants, rather than the other (correct) way around. By restricting the randomization to numbers already logged for each species, two important components of their natural history are retained: the set of resources used and niche breadth (because we have not changed the numbers, but simply reordered them, for each species). So *P. platyrhinos* becomes a beetle specialist rather than an ant specialist, but a specialist just the same. We then rerun the overlap analysis with such shuffled numbers—but do so thousands of times, for thousands of pseudocommunities, to answer our question of whether our real community differs from all possibilities in terms of pattern and organization. Inevitably, some of the structure of the prototype is mirrored in these randomized replicates, though interestingly, similarity among consumers is usually higher and more homogeneous, and guild structure less evident, in randomly constructed models than in real systems. To put this another way, real species typically differ more from each other in their ecological attributes than do hypothetical species in pseudocommunities, indicating that something has caused them to diverge in their use of resources. Random sampling from among available resources does not apply in the real world. (Indeed, if all lizards in a particular place did use resources randomly, not only would most of the interesting questions become trivial, but we would have to reevaluate the theory underlying most of ecology!)

The community analysis discussed thus far is basically descriptive, even though substantial data collection and computations are involved. We are simply describing quantitatively what most good field naturalists observe qualitatively. Indeed, the qualitative "feel" for how the natural world works is what drives such analyses, which nowadays are aided by powerful computers that allow us to do things we could only dream of a few short years ago.

All lizard community analyses show that lizard assemblages are structured (species do not use resources randomly) and that at least some guilds exist (some sets of species within the assemblage do things in a similar way).

This is so because a host of variables influence each species differently, including species interactions (competition, predation, and parasitism), biogeography, phylogenetic relatedness, body size, and morphology. As a result, lizard community studies are challenging, but also illuminating.

With the exception of very simple systems, natural communitywide experiments are logistically impossible. Laboratory- or computer-based experimental studies in which a single variable is changed while everything else remains constant therefore offer perhaps the best opportunity to examine community dynamics. One approach to seeing how change affects an assemblage is to insert species data from other assemblages into the one of interest. This approach is useful for examining questions such as, How well do resource utilization patterns observed among sympatric lizards in a given system mesh? Or, Can evidence be found for ecological adjustments among coexisting species?

Feeding nearly exclusively on ants, the Amazonian thornytail, *Uracentron flaviceps* (Iguanidae), lives in social groups in hollow trees.
(Laurie Vitt)

To address such questions, Eric Pianka (1986) under-took a series of artificial "removal-introduction experiments" using data on lizard diets. Resource matrices for diets were assembled and analyzed to estimate preferences, or "electivities," of consumer species on prey resources over many different study sites. The removal-introduction part of the procedure was relatively simple: At each study site (thirty in all, though the number for any given species varied), a selected lizard species was represented by an observed "resident" population. Residents were then systematically replaced by "aliens"—individuals of the same species taken from each of the other study sites. A moderately large number of such "transplants" was possible: for a ubiquitous species found on ten different study sites, for example, nine alien "introductions" could be made on each area, allowing a total of ninety alien-versus-resident comparisons. Meanwhile, resource utilization spectra of all other resident species remained unaffected. The "goodness of fit" among species was estimated by the amount of niche segregation observed.

Each resident species' observed overlap with all other members of its community was compared to the overlap a transplanted alien experienced. The result was less overlap for residents of most species than for aliens. Among all 90-odd species over all thirty study areas, residents outperformed aliens in 1,871 out of 3,014 cases, or 62 percent of the time. This trend is more pronounced when expanded food resource matrices are used: in the Kalahari semidesert, when 46 prey categories (including termite castes) are recognized instead of only 20, residents outperform aliens in 810 out of 1,056 cases (76.7 percent); in a comparable analysis for two Australian study areas using 200 prey resource states, residents outperformed aliens in 39 of 52 cases (75 percent). These results strongly suggest that compensatory interactions to reduce overlap occur among members of naturally coexisting lizard assemblages.

Community ecologists need ways of reducing complex, multidimensional systems to a simple graphical form in which structure and organization can be appreciated. Working on Neotropical fish assemblages in South and Central America, Kirk Winemiller, now a professor at Texas A & M University, developed a promising hybrid approach to compare aquatic systems with desert ones (Winemiller and Pianka 1990).

Raw data for these analyses were formatted as resource utilization matrices, a process that involves considerable tedious work; if a system is changing over time, statistical sampling of all species in the system needs to be done both quickly and repeatedly, at various times. Entries in a resource matrix are used to calculate probabilities that a given consumer species will use particular resources (with probabilities ranging from 0, indicating no use, to 1).

Next, resource utilization needs to be standardized in terms of relative availabilities (Ivlev 1961), since simple dietary proportions weight uncommon or very abundant resources disproportionately. Resource availability, however, is not easily measured in the field. Insects can be sampled with sweep nets, DeVac vacuum cleaners, tanglefoot sticky traps, pit traps, or burliese funnels, with each technique yielding a different result depending on ease of capture of specific insects. Jaime Pefaur and William Duellman (1980) faced this problem while studying Andean reptiles and amphibians. After fencing study plots and collecting all frogs and squamates, they proceeded to collect all conspicuous insects in the plots to compare against stomach contents. Incredibly enough, they succeeded in finding only about 10 percent of insects eaten by reptiles and amphibians (W. Duellman, pers. comm.)!

Resource availabilities, of course, depend to some extent on each species' use of space and time as well as on its particular sensory capacities and foraging mode. We know from many community studies on desert and Neotropical lizards that different species do different things, segregating on the basis of preferred food, time of activity, and microhabitat (Lawlor 1980a,b; Pianka et al. 1979; Pianka 1986; Vitt 1995; Vitt and Zani 1998a,b; Vitt, Sartorius et al. 2000; Winemiller and Pianka 1990). The long-standing paradigm is that ecological differences that allow coexistence among lizard species (i.e., the reason for structure) are rooted in interspecific competition. The logic is quite simple: if species use particular resources better than other species (competitors), they will, all else being equal, increase more rapidly in population size and ultimately eliminate or displace competitors; thus, if species are to coexist, they must use different resources. Not only does a great deal of empirical evidence confirm this notion, but computer simulations support it from a theoretical viewpoint as well. Lizard assemblages may contain separate sets of species so drastically divergent evolutionarily that they have never interacted. The ability of any one of these species to enter a particular assemblage may result from this evolutionary difference, combined with the failure of other species to exploit particular resources, even in the absence of competitors.

A lizard fauna of the Amazonian rain forest, comprising nineteen relatively common species and several rare ones, provides insight into such an assemblage. Five families are represented: Iguanidae (three subfamilies), Gekkonidae, Teiidae, Gymnophthalmidae, and Scincidae. Each family has experienced independent evolutionary histories, such that generalizations about each can be made regardless of locality. For example, teiids, regardless of where they occur, are active foraging heliothermic lizards, whereas gekkonids are sit-and-wait foragers and not heliothermic (or at least much less so than teiids). A pseudocommunity analysis of diet and microhabitat data reveals structure in this assemblage (Vitt et al. 1999). However, when the lizards' evolutionary histories are considered, a significant portion of observed structure is tied to phylogeny: species that are more closely related are also more similar in ecological characteristics. Indeed, the most prominent guild in this assemblage has two congeneric species, *Plica plica* and *P. umbra,* that are ant specialists. Ants, however, are relatively rare in diets of the other seventeen species. Did ant specialization result from interactions with the other species, or were the two ant specialists able to enter the assemblage because they were already specialized on ants but none of the other species were?

Answers to these questions are embedded in tropidurine phylogeny (fig. 7.4). The sister taxon to all other tropidurines is *Uranoscodon,* a shade-seeking rain forest lizard that rarely eats ants. The other lineage, which ultimately produced *Plica plica* and *P. umbra,* has its center of diversity in open habitats. Moreover, all tropidurines eat some ants, but the arboreal Amazonian species, which are farthest out on the evolutionary tree, specialize on ants. Historically, then, ant specialization was evolving prior to invasion of lowland rain forest.

An increasing tendency toward arboreality also occurred historically within tropidurines. Most likely, when *Plica plica* and *P. umbra* or their ancestors entered Amazonian rain forest, they were already arboreal—the smaller-bodied *P. umbra* living on trunks of moderate-sized trees, and the large-bodied *P. plica* living on trunks of the largest trees—and fed on ants. They faced no competition: No teiid at this site is arboreal (one, *Kentropyx calcarata,* forages off the ground on low shrubs as well as on the ground), none specializes on ants, and the only lizards that

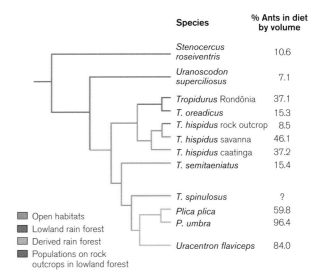

Figure 7.4 Evolution of ant-eating in Amazonian lizards.

use tree trunks at all are several small-bodied *Anolis* and the small gecko *Gonatodes humeralis.* Only the latter is common on trunks and restricted trunk bases, rarely if ever used by tropidurines, which prefer broad surfaces. Gymnophthalmids, meanwhile, are all terrestrial or semiaquatic, living in leaf litter or on stream banks; no ant specialists exist in this group either.

Does all this mean that species interactions, particularly competition, are not important determinants of community structure? No. Rather, it suggests a historical component to development of community structure, tied to the evolutionary history of interacting species.

FOOD WEBS

Any community of organisms can be represented by a food web describing trophic relationships; the food web is in turn generally composed of many food chains, each representing a single pathway up the food web and seldom consisting of more than five to six links, usually only three or four. The perentie, *Varanus giganteus,* for example, is at the top of the partial Australian desert lizard food web shown in figure 7.5. Although all the lizards depicted here eat prey other than lizards and all are eaten by predators other than lizards, making this but a very small subset of a much larger (community) food web, this illustration demonstrates the complexity of natural systems even

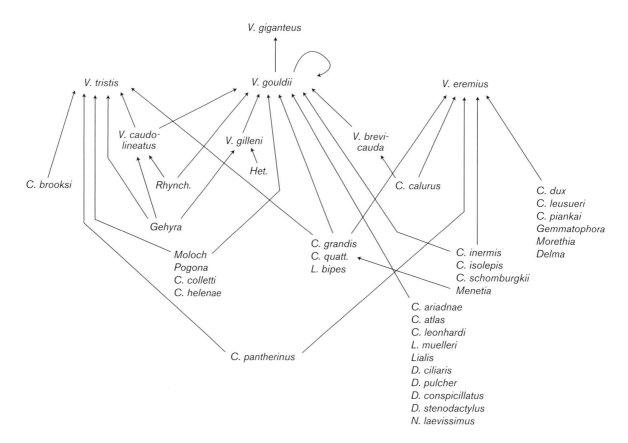

Figure 7.5 Part of the food web in the Great Victoria Desert of Australia. Each name represents a different species of lizard. Abbreviated genera: *V.* = *Varanus, C.* = *Ctenotus, D.* = *Diplodactylus, L.* = *Lerista, N.* = *Nephrurus, Rhynch.* = *Rhynchoedura,* and *Het.* = *Heteronotia.* The "top" predator, a monitor lizard, *Varanus giganteus,* eats another monitor, *Varanus gouldii,* which consumes twenty-seven species of lizards, including its own and three species of pygmy monitors *(V. caudolineatus, V. gilleni,* and *V. brevicauda). Varanus tristis* is arboreal and eats other lizards and some insects, but also feeds on baby birds and bird eggs. Another terrestrial pygmy monitor, *V. eremius,* preys on fourteen other species of lizards. Lizards lower down in this food web feed on various sorts of arthropods, which prey on other insects or eat plants. Like all food webs, this one is incomplete; many more links between species doubtless occur. A more satisfactory food web would include all members of the community concerned, including plants and insects, species by species, and would indicate actual rates of flow of energy and matter for each link in every food chain.

when only lizards are considered. *Varanus gouldii,* for example, which is eaten by *V. giganteus* and by other (larger) *V. gouldii,* eats nearly every other lizard species. As a group, therefore, all seven species of *Varanus* have a significant impact on other lizard species as predators.

The number of species in a food web *(n)* directly determines the number of possible connections, according to the formula [*n(n* − 1)]/2. Thus a food web with 4 species has 6 possible connections, one with 5 species has 10, and one with 6 species has 15. As species increase in number, the possible connections likewise rapidly multiply: 190 for a community of 20 species, 780 for 40. Of course, in real food webs not all species are connected, so the actual number of connections is much smaller than the possible number. To compare communities, therefore, ecologists refer to "connectance," defined as the actual number of connections divided by the total possible number of connections. This simple numerical index of food web complexity is thought to relate directly to community stability. Higher connectance means that more checks and balances are in place to regulate populations, possibly making them more stable.

Stable natural communities, however, are quite inefficient, with only about 10 to 15 percent of the energy at any given trophic level available to the next higher trophic level (an "ecological efficiency" of 0.10 to 0.15). Why is this so? One answer lies in natural selection. When it operates on individual predators in homogeneous predator-prey systems, selection favors food-capturing ability, which in turn renders energy flow up through trophic levels more efficient; yet ultimately, the system is left more unstable as overexploitation, the extinction of prey, and starvation of the predators themselves result. The reverse is seen when selection operates on individual prey in heterogeneous habitats offering more hiding places, for then escape ability is favored, reducing energy flow and enhancing stability. In the coevolutionary arms race between a predator and its prey, the prey must always remain a step ahead of their predators; only then will a community be stable.

As a corollary, community-level properties of ecological efficiency and community stability may in fact be inversely related precisely because natural selection operates at the level of *both* predators and prey. Thus, the apparent constancy and low level of ecological efficiency observed in natural ecosystems could be a result of the "compromise" that must be reached between coevolving prey and their predators.

In this chapter, we have highlighted lizards as model subjects for basic ecological studies. Lizards can be exploited to learn how natural communities are organized as well as why ecosystems follow apparent rules. Such understanding is vital for humans who must manage simpler artificial (man-made) ecosystems.

Gonocephalus chamaeleontinus
in a dark forest. (L. Lee Grismer)

Part Two

LIZARD DIVERSITY

CHAPTER 8

IGUANIANS

The agamid *Lophognathus longirostris* occurs in arid Australia. This one is in the Great Victoria Desert in Western Australia. (Eric Pianka)

Iguania is the sister taxon to all other lizards and contains three families: Iguanidae, Agamidae, and Chamaeleonidae. Lizards with dorsal crests, colorful dewlaps, and seemingly endless sets of push-ups or head bobs—these are what come to mind when one thinks of iguanians. Morphological, ecological, and behavioral diversification is strikingly similar between iguanids (primarily New World) and agamids (strictly Old World), with numerous examples of convergent evolution. Chameleons conjure up a somewhat different image, largely because of their zygodactylous toes;

coiled, prehensile tails; and turretlike eyes. Nevertheless, chameleons are adorned with often striking morphological ornamentation, and their behavioral antics are right in line with other iguanians. All Iguania exhibit an ancestral condition shared with the tuatara: a fleshy tongue and lingual prehension of small prey (Schwenk and Throckmorton 1989).

Nomenclatural history of the Iguania has been in flux in recent years. For a long while the above three families prevailed. In 1989 Darrel Frost and Richard Etheridge elevated eight presumably monophyletic groups within "Iguanidae" to family-level status and restricted use of the family name Iguanidae to a much smaller group (formerly a subfamily) of nine genera—31 species in all—of large herbivorous lizards found primarily in the New World. This change has been controversial, partly because many details of evolutionary relationships among groups within Iguania remained unresolved (Schwenk 1994a). Even more recently, additional former subfamilies within "Iguanidae" have been elevated to family level (Frost, Etheridge, et al. 2001). Whether these are called families or subfamilies is arbitrary. We adopt a conservative approach, retaining Iguanidae as a family, with clades recognized by Frost, Etheridge, et al. as subfamilies. Whereas Chamaeleonidae and Iguanidae now appear to be monophyletic, Agamidae is not (Macey et al. 1997) because chameleons evolved from agamids, though they have traditionally been excluded (a problem easily resolved by recognizing a clade called Acrodonta, which includes both agamids and chameleons).

The north African agamid _Laudakia stellio._ (R.D. Bartlett)

During the Middle Jurassic, about 180 M.Y.B.P., the northern Laurasian plate separated from the southern Gondwanan plate. Gondwana contained primitive Iguanian lizards (Estes 1983). Later Gondwana itself fragmented, isolating Iguanians on the three large southern landmasses, Africa, South America, and Australia. These ancient lizards subsequently underwent multiple massive adaptive radiations, giving rise to three groups currently recognized as families, which dispersed into the Northern Hemisphere as well. Members of these three families, constituting about 1,230 extant species, forage visually, primarily by sitting and waiting for prey to move within range (herbivorous lizards are exceptions), and do not appear to discriminate prey on the basis of chemical cues. Most striking are radiations within one family, Iguanidae (eleven subfamilies are recognized). We now consider each major iguanian clade.

AGAMIDS

Agamids are small, medium, to large diurnal lizards found across central and southern Asia, Southeast Asia, New Guinea, Australia, the Solomon Islands, and all of continental Africa (they are missing from Madagascar). All have movable eyelids. Scales are irregularly shaped and are rough, spiny, or keeled in many but not all species. Enlarged scales are scattered across the dorsal surface in many species, sometimes forming dorsal crests. Many species possess intrascalar pores (Witten 1993). Head scales are usually smooth and seldom enlarged. The external ear opening with its tympanic membrane is conspicuous in most agamids (with the exception of some Australian _Tympanocryptis_ and Indian and Sri Lankan _Otocryptis_). Many agamids exhibit striking sexual dimorphism in body size as well as color pattern (dichromatism). Many are territorial, though some, such as Australian _Pogona,_ do not appear to defend territories. All agamids lay eggs except for a few species of the high-latitude northern Eurasian genus _Phrynocephalus,_ which are live-bearers (as may be _Cophotis,_ of Sri Lanka). Small species have small clutch sizes, but some larger species lay more than a dozen eggs. Some agamids lay multiple clutches of eggs during a growing season.

A unique derived feature that ties the Agamidae with its sister group Chamaeleonidae is "acrodont" (modified pleurodont) teeth, which are fused to the top of the jawbones and are not replaced after they are formed (Estes

The Australian bearded dragon *Pogona barbatus* with its "beard" erected. (C. Ken Dodd)

et al. 1988); rather, as a lizard grows, new teeth are added from the rear (Witten 1993). Agamids also have caniform (sometimes fanglike) pleurodont teeth anteriorly, which are replaced continuously.

Agamids range in appearance from commonplace to bizarre to spectacular. One of the most striking is tropical Australia's frilled lizard, *Chlamydosaurus* (= cloaked lizard), which has evolved a large, colorful fanlike frill around its neck that is spread out during a threat display. Two modified elongate hyoid bones form rods, which are used to expand the frill. Like many long-legged lizards, frilled lizards can run bipedally.

Seven species of "bearded dragons," genus *Pogona* (= bearded), live in Australia. Although any given area will host but a single species, the geographic distribution of all seven includes most of the continent. These large lizards are semiarboreal. *Pogona* have a distinctive throat pouch, which is extended in defensive displays, but they do not appear to defend territories. They lay many small eggs.

An interesting genus of "flying" lizards that ranges from

India to Southeast Asia, *Draco* (= dragon), has evolved long movable ribs that support a thin winglike, often colorful, membrane of skin called a pterygium. Wing colors often differ between sexes. With their "wings" extended, these delicate, slender agamids glide gracefully between trees, losing altitude along the way; steering with their long tails, they can travel as far as 50 m in this manner, raising up and stalling at exactly the right moment to make a gentle landing, head up, on a tree trunk. They then scamper upward, gaining altitude in preparation for their next flight. When on the ground, *Draco* are clumsy and vulnerable to predators. About 40 species range from Thailand and Singapore through Borneo to various islands of Indonesia; 1 species occurs in India. *Draco* vary greatly in body size—because smaller species do not lose altitude as fast as large ones, they can glide farther. Many exhibit pronounced sexual dimorphisms in size and color, and in some, females are more colorful than males (J. A. McGuire, pers. comm.). Males defend territories, courting females by extending their brightly colored throat appendage (Daniel 1983), much as anoles do in the New

The agamid *Hypsilurus spinipes* lives in rain forests of southeastern Queensland and northeastern New South Wales, Australia.

(Dr. Hal Cogger)

World. *Draco* consume large numbers of ants and termites. Their unusual eggs are elongate and spindle shaped with dense calcium carbonate "caps" at each end, the function of which has not yet been studied (C. H. Diong, pers. comm.).

The agamid genus *Ctenophorus* (= comb bearing) has undergone an adaptive radiation in Australia, where about 20 species are found. Some are rock dwellers, others live on the ground in open sandy desert habitats. *Ctenophorus clayi, C. inermis,* and *C. reticulatus* have blunt snouts and short legs; others, such as *C. cristatus, C. isolepis,* and *C. scutulatus,* are long-legged fast runners (Pianka 1971b). Males are usually much more colorful than females.

An impressive arboreal Australian lizard, *Lophognathus* (= crested mouth) *longirostris* (= long nose), sports a tail three times its snout-vent length. Arboreal lizards often have long tails, used as counterbalances in climbing. These uncommon large, agile, and fast agamids are never found far from trees and appear to be associated with sandridges in desert habitats. They can attain speeds of 24 km/h running bipedally on their hind legs.

Another interesting agamid is *Hypsilurus* (= upsilon tail), formerly included within the Southeast Asian genus *Gonocephalus.* These large arboreal lizards live in rain forests along the east coast of Australia and in New Guinea and are somewhat reminiscent of Neotropical

Iguana (though they are apparently carnivorous rather than herbivorous). *Hypsilurus boydii* has a conspicuous dorsal crest of enlarged scales as well as a dewlaplike throat extension. They are seldom seen, as they rely heavily on their cryptic coloration, usually remaining motionless to avoid detection when faced with danger.

The so-called water dragon *Physignathus* (= bladder jaw) occurs in Australia and New Guinea, almost always near water. These large, long-lived lizards escape by jumping into water, and, like the New World corytophanine *Basiliscus,* they sleep in bushes and trees overhanging water. Their tails are compressed laterally, making them good swimmers, and nostrils open dorsally, probably facilitating breathing while in the water. *Physignathus* can stay submerged for extensive periods (over an hour!); heart rate slows during diving, conserving oxygen, and carbon dioxide is released through the skin. Water dragons maintain somewhat lower body temperatures (about 30°C) than most other agamids.

An impressive tropical, herbivorous, aquatic agamid, found in the Indo-Australasian archipelago, including New Guinea and the Philippines, is aptly named *Hydrosaurus* (= water lizard). In Southeast Asia, *Hydrosaurus* has a large sail-like fin on the dorsal side of its tail, reminiscent of some fossil reptiles *(Dimetrodon);* the function of this fin remains to be studied. Like other herbivores, these lizards harbor intestinal endosymbiotic microbes that produce cellulases, aiding in digestion of plants. Expert swimmers, *Hydrosaurus* take refuge in the water, and, like *Basiliscus* and *Physignathus,* they spend most of their time in branches overhanging water.

The genus *Gonocephalus* (= seed head) has undergone an adaptive radiation in southeastern Asia. These striking, large, semiarboreal agamids are laterally compressed and have very prominent ridges above their eyes. Males sport a prominent dorsal crest of enlarged scales extending from the nape of their neck down the center of the back, as well as a large throat flap. Four species in a similar, closely related genus, *Acanthosaura* (= spiny lizard), have discontinuous throat flaps. Both *Gonocephalus* and *Acanthosaura* rely heavily on camouflage. Usually found near water, *Gonocephalus* take refuge in the water. Adept swimmers and climbers, they can undergo quite dramatic color changes. In courtship displays, males extend their gular pouch, head-bob, and move their heads from side to side several times. Sometimes, their entire body moves up and down and their tail is waved from side to

In eastern Australia, the water dragon *Physignathus lesueuerii* perches on trees above water, dropping into water to escape.
(Dr. Hal Cogger)

side. Copulation is brief. Females lay from three to six large eggs.

Members of two genera of tropical arboreal agamids, *Ceratophora* (= horn carrying) and *Harpesaurus* (= sickle lizard), have evolved unusual appendages on their snouts. Horns are larger in males than in females and in some species may be entirely missing in females. The rostral appendage of one species, *Ceratophora tennentii,* is covered with scales and shaped like a leaf, flattened from side to side and coming to a point. Another species, *Ceratophora aspera,* has a snout appendage reminiscent of a small pine cone. Male *Ceratophora* expand an erectile sail-like crest on the back of their necks and backs when displaying. These slow-moving chameleonlike lizards are sit-and-wait ambush predators, relying on camouflage to evade enemies. Several species of *Ceratophora* occur in Sri Lanka, where they are endangered due to habitat loss. Snout appendages of *Harpesaurus* curve forward and upward like the horns of a rhinoceros. Several species of *Harpesaurus* occur in montane rain forests on Java and Sumatra. Little is known about these rare and endangered lizards.

The great anglehead lizard, *Gonocephalus grandis,* of Malaysia can change from brilliant green to jet black in minutes. (C.H. Diong)

This weird Sri Lankan knucklehead lizard, *Ceratophora tennentii,* easily remains undetected in its rain forest home.
(Christopher Austin)

Lyriocephalus (= lyre head) is another agamid with a prominent scaled protuberance on its nose. These slow-moving green arboreal lizards, which are found in second-growth sunny montane forests in Sri Lanka, have a neck flap and a light yellow throat flap. Males are territorial. *Lyriocephalus* relies on cryptic coloration to avoid detection. When threatened, it opens its mouth wide, sometimes leaving it open for several minutes; the mouth is lined with bright purple epithelium (R. Etheridge, pers. comm.). In captivity, earthworms are favored food, but little else is known about these lizards.

Another interesting Sri Lankan agamid is *Cophotis* (= deaf ear) *ceylanica,* which are covered with very large, rough, overlapping keeled scales. No tympanum is evident. In these gray-brown lizards, a prominent broad white stripe extends from the lip to the shoulder. Adept climbers with prehensile tails, they prefer trees with rough bark but also cling to thin twigs. They can also swim, dropping into water at the slightest provocation.

Leiolepis (= smooth scaled) of Southeast Asia have bodies covered with small, smooth body scales, although their tail scales are keeled. Some systematists recognize an agamid subfamily Leiolepidinae, containing *Uromastyx* and *Leiolepis.* Commonly called butterfly agamas, these handsome and agile lizards dig deep burrows. Elongate ribs allow *Leiolepis* to flatten their bodies; they are even reputed to be able to parachute, although this requires confirmation. In Bangkok, Thailand, *Leiolepis* are regularly sold as food for humans.

Several species of large spiny-tailed herbivorous agam-ids in the genus *Uromastyx* (= spiny tail) thrive in the Sahara, the Middle East, and deserts and steppes of central and western Asia. They retreat to extensive burrows or rock crevices when threatened, blocking the entrance with their spiny, sharp, muscular tails. If pursued, they thrash their tails vigorously from side to side—a few blows from this sharp spiny club are enough to discourage most small enemies. In the same geographic region, in sandy habitats, considerably smaller "toad-headed" agamids in the genus *Phrynocephalus* dig short tunnels.

Diversity of agamids is low in Africa south of the Sahara, possibly because they arrived relatively recently. Nevertheless, a few interesting species have done well. The two enlarged fanglike canine teeth of the Kalahari *Agama aculeata* are large enough to draw blood. During the breeding season, males are exceedingly colorful with blue head, bright red nape, and yellow shoulders. Females lay large clutches of up to a dozen small eggs. Local myth has it that these agamas climb trees or posts to scan the horizon for rain. Locals also warn that these lizards are venomous, explaining that they do not make their own poison but obtain venom by "milking" cobras. When questioned further, Afrikaners explain that they are often seen with their head inside a cobra's mouth extracting cobra venom. Of course, a more likely and simpler explanation for such an observation is that cobras eat them!

Another South African agamid, *Agama planiceps,* is found on rocks at the edge of the Namib Desert in Namibia. Females are drab brown with yellow markings, blending almost perfectly with the rocks on which they

Living in mountains of Sri Lanka, *Lyriocephalus scutatus* varies considerably in color, with some individuals being brilliant yellow. (John Murphy)

Uromastyx acanthinura **is in a clade of herbivorous agamids.** (L. Lee Grismer)

Large male ***Agama agama*** **are brilliantly colored during breeding season.** (Walter Hödl)

Calotes versicolor **of southwestern Asia are often called "bloodsuckers" because of their red throats.** (C.H. Diong)

Calotes, also arboreal, are bright green. A closely related long-tailed climbing lizard genus, *Bronchocela,* is more slender, laterally flattened, and with smaller scales than *Calotes.* In a terrestrial relative found in India, *Sitana ponticeriana,* males have evolved a large blue and bright red fan on their throat that extends all the way down to the chest region; it is opened and closed in courtship displays toward females (Daniel 1983).

Australia is home to many magnificent animals, but one of the most bizarre of all is a spiny lizard known as the thorny devil or mountain devil. John Gould first exhibited a thorny devil in London in 1840, and John Gray illustrated this spectacular specimen in 1841, giving it its scientific name, *Moloch horridus,* named after Milton's Moloch, in turn based on a fearful Canaanite god called Moloch, a horrid king smeared with the blood of human sacrifice. The specific name *horridus* means rough or bristly, or secondarily, dreadful. Close-ups of thorny devils are often shown in movies filmed in the Australian interior, giving many people an image of these lizards as gigantic monsters. In fact, thorny devils are moderate-sized lizards about four to six inches long.

British biologist William Saville-Kent was among the first to keep thorny devils in captivity in Australia, observing their behavior in detail (Saville-Kent 1897). He discovered that they fed almost exclusively on ants, and noted the striking morphological similarity between the thorny devil and North American horned lizards, genus *Phrynosoma.* On the basis of this similarity in body form, Saville-Kent predicted that horned lizards would be found to be ant eaters. His prediction proved correct. A multivariate morphometric analysis demonstrates that thorny devils and horned lizards are actually anatomically more similar to each other than either species is to other members of its own clade (Pianka 1986, 2001).

Such organisms that fill similar ecological niches in different, independently evolved biotas are known as "ecological equivalents." Convergent evolution of this sort is interesting because it implies that similar selective pressures have resulted in the same solutions to particular environmental situations. The very existence of convergent species pairs indicates that evolutionary pathways can be predictable and repeatable.

Thorny devils occur through most of arid Australia, particularly on sandy soils, but seldom on stony soils in two quite different habitats: spinifex-sandplain and -sandridge deserts of the interior and the mallee belt of southern

live, but males sport bright red heads and tails against a blue-black body. Dorsoventrally flattened, they take refuge in narrow rock crevices. Both sexes defend territories, with male territories being several times larger than those of females (as in many Iguanians, a male mates with several females that reside inside his territory).

Some common agamids in the genus *Calotes* in India are called "bloodsuckers" because males develop bright red heads during the breeding season. In other species of *Calotes,* males have blue or green heads. Females are much duller gray and brown. These lizards have a prominent dorsal crest of enlarged scales. Agile climbers, they are adept at hiding themselves behind branches. During the breeding season, the head, shoulder, and parts of the forelegs of males become brightly colored, often with black patches on the sides of the throat as well. Males defend mating territories, standing up erectly, facing off, and flattening their bodies from side to side, extending their gular pouches. If one male does not back away, a serious biting fight ensues. Males court females using push-ups and head-bob displays (Daniel 1983). Other species of

South Australia and southwestern Western Australia; in those areas, their geographic distribution corresponds more closely to the distribution of sandy and sandy loam soils than to any climatological feature.

Thorny devils are obligate ant specialists, eating virtually nothing else (Pianka 1997; Pianka and Pianka 1970). They consume several species, especially very small *Iridomyrmex* ants, particularly *Iridomyrmex flavipes.* Feeding rates have been estimated at 24 to 45 ants per minute. Occasional objects such as small stones, sticks, tiny flowers, and small insect eggs are also ingested, although these are probably objects being carried by ants and eaten only accidentally. Large numbers of tiny ants are eaten per meal by an individual thorny devil (estimates range from 675 to 2,500).

Fecal pellets of thorny devils are very distinctive: black, glossy, perfect prolate spheroids, often found in neat piles in the open or among sparse vegetation. Individuals have specific defecation sites—"toilets" separate from their basking and feeding sites. Tracks and accumulations of fecal matter indicate that thorny devils often return to such spots several days in succession.

These lizards have a hygroscopic system of grooves in their skin that leads to the corners of their mouth. They take up water via capillary action via these grooves, using a gulping oral mechanism to move water along the grooves and into their mouths (Bentley and Blumer 1962). They can actually drink water from dew that falls on their backs, gaining as much as a gram of water during a rainstorm (Pianka et al. 1996).

They also have a curious knoblike spiny appendage on the back of their necks, which has sometimes been likened to a false head. When threatened, these lizards tuck their real heads down between their forelegs, leaving this false head in the position of the real head. This makes them difficult for most predators to swallow, even snakes. Once Pianka caught a thorny devil that had a damaged knob, which looked as if a mammalian predator had chewed on it. When disturbed, they inflate themselves with air like little puffer fish, presumably another antipredator defense.

Thorny devils walk slowly with a jerky motion, often freezing in place while walking, with one foot off the ground. This probably helps to conceal a lizard caught out in the open. They also change color rapidly: when warm and active, they are usually a pale yellow and red, but when alarmed or when they are cold, they are dark olive drab. Despite their camouflage and thorny spines, thorny devils are not immune to predation. Australian aborigines and bustards, both visual predators, prey on thorny devils.

Individuals are usually heavily parasitized, with almost all specimens having many nematode worms in their intestinal tracts. These parasites probably exploit ants and termites as intermediate hosts and various species of lizards as their final hosts. In addition, a small species of tapeworm, *Oochoristica piankai,* was described from the small intestines of thorny devils (Bursey et al. 1996).

Thorny devils are heliotherms, relying on solar energy to raise their body temperatures above ambient temper-

Indonesian *Bronchocela jubata* are brilliant green when in sun.
(Rafe M. Brown)

atures. Body temperature varies directly with air temperature. In one study of eighty-three active thorny devils, average body temperature was 33.3°C (Pianka and Pianka 1970).

Adult female thorny devils are larger and stouter than adult males. They range from 80 to 110 mm in snout-vent length and weigh from 33 to 88.7 g, whereas adult males are all under 96 mm in SVL and never weigh more than 49 g. Hatchling males and females do not differ in size. An Australian naturalist, C. C. Sporn, kept captive individuals for five to eight years and showed that females and males grow at about the same rate during their first year, but after that, females grow faster than males, continuing to grow until at least their fifth year (Sporn 1955, 1958, 1965). Sporn considered it highly likely that thorny devils live up to twenty years in nature. Someone should undertake a detailed long-term mark-recapture study of these interesting lizards.

Thorny devils have a bimodal seasonal pattern of activity. They move little during the coldest winter months (June and July) or during the hottest summer months (January and February), with most activity occurring during a three-month austral autumnal period (March–May) and during a five-month activity period that spans late winter, spring, and early summer (August–December), when mating and egg deposition take place. During hot summer days they are inactive, retreating into shallow underground burrows that they have dug.

Thorny devils are quite sedentary during summer and autumn, with an individual's movements generally being restricted to an area less than 10 m on a side. Within this area are one or more ant trails, a defecation site, and several small bushes with scattered dead leaves and/or loose *Triodia* spinifex tussocks beneath them. The lizards spend the night and heat of midday within the protective cover of such shrub complexes (sometimes actually within a loose spinifex tussock), and take refuge in them upon the approach of a potential predator. Tracks and observations on several focal individuals suggest the following general pattern of daily activity: A lizard first walks a short distance out into the sun from the shrub(s) under which it has spent the night. After basking and bellying

MEET YOU UP BY THE *THRYPTOMENE* BUSH NEXT SPRING

One day in late August 1967, my ex-wife Helen discovered a small aggregation of thorny devils—two males and two females—in an area immediately adjacent to and beneath a large *Thryptomene* shrub. Backtracking them, we found that all four had traveled varying distances (45–75 m) toward and converged upon the same shrub, all in less than twenty-four hours (strong prevailing unidirectional winds had erased all tracks the day before). What's more—and this was especially interesting—each had come from a different direction. Earlier that same month we had observed evidence of a similar gathering at another study site, although we could find only two of these participants, both males. Tracks indicated the presence of two other much larger individuals (probably females) and showed that considerable mingling had oc-

curred. Despite concerted efforts, we found no evidence of such aggregations at any other time during our 18-month expedition.

These two cases, which were almost certainly mating contacts, are striking because of both their rarity and their simultaneity. The fact that these lizards converged from all points of the compass despite a prevailing unidirectional wind suggests that they did not rely on airborne chemical scent plumes to find each other. Indeed, one could speculate that these thorny devils had a pre-arranged rendezvous.

In mid-September 1995 my research group introduced a male thorny devil into an enclosure with a female. Neither lizard appeared to be aware of the other for long periods lasting from fifteen to sixty minutes. Then the male would suddenly move toward the female and rap-

idly bob his head a number of times. If the female remained stationary, the male would attempt to mount her from behind, with one hind leg still on the ground. The unreceptive female invariably threw the male off her back with a quick sideways roll and moved quickly away from him.

At about the same time in that austral spring of 1995 we glued small transmitters to the backs of nine thorny devils—three male and six female—to conduct radiotelemetric studies. We then followed their movements twice daily from mid-September through early November 1995. Two males disappeared suddenly, probably eaten by predators—most likely Australian bustards, which were abundant in the area. One female and one male undertook long-distance cross-country treks, traveling a kilometer or so over a period of down in the sand, the lizard attains an active body temperature and begins to move. Often an individual then walks 5–10 m to its defecation site, relieves itself, and returns by nearly the same route to the shrub complex, stopping to feed at an ant trail somewhere along the way. Like horned lizards, thorny devils require a steady diet of ants. Thorny devils have been successfully held captive for long periods in Australia, where there is an abundance of *Iridomyrmex* ants, but all those that have been exported soon sickened and died.

Eggs are laid from mid-September through late December. Usually only a single clutch is laid per year. Clutch size in thorny devils varies from three to ten, with a mode of eight eggs per clutch. In one study, seven clutches had incubation times of from 90 to 132 days (mean 118 days) (Pianka et al. 1996).

Mating has been observed in autumn, which suggests that thorny devils may have a mechanism of sperm storage. In contrast to the relatively sedentary summer-autumn existence, thorny devils move over much greater distances during August and September, when most mating occurs. One encounters many fresh cross-country tracks in early spring, relatively straight tracks extending for linear distances of over a hundred meters. Such increased movements doubtlessly increase the probability of contact between individuals.

CHAMELEONS

Chameleons are an ancient Old World family now found across much of Africa, Madagascar, southern Spain, the southern Arabian peninsula, and southern India as well as on many islands surrounding those regions (Crete, Cyprus, Malta, Mauritius, Seychelles, Sicily, Socotra, Sri Lanka, etc.). Fossils have been found in central Europe and China, an indication that they were once much more widespread. About half of the 150-odd species occur on Madagascar. Two subfamilies, Chamaeleoninae and Brookesiinae, are currently recognized, although these are probably not monophyletic entities (Klaver and Böhme 1997; Necas 1999). Chamaeleoninae includes two currently recognized genera: *Bradypodion* and *Chamaeleo*

---

several days. The other thorny devils were fairly sedentary, moving only short distances (Pianka et al. 1998).

While wearing radio transmitters, three female thorny devils excavated nest chambers and laid clutches of eggs (19 Sept., 3 Oct., 12–13 Nov.). All three nesting burrows were located on the southern side of the same sand ridge. (Two females crossed over the ridge from the north side prior to digging.) Females spent several days excavating their nesting burrows, using their forefeet to toss sand back to their hind feet, which kick the sand out of the tunnel. Their nesting burrows went down about 20 cm and then took an abrupt right-angle turn. After laying eggs, females carefully backfilled these tunnels, smoothing over the sand and leaving no trace of a burrow. The three females invested heavily in their clutches, with relative clutch masses (weight of eggs divided by female weight before oviposition) of 34.2 percent, 41.7 percent, and 40.9 percent, respectively (Pianka et al. 1996).

My daughter Gretchen, who was acting as an assistant, erected wire enclosures around the first two nests to exclude predators and to hold in the hatchlings so they could be weighed, measured, and photographed. The third nest fell victim to predation, dug up probably by a sand goanna, *Varanus gouldii*. I checked nests daily until early February. Hatchlings emerged from the first clutch on 20–21 January 1996 and from the second clutch on 7 February 1996. Hatchlings, which emerged from a single foot-long diagonal tunnel to the surface, weighed an average of 1.8 g and measured from 63 to 65 mm in total length (snout to tail tip).

Shortly after the eggs hatched, I carefully excavated both nesting burrows: each had an air-filled chamber about 8 x 13 x 15 cm in dimension, some 25 cm below the hard-packed sand's surface, with an air temperature of about 31°C. I was surprised to find no eggshells, and concluded that hatchlings had eaten their egg cases—which could add to their body mass and provide hatchlings with calcium and other materials to support early growth. I know of no other squamate reptile that deposits its eggs in an air-filled chamber or consumes its own eggshells. Eggshell consumption could, in fact, be facilitated by the air chamber: since other squamates cover their eggs with damp sand, it may be nearly impossible for a hatchling to find and eat its own eggshell. *(PIANKA)* (some recognize *Calumma* and *Furcifer*). Two genera of Brookesiinae are also recognized: *Brookesia* and *Rhampholeon*. *Brookesia* are found on Madagascar and may have evolved from an African *Rhampholeon*-like ancestor (Martin 1992). Differences exist among species in relative tail length, skull structure, and pigmentation in the peritoneal region, suggesting that *Rhampholeon* is a metataxon (Bauer 1997).

These very distinctive lizards are famous for their ability to change color rapidly. Chameleons move their eyes independently and have exceedingly long projectile tongues used to capture prey (in some species, the tongue is longer than the lizard's SVL). Another unique adaptation of chameleons concerns their feet, which have two grasping pads rather than the five distinct toes evident in most lizards. Two digits are fused to form one pad and the other three digits are fused to form the other pad on each front and hind foot, a condition called zygodactyly (= yoke toed; see below). Chameleons vary greatly in size, from tiny inch-long lizards (*Brookesia minima,* for instance, is only about 30 mm in snout-vent length) to giants two feet and more in length. Large chameleons are arboreal and have very muscular prehensile tails, which are used as a fifth leg in climbing. Radiations of dwarf chameleons have evolved in Madagascar *(Brookesia)* and in Africa *(Bradypodion* and *Rhampholeon)*. Tails of nearly all *Brookesia* and *Rhampholeon* are short and do not appear to be prehensile (these are known as "false" or "stump-tailed" chameleons). The exception is *Rhampholeon marshalli,* which has

Dwarf chameleons, like *Rhampholeon spectrum* of Africa, are tiny compared with most African *Chamaeleo.* (Chris Mattison)

a tail often exceeding 50 percent of body length, compared to less than 33 percent in other *Rhampholeon* (Broadley 1971; Hillenius 1986). All chameleons are slow-moving lizards that ambush their prey, and all rely on camouflage to avoid predators. Chameleons have no external ear opening. Most chameleons are compressed laterally. Some display pronounced sexual dimorphisms in color, size, and shape. Males of some species are exceedingly colorful and may have crests, horns, or casques on their heads for use in combat with other males and in courting females. Some tiny *Brookesia* and *Rhampholeon* are colored to match twigs, rendering them highly cryptic. Most chameleons lay eggs, although some bear living young. Relatively high fecundity is common. Both eggs and neonates are small. The tiny eggs of *B. minima,* for instance, are only 2.5 mm long, and the hatchlings are only about 15 mm in total length (Glaw and Vences 1992).

Chameleons move differently from all other lizards. Their bound, zygodactylous toes wrap around twigs from opposite directions, providing exceptionally stable support. Limbs move slowly and methodically, with one front limb reaching forward, grasping a branch, and pulling the body forward while the other front limb begins to stretch forward in slow motion. Meanwhile, the opposite hind limb disengages and moves slowly forward, changing the position of its grasp on the branch. Rather than moving its legs forward in the sprawling manner used by most other lizards, a chameleon moves its legs nearly directly under its body. Lateral forces on the body experienced by lizards with sprawling locomotion are thereby reduced, such that the body does not appear to undulate. A moving chameleon's center of gravity, moreover, remains nearly directly over its grasping feet. The stability achieved by a combination of grasping feet and a stabilized center of gravity allows chameleons to stop in midstride when disturbed and balance on a branch with one or two legs having no contact with the substrate. When placed on flat surfaces, most chameleons are remarkably clumsy because there is nothing for them to hold on to!

Nearly all lizards and other nonmammalian tetrapods are capable of independent eye movement (Wall 1942; Kirmse 1988). Chameleons have carried this to the extreme: a chameleon's left eye can be looking above and behind, while its right eye looks forward and down. Each eye covers a full 180 degrees horizontally and 90 degrees vertically (Wall 1942), a range of movement that allows a

Slowly and methodically, a _Bradypodion pumilum_ reaches for a twig in Africa. Note its zygodactylous toes and prehensile tail.

(David M. Hillis)

chameleon to survey its entire surroundings without moving its head. Not only can it see behind itself, but its only blind spots are close in, directly above the lizard's back and directly beneath it. When its eyes are looking at different visual fields, chameleons move their eyes rapidly in a stop-go fashion. Only one eye sends signals to the brain at a time, with eyes alternating at approximately one-second intervals (Ott et al. 1998). Fused eyelids covered with tiny scales cover most of the eyeball except for a tiny round hole in the center, just above the pupil.

Chameleon eyes enable exceptionally effective detection and regulation of light (Necas 1999). A chameleon's lens, which focuses extremely rapidly, acts like a telephoto lens, enlarging visual images (Land 1995; Ott and Schaeffel 1995). Moreover, a chameleon's eye contains as many as 756,000 visual cells (cones) per square millimeter, vastly more than a human eye (Wall 1942). A very large image is projected on the large and exceedingly sensitive retina, greatly aiding prey detection and range finding. Their eyesight is so good that chameleons can spot dangers many dozens of meters away. When a potential food item such as an insect is detected, both turretlike eyes are trained on it binocularly and both eyes send information to the brain. Chameleons assess distances to prey with acute accuracy, projecting their tongues to precisely the distance needed for prey capture.

Binocular vision is not a requisite for accurate estimation of distance by chameleons. A chameleon can learn to perform almost as well with only one eye as it can using both eyes. Clever experiments by Lindsay Harkness (1977) showed that chameleons measure distances by accommodation (focusing). Using _Chamaeleo jacksoni,_ which possess three horns ideally positioned to support eyeglasses, Lindsay fitted spectacles that produced images of prey appearing to be either closer or farther away than the real prey. Chameleons projected their tongues to the distance of the optical image, undershooting apparent close prey and overshooting apparent distant ones. The speed of accommodation among chameleons is the fastest ever measured in any vertebrate. One of the greatest remaining enigmas concerning chameleons is determining how their brain processes visual information.

Many anatomical precursors of the complex projectile tongue system of chameleons are present in their sister group, the Agamidae (Schwenk and Throckmorton 1989). These include the hyoid apparatus, an extended tongue process, a thick sticky tongue pad, and an elongate circular muscle. Other iguanians and _Sphenodon_ often use their tongues to capture and subdue small prey. The agamid *Phrynocephalus helioscopus* exhibits an incipient form of tongue projection (Schwenk and Bell 1988).

Chameleon tongues are hollow, encasing a long, tapering spike of cartilage, the hyoid horn, which is attached to the U-shaped hyoid apparatus. At rest, this entire tongue assembly—which consists, in addition to the flexible hyoid horn, of a sticky tip, an accelerator muscle (a sphincter that squeezes against the hyoid horn; Martin 1992), retractor muscles, and the stiff lingual process—sits in the bottom of the mouth, with its base well down into a lizard's throat behind its head. Before striking, a chameleon shifts the entire assembly upward and forward toward the front of its mouth (a movement that has been likened to "rolling a cannon forward"). Muscles contract, raising the hyoid apparatus above the lower jaw. Aiming with its entire head, a chameleon then opens its mouth and, immediately before it is ready to fire, slowly extrudes the tip of its tongue. Because the hyoid horn tapers toward its tip, when the accelerator muscles contract, the entire tongue is literally hurled out of the mouth. Cavities between muscles, cartilage, and bone are lubricated, minimizing friction. Chameleons can "shoot" their tongues at temperatures as low as 8°C.

Once the accelerators fire, a chameleon can control only the length of its strike. This is accomplished by means of a strong tendon attached to the tip of the hyoid horn, which, in concert with other tendons, stops the tongue in midflight, shielding the shock. Tongues accelerate very rapidly, extending to one to two body lengths in just a fraction of a second—but decelerate almost instantly. Maximal acceleration is nearly 500 m/s², with speeds higher than 20 km/h reached. Chameleons often snatch flying insects out of the air using their deadly accurate tongues. The sticky tip has flaps that wrap around a prey object grasping it. Adhesion is promoted both by interlocking filamentous papillae and by wet adhesion, and prey items are drawn deeply into tongues, suggesting that suction may also be involved (Schwenk 2000). Chameleons can lift about half their own body weight with their tongues, with some very large species occasionally eating small birds and mice (Dischner 1958).

Some chameleons possess an interesting temporal gland at the corners of their mouth. These dermal pouches function as a primitive holocrine gland, secreting an odoriferous material. In some species, these pouches are turned outward when the mouth is opened, apparently a defense tactic, though studies are needed to ascertain whether pouch secretions in fact repel potential predators. Chameleons sometimes rub these temporal pouches against substrates, attracting insects that are then captured by the same chameleon, suggesting that the smelly secretion can also act as a lure (Necas 1999). The secretions may be used to mark territories as well, although chameleons have rather feeble olfactory capabilities (Jacobson's organs are vestigial or completely lacking in chameleons; Martin 1992); nevertheless, chameleons have been observed licking marked branches. Functions of these dermal pouch secretions certainly merit further study.

When threatened, chameleons change color suddenly, usually becoming much darker. Often they hiss and gape. Some species have bright colors (often yellow) inside their mouths, which are used in defensive displays. When disturbed, some small species simply fall off their perches, coiling up as they fall to the ground, where they remain motionless.

Many factors influence a chameleon's color, including temperature, light, time of day, season, health, nutritional and reproductive status, danger, and mood swings (stress etc.). Chameleons use their colors to communicate with other chameleons. Females signal their reproductive state, and males can provoke combat or signal submission just by changing color. Male chameleons are very aggressive toward one another, sometimes engaging in pugnacious battles. In species with horns, males ram their opponents with their horns. Males also use their colors when courting females. Differences between species in colors and displays constitute effective reproductive isolating mechanisms.

IGUANIDS

Iguanidae contains eleven presumably monophyletic radiations (considered families by Frost, Etheridge, et al. 2001), each of which we consider separately. Their global distribution is interesting. Most are restricted to the New World, including the West Indies and Galápagos Islands. Two species occur on Pacific islands, Fiji and Tonga. One subfamily is restricted to Madagascar and the Comores.

IGUANINAE

These lizards occur from the southwestern United States (*Dipsosaurus, Sauromalus*) through Mexico (*Ctenosaura,* which includes the formerly recognized genus *Enyalio-saurus*) into central and South America (*Iguana*). One genus, *Brachylophus,* occurs on the Fiji and Tonga island groups as well as Wallis and Futuna, presumably having rafted there from the Americas on equatorial currents (Gibbons and Watkins 1982). All iguanines are herbivorous (Burghardt and Rand 1982). All are egg layers, and some have very large clutches. Females of many species nest communally, presumably due to a shortage of suitable nesting sites. Males of most species tolerate females but defend territories against other males, presumably thereby acquiring copulations with several females and enhancing their reproductive success.

The evolutionary history of iguanines is better known than for most groups of lizards. The ancestor of iguanines was most certainly herbivorous and relatively large in body size. The Fijian iguana, *Brachylophus,* is the sister taxon to all other iguanines except *Dipsosaurus* (Sites et al. 1996; Wiens and Hollingsworth 2000); it must have rafted out from South America on the equatorial current many millions of years ago. Marine and land iguanas (*Amblyrhynchus* and *Conolophus*) on the Galápagos are most closely related to mainland spiny-tailed iguanas (*Ctenosaura*), whereas chuckwallas (*Sauromalus*) may be most closely related to green iguanas (*Iguana*), though their relationships remain uncertain (Wiens and Hollingsworth 2000). Even though the desert iguana (*Dipsosaurus dorsalis*) lives in southwestern North American deserts alongside chuckwallas, the two are only distantly related.

Relatively little is known about Fijian iguanas, *Brachylophus,* except that they are moderately large (about 1 m) green arboreal lizards that eat leaves. Two species are known; both have small scales and a low crest on their necks and down their backs. Male *Brachylophus vitiensis* have a dewlaplike gular pouch (Gibbons and Watkins 1982).

The desert iguana of the southwestern United States and northern Mexico, *Dipsosaurus dorsalis,* is a medium-sized pale-colored lizard, restricted to the Sonoran and Mojave Deserts. These lizards climb up into creosote bushes (*Larrea divaricata*), often as high as a meter above ground, and eat flowers during the spring blooming season. Leaves of other plants and fecal pellets of small mammals are also eaten (Norris 1953). These lizards are very alert and quickly retreat to their burrows at any threat. Males are strongly territorial. *Dipsosaurus* are active during the hottest periods of the day (perhaps as an anti-

This juvenile *Iguana iguana* from the Brazilian Amazon can reach some 2 m in total length as an adult. (Laurie Vitt)

predator tactic) and maintain very high body temperatures. Desert iguanas infected with parasitic microbes maintain higher body temperatures than they do when uninfected. Moreover, if infected lizards are prevented from attaining higher temperatures, microbes prosper and infected lizards suffer, suggesting an adaptive advantage to fever (Vaughn et al. 1974).

Chuckwallas (*Sauromalus*) are medium to large rock dwellers in the American desert southwest. When threatened, these lizards take refuge deep in rock crevices and bloat themselves up with air, making it extremely difficult to extract them. Males are highly territorial and have been called "tyrants"; they display from atop their rock pile, their head-bobbing having an individually distinct "signature" (Berry 1974). Juveniles and females are tolerated but other males are not.

Several very large species of chuckwalla occur in Baja California and on islands in the Gulf of California. Although gigantism is often attributed to evolution of large body size on islands, several mainland chuckwallas (*S. obesus, S. klauberi,* and *S. slevini*), which are smaller than island species (*S. hispidus* and *S. varius*), are derived from island species, and they became smaller once they dispersed to the mainland (Hollingsworth 1997).

Adult body size varies considerably among populations of chuckwallas even when island populations are not considered. In a preliminary study, Ted Case (1976) found that body size appeared to be related to climatic conditions that influence resource availability: chuckwallas in places with high resource availability grew faster and reached larger body sizes. However, this relationship was confounded by

A chuckwalla, *Sauromalus hispidus,* basks on a rock on Isla Smith, Baja California, Mexico, while viewing the Gulf of California.

(L. Lee Grismer)

data on stored fat: chuckwallas at an extreme site, presumably with low resource availability (Amboy, in the Mojave Desert), had larger bodies, and more body fat, than chuckwallas at a less extreme site (Little Lake, also in the Mojave). This discrepancy led Case to speculate that genetic divergence might have occurred. More recently, Chris Tracy (1999) confirmed that differences in body size among chuckwalla populations has a genetic basis, at least in part. He performed a carefully designed "common garden" experiment, bringing juvenile *Sauromalus obesus* from various sites into the laboratory and treating them all exactly the same. He predicted that if genetic differences exist, chuckwalla body sizes would differ among populations after a period of time—and they did. Yet Tracy also found in field studies that resource availability plays a role in determining how chuckwallas partition their energy. In contrast to Case's conclusion, Tracy suggested that the duration of time that energy was available, rather than absolute resource availability at any given time, contributed most to nongenetic differences in energy use, and thus to body size. Chuckwallas with a genetic predisposition for large size will be large as long as the growing season is long enough: a classic example of the interaction between proximate (resource availability) and ultimate (genetics) factors leading to differences in body size.

Spiny-tailed iguanas, *Ctenosaura* (about a dozen species), can be terrestrial, arboreal, or rock dwellers (saxicolous). Some species have extremely spiny tails. They appear to be active all year long, grow rapidly, and mature early, and some are very fecund. In Central America, these lizards are a staple food item for humans.

Ctenosaura similis is common in dry forests of western Central America. Males are larger than females and have relatively larger heads. At Palo Verde, Costa Rica, females dig permanent burrows in open areas; these burrows constitute a key resource for females, and much of their activity takes place there (Gier 1997). *Ctenosaura similis* feed on vegetation, but localized seasonal abundance of flowers from certain preferred trees results in additional activity concentrated near food. Lizards often climb into mango trees to take flowers directly from the plant. Males form a dominance hierarchy based on locations of female burrows, with dominant males establishing territories that include female burrows and subdominant males remaining at the periphery of territories of dominant males. Males with high social status acquire most matings, indicating that establishing territories containing burrows pays off.

Some Mexican spiny-tailed lizards hide in tree hollows. In the Tepalcatepec Valley of Michoacán, the arboreal *Ctenosaura clarki* (formerly referred to as *Enyaliosaurus*) is restricted to open arid shrub forest habitat (Duellman and Duellman 1959). When disturbed, these lizards enter cavities in trees, blocking the entrance by arching their muscular spiny tails. As they grow, they select larger and larger cavities.

In Central and South America, green iguanas *(Iguana iguana)* live high in the canopy, seldom descending to the ground, except when females dig burrows to lay eggs. On Barro Colorado Island in Panama, females swim to a small island and nest communally (Rand 1968). After hatching, juveniles swim back to the mainland, where they establish themselves lower down in the canopy than adults. Juveniles eat leaves on which droppings of adults have fallen, thus acquiring intestinal endosymbionts that synthesize cellulases, aiding digestion of plants. *Iguana* grow fast, and both their eggs and flesh are highly edible to humans. One species, *Iguana delicatissima,* was named for its palatability to humans (a cruel joke).

On some Caribbean islands, eight species of *Cyclura,* which boast a prominent dorsal crest of scales, are on the brink of extinction. Surprisingly little is known about these large (one species, *C. nubila,* reaches a snout-vent length of 750 mm), conspicuous, ground-dwelling, lizards. The most common species, *C. carinata,* occurs in the Caicos and Turks Islands of the West Indies, where it can reach densities of 31 individuals per hectare (Iverson 1979). Others occur on the Cayman Islands, Cuba, and Puerto Rico, though original Puerto Rican species are extinct. Adult males are territorial, defending sites with quality food. These long-lived lizards have a mean generation time of fourteen years, which, in conjunction with pressures from humans, may account in part for their rapid disappearance (see chapter 15).

The monotypic *Amblyrhynchus cristatus* and *Conolophus subcristatus* are, respectively, the marine iguana and the land iguana of the Galápagos. Their ancestors rafted there from mainland South America long ago. *Amblyrhynchus,* which dives and eats marine algae, is the only modern lizard that utilizes the ocean as habitat. (Cretaceous mosasaurs were also highly aquatic, somewhat sharklike, predatory marine lizards.) Marine iguanas secrete excess sodium via enlarged nasal salt glands. Instead of having burrows, these large dark lizards pile up at night on the rocky shore in groups of as many as fifty; this conserves body heat at night, presumably enhancing digestion. The large land iguana *Conolophus*—endangered because of predation on juveniles by introduced cats and dogs—eats cactus flowers, pads, and fruits. Female land iguanas copulate with multiple males and store sperm, presumably increasing genetic variation among their offspring (Werner 1983). Females make long journeys up volcanoes and into calderas to build nests and lay their eggs.

OPLURINAE

Endemic to Madagascar plus some small adjacent islands where no other agamids or iguanians are found, oplurines are known as spiny-tailed lizards or Madagascar swifts. Only two genera, *Oplurus* and *Chalarodon,* are known. *Oplurus* (6 species) are rock dwellers or arboreal lizards with strongly keeled scales (some have spiny tails). The small monotypic terrestrial genus *Chalarodon* has smooth scales and lives in sandy areas. All lay eggs and all are diurnal heliotherms, living in hot, dry, open habitats. Males assume bright colors during breeding season and defend territories.

PHRYNOSOMATINAE

Phrynosomatinae is a North American subfamily (though a few representatives reach Central America) consisting of 10 genera: horned lizards *(Phrynosoma),* sand lizards *(Callisaurus, Cophosaurus, Holbrookia,* and *Uma),* swifts *(Sceloporus),* and their allies *(Petrosaurus, Sator, Urosaurus,* and *Uta)*—about 120 species in all. Most lay eggs, although viviparity has arisen in both *Phrynosoma* and in *Sceloporus.*

In North American deserts, active zebra-tailed lizards *(Callisaurus draconoides)* are usually found in open sun (Pianka and Parker 1972). When approached, these fleet-

Adult male _Uta nolascensis_ in Baja California are bright blue during breeding season. (Chris Mattison)

footed lizards curl their tail up over their back, revealing its black-and-white-banded underside, and coyly wag it from side to side—until you get too close, when they take off running in zigzag fashion across the desert (they have been clocked at 24–32 kph). _Cophosaurus texanus,_ commonly known as "dog-tailed" lizards in reference to tail wagging, behaves similarly. Such tail waving not only distracts a predator's attention from a lizard's body, but may also signal that the lizard knows a predator is there and will be able to escape. This could encourage a predator to continue on its way, saving time and energy for both. Tail waving in lizards with patterns on the underside of their tail no doubt also serves as a social signal between individuals; in addition, it might distract the attention of alert prey, providing a lizard with a meal of its own.

The phrynosomatines _Sceloporus magister_ and _Urosaurus graciosus_ are found in trees at some distance above ground. While _Urosaurus_ frequents shrubs or smaller branches in the tree canopy, where it captures most of its insect prey, _Sceloporus_ uses Joshua trees or tree trunks as perches from which to make forays to the ground to feed. Such climbing ground feeders (a more appropriate label than "arboreal" or "semiarboreal") exploit a distinctly different microhabitat than true ground-dwelling species, which forage at considerable distances from trees; hence competition for food is likely be reduced by this differential use of space. _Urosaurus graciosus_ and _Sceloporus magister_ display narrower variances in body temperature than do terrestrial species as well. Arboreal habits often

facilitate efficient, economic, and rather precise thermoregulation. Climbing lizards have only to shift position slightly to be in sun or shade or on a warmer or cooler substrate, and normally do not move through a diverse thermal environment (Huey and Pianka 1977c). Moreover, arboreal lizards need not expend energy making long runs as do most ground dwellers.

Sceloporus has undergone an extensive adaptive radiation, occurring in Canada, most of the continental United States and Mexico, and Central America. All 80 or so species are diurnal heliothermic lizards. Some species are terrestrial (e.g., _S. scalaris_ and _S. virgatus_), others arboreal (e.g., _S. clarki_), and still others are saxicolous (e.g., _S. poinsettii_). Most _Sceloporus_ lay eggs, but viviparity has arisen several times (Blackburn 1982).

One of the first detailed population studies of any lizard was conducted on the rusty lizard, _Sceloporus olivaceus,_ in central Texas (Blair 1960). Over three thousand individuals of these climbing lizards were toe-clipped and observed over a five-year period. Survivorship of juveniles was very low. Lizards reached sexual maturity one year after hatching. Females laid multiple clutches per season (as many as four or five), and some lived for several years. Clutch size increased with female size (and age): yearling females laid clutches of about eleven eggs, two-year-olds had about eighteen eggs per clutch, and three-year-old and older females laid clutches of about twenty-four eggs.

Another similar study was undertaken on the considerably smaller side-blotched lizard, _Uta stansburiana,_ in western Texas (Tinkle 1967). These terrestrial lizards, which tend to be found under shrubs but also climb rocks when available, laid at least three to four clutches of four eggs each season, producing an average of about fifteen young per female per season. Both sexes defended territories. Survivorship was low, with very few lizards living to reproduce in their second year, making these lizards essentially annuals. Survivorship was slightly higher among females than it was in males.

In southwestern North America, _Uma_ (3 species) forage in sandy areas by sitting and waiting in the open. They eat a fairly diverse diet of various insects, such as sand roaches, beetle larvae, and other burrowing arthropods. _Uma_ listen intently for insects moving in the sand, then dig them up. Sometimes these lizards dash, dig, and paw through a patch of sand and then watch the disturbed area for movement. They also eat considerable quantities of plant material, including flowers and leaves. Plant material is most important in spring, even though arthropod abundance is high then, indicating that ingestion of plants is not accidental (Durtsche 1995). Fringed toe lamellae assist these lizards in running on top of and moving through loose sand. Their lower jaws are countersunk, and they have flattened, duckbill-like, shovel-nosed snouts, which enable them to make remarkable "dives" into the sand even while running at full speed. These lizards then wriggle along under the surface for a meter or more. One must see such sand diving to appreciate its effectiveness as a disappearing act. *Uma* also have special modifications of their eyelids and nostrils (Norris 1958) and exceptionally sensitive hearing, with a very good directional location ability (Stebbins 1944). All *Uma* had restricted geographic ranges before human encroachment, and some are now threatened due to extensive loss of suitable habitat (though only one species, *U. inornata,* is federally listed).

Erroneously called "horny toads," horned lizards are bizarre, spiny, ant-eating lizards unlike any other lizard in North America. Fourteen species are recognized, eight of which are found within the continental United States (one reaches southern Canada); the remaining six little-known species are restricted to Mexico (one reaches Guatemala). The generic name for horned lizards, *Phrynosoma,* is Greek for "toad-bodied."

Horned lizard species are easily distinguished from each other by the arrangement of occipital and temporal horns on the head, as well as by such features as the number of rows of lateral, abdominal fringe scales and dorsal scale patterns. These lizards are extremely variable in color and color patterns, tending to match sand or rocks in their local environment. Geographic ranges of most species of *Phrynosoma* have been severely reduced, and several species are potentially threatened due to habitat loss. One Mexican short-horned species, *P. ditmarsi,* was for many years thought to be very rare; however, when Vincent Roth, former director of the Southwestern Research Station near Portal, Arizona, went to where ants found in the stomachs of museum specimens were known to occur, he found a thriving population in northern Sonora.

Although the ancestral state of horned lizards is oviparity, one clade of mostly high-altitude species (*P. braconnieri, boucardi, ditmarsi, douglassi, hernandezi, orbiculare,* and *taurus*) is live bearing, indicating that viviparity arose only once in the genus. The fact that these species are all montane also supports the idea that drier and colder mountain climates demand that progeny be retained internally until birth; egg-laying is a much less viable strategy for that habitat.

Horned lizards are a rather fecund group, laying many eggs or giving birth to many more offspring than other similar-sized lizards. The median clutch size for *P. cornutum* is twenty-five (one specimen laid forty eggs!), *P. asio* lays seventeen on average, and *P. hernandezi* bears up to sixteen live young. Relative clutch mass among horned lizards ranges from 13 to 35 percent (that is, a clutch or litter constitutes 13–35 percent of the mother's weight). Some species also reproduce twice in a season. This large investment in offspring throughout the active season weighs down females and makes them vulnerable to predators. Because the tiny babies are easy prey for a multitude of predators, horned lizards would become extinct without such high fecundities.

Horned lizards have evolved a variety of mechanisms to fend off predators, which include loggerhead shrikes, hawks, roadrunners, a variety of snakes, coyotes, and foxes. Their first line of defense is to remain cryptically hidden from a predator's sight. This is accomplished in three ways: by matching substrate background color; by remaining motionless when approached; and by means of their various spines and fringes of scales, which diffuse their shadow. Second, their formidable body armor of spines and horns pose a significant threat to many pred-

By tracing the present distribution of ants found in stomachs of preserved *Phrynosoma ditmarsi,* researcher Vincent Roth "rediscovered" this horned lizard in northern Sonora, Mexico.
(Eric Pianka)

ators, as evidenced by snakes and birds found dead with lizards' horns projecting through their throats. Horned lizards can further capitalize on their armor by inflating their bodies with air until they look like spiny balloons. At least five species of horned lizards, *P. asio, coronatum, cornutum, orbiculare,* and *solare,* squirt blood from their eyes when attacked, especially by canid predators such as foxes and coyotes. A canid will drop a horned lizard after being squirted and attempt to wipe or shake the blood out of its mouth, clearly indicating that the fluid has a foul taste (Middendorf and Sherbrooke 1992). As pointed out in chapter 3, the blood of some horned lizards may contain toxins derived from ants they eat. Horned lizard phylogenies suggest either that blood-squirting behavior has evolved independently several times, or if it evolved only once, it must have been lost in some subsequent lineages (Pianka and Hodges 1998).

Various features of horned lizard anatomy, behavior, diet, temporal pattern of activity, thermoregulation, and reproductive tactics can be profitably interrelated to provide an integrated view of the ecology of these interesting lizards (Pianka and Parker 1975b). A defining characteristic of horned lizards is that they are ant specialists.

While some species eat almost only ants, others eat a variety of insects in addition to ants. In a dozen species surveyed by Pianka and Parker (1975b), percentage by volume of ants in diets varied from 11.3 percent in *P. ditmarsi* to 88.8 percent in *P. solare*. Several features of skulls and lower jaws of these twelve species vary with degree of myrmecophagy (Montanucci 1989): ant specialists have shorter epipterygoid bones and longer tooth rows than do horned lizards that consume fewer ants.

Because ants are small and contain much indigestible chitin, large numbers of them must be consumed. Hence, an ant specialist must possess a large stomach for its body size. Consider, for example, the desert horned lizard, *Phrynosoma platyrhinos,* whose stomach occupies about 13 percent of its overall body mass—a considerably larger percentage than in any sympatric desert lizard species, including the herbivorous desert iguana, *Dipsosaurus dorsalis* (even though herbivores typically have larger stomachs than carnivores, because their assimilation rates are lower). Such a large gut necessitates a tanklike body form, which in turn reduces speed and decreases a lizard's ability to run from predators. As a result, natural selection has favored protective spines and cryptic behavior.

The collared lizard *Crotaphytus dickersonae* occurs along a small area of coastline and on Isla Tiburón in Sonora, Mexico. (Chris Mattison)

Specialization on ants is feasible only because these insects tend to be spatially clumped, thus constituting a concentrated food supply. By trap-lining from ant mound to ant mound, horned lizards are able to find and eat more than two hundred ants a day. This requires time, however, which means that *Phrynosoma platyrhinos* is active over a longer time interval than sympatric lizards. While active, moreover, it moves about relatively little, partly to avoid detection by predators, and it tends not to traverse from sun to shade (the vast majority are in the open sun when first sighted). This decreased movement contributes to a variance in body temperature significantly greater than in sympatric lizards (Pianka 1966).

The high reproductive investment of adult horned lizards is probably a direct consequence of their robust body form as well. Lizards that must move rapidly to escape predators, such as whiptail lizards *(Cnemidophorus)*, would hardly be expected to weight themselves down with eggs to the same extent as animals like horned lizards that rely almost entirely on spines and camouflage to avoid enemies (Vitt and Congdon 1978).

CROTAPHYTINAE

Crotaphytinae is a distinctive small clade that contains just two monophyletic genera, *Crotaphytus,* or collared lizards (9 species), and *Gambelia,* or leopard lizards (3 species) (McGuire 1996). These usually colorful, egg-laying lizards are restricted to central and western North America and northern Mexico (including Baja California). All species are sexually dimorphic, with males larger than females in some but females larger than males in others. Females develop special red or orange spots when carrying eggs.

Large and voracious predators, these lizards feed on large insects and other lizards. All but one of the collared lizards are associated with rocky habitats, the sole exception being *Crotaphytus reticulatus,* which, like leopard lizards, occurs in open flatland areas. Geographic distributions of some species of crotaphytines, such as *C. collaris* and *Gambelia wislizeni,* are extensive, but other species, such as *C. antiquans, C. dickersonae, C. grismeri, C. reticulatus,* and *G. silus,* have very restricted ranges. Due to extensive habitat loss, *G. silus* is a federally listed endangered species.

Gambelia wislizeni, known as the long-nosed leopard lizard, ranges from eastern Oregon and southern Idaho through Nevada, southern California, Utah, Arizona,

northern Baja California, and Sonora, with scattered records from New Mexico, Texas, Chihuahua, and Coahuila. In the southern parts of their range these lizards are quite large (105–115 mm SVL) and eat vertebrate prey (especially other lizards), whereas in the north they are smaller (80–95 mm SVL) and feed primarily on large insects (Parker and Pianka 1976). In sandy habitats, tracks of leopard lizards are distinct from those of other lizards: they look like long-fingered handprints.

LIOLAEMINAE

Liolaeminae are restricted to southern South America, extending in the Andes through Bolivia, Peru, and coastal Chile, with one species endemic to beaches of Rio de Janeiro. It consists of three genera: *Liolaemus, Ctenoblepharys,* and *Phymaturus.*

Liolaemus is by far the most speciose genus of liolaemines. About 150 species are known, of which more than 60 are found in Chile alone (Donoso-Barros 1966; Etheridge and Espinoza 2000). Many *Liolaemus* live at high elevations, but they can be found in nearly every habitat of southern South America. Like many other iguanians, *Liolaemus* bask in sun to gain heat (Marquet et al. 1989). Reproductive tactics are split almost evenly between oviparity and viviparity—the latter being the strategy of many high-elevation species. One liolaemine, *L. magellanicus,* is the southernmost lizard in the world, living in coastal shrub and coastal steppe of Tierra del Fuego in southern Chile (Jaksíc and Schwenk 1983). In much of southwestern South America, species of *Liolaemus* dominate the lizard fauna.

Detailed ecological studies have been conducted on some species, and numerous natural history observations have been made on others (Jaksíc and Núñez 1979; Jaksíc et al. 1980; Etheridge and Espinoza 2000). In the Mediterranean climate region of southern Chile, three to four species of *Liolaemus* always occur together and dominate the lizard fauna, with species composition varying according to elevation (Fuentes 1976; Fuentes and Jaksíc 1980). In a coastal sage habitat, the lizard fauna is represented by *L. lemniscatus, L. fuscus,* and *L. chilensis*—a combination that is similar ecologically to the lizard fauna at Camp Pendleton, a coastal sage habitat in southern California, composed of two phrynosomatines and a teiid. Likewise, a Chilean chaparral site has *L. lemniscatus, L. fuscus, L. monticola, L. tenuis,* and a teiid, *Callopistes macu-*

latus, while a similar chaparral site in southern California has an ecologically similar fauna consisting of four phrynosomatines and one teiid. The same holds true for montane chaparral sites: one in Chile has three liolaemines and one leiosaurine, *Liolaemus schroederi, L. nigroviridis, L. tenuis,* and *Pristidactylus* (formerly *Urostrophus*) *torquatus,* while an analogous southern California site has three phrynosomatines, one anguid, and a skink.

Species of *Liolaemus* drop out rapidly to the northeast in southern South America, with the distribution of the genus ending in coastal regions of extreme southern Brazil, with one exception: isolated by hundreds of miles, *L. lutzae* lives on beaches along the Atlantic coast near the megacity of Rio de Janeiro (Rocha 1992; Rocha and Bergallo 1992). Its habitat is limited to a 200-km-long strip of sandy coastal ridges 50–100 m wide and containing herbaceous and shrubby vegetation. This endemic lizard has been severely impacted by human activities.

Phymaturus (= swollen tail) are robust, dorsoventrally flattened rock lizards. Most of the ten described species (others remain undescribed) are found at mid to high elevations to 4800 m in Argentina, but one species enters Andean regions of eastern central Chile (Donoso-Barros 1966; Cei 1986, 1993; Etheridge and Espinoza 2000). The species are divided into two main clades: a primarily northern clade, members of which are distributed at high elevations, and a primarily Patagonian clade, members of which live at high latitude but moderate elevations (rarely exceeding 2000 m). All *Phymaturus* are herbivorous, heliothermic, presumably territorial, and bear living young. One species, *P. patagonicus,* either is a facultative parthenogen or has the longest sperm retention ever recorded for a squamate (Chiszar et al. 1999). They have converged on a morphology and ecology similar to those of chuckwallas.

LEIOCEPHALINAE

Leiocephalus (= smooth head) consists of 21 species of terrestrial lizards endemic to the West Indies (Pregill 1992). (The species list has recently shrunk by two, due to extinctions attributable to humans.) All but one species, *L. carinatus,* are endemic to single islands; *L. carinatus* occurs on Cuba, Grand Cayman, Little Cayman, Cayman Brac, and Little Bahama Banks. The distribution of these lizards within islands is spotty, and in some cases densities are high. Hispaniola is home to the most species, 12, but some islands have only a single representative. As a group, *Leiocephalus* are considered relictual in the West Indies because their present distribution within the islands is much less than it was in the past. Jamaica once had a single species, known now only from fossils. Putative fossil specimens from Nebraska, Wyoming, and Florida cannot be unambiguously assigned to *Leiocephalus;* in any case, their occurrence in those areas makes little biogeographic sense. Since there are no North or Central American leiocephalines, colonization of the West Indies by these lizards probably occurred not from the north but from South America.

Locally, some *Leiocephalus* are known as curly-tailed lizards because of their habit of curling their tail up over their body like a coiled watch spring when disturbed. Some species are only 55 mm in snout-vent length, whereas others are quite large, reaching 140 mm SVL. Fossil species were even larger, reaching at least 200 mm SVL. Males are larger and have relatively larger heads than females for all species studied (e.g., Smith 1992; Schreiber et al. 1993). All species are terrestrial, and all live in dry habitats. *Leiocephalus* feed on a variety of insects, other lizards, and buds, flowers, and fruits of plants (Sampedro Marin et al. 1979; Schoener et al. 1982). Females produce small clutches (1–3 eggs) of very large eggs (Smith and Iverson 1993).

TROPIDURINAE

Tropidurus and its relatives have diversified on mainland South America and in the Galápagos Islands. *Uranoscodon superciliosus,* a shade-seeking arboreal lizard living along edges of waterways in Amazon rain forest (Howland et al. 1990), is the sister taxon to the *Tropidurus* group of tropidurines. Known in the popular pet trade as "mop-headed iguana," because its skull is highly elevated in the orbital region, *Uranoscodon* feeds on living and dead invertebrates that wash ashore in flotsam along rivers and streams. As rising wet season rivers flood the forest, *Uranoscodon* move with the edge so that they are nearly always at the land-water interface. A green alga grows on their skin, giving them a green sheen and making them cryptic on tree trunks and vines covered with algae. In Brazil, they are often called *calango cego* (blind lizard) because they rely so heavily on crypsis that they are easily approached by an observer.

Remaining tropidurine genera are *Eurolophosaurus,*

Found on Bahama, Cayman, and Cuba, *Leiocephalus carinatus* curls its tail to signal predators that the lizard is aware of their presence and likely to escape. (C. Ken Dodd)

One of the flattened rock lizards, *Tropidurus helenae* occupies narrow crevices in rock outcrops in northeastern Brazil. (Ivan Sazima)

Microlophus, Plica, Stenocercus, Strobilurus, Tropidurus, and *Uracentron.* All of these bask to gain heat, even though some, such as *Stenocercus roseiventris,* live in undisturbed rain forest. Species of *Stenocercus* are spread from Colombia across Ecuador, extending in a southeasterly arc to coastal Argentina. Most *Microlophus* occur on the Galápagos Islands and in South America west of the Andes. Most *Microlophus* species occur on rocks and are highly visible. A few have spectacular male coloration. One, *Microlophus koepkeorum,* occurs only along the eastern portion of the Sechura Desert in Peru. Excluding the Amazon and Orinoco basins, *Tropidurus* occur in most of South America east of the Andes. Diversity within this genus is only beginning to be appreciated (Rodrigues 1987; Frost, Rodrigues, et al. 1992, 2001). In open habitats such as caatinga and cerrado, they occur on termite nests *(T. hispidus),* on expansive *lajeiros* (large granitic slabs) *(T. hispidus* and *T. semitaeniatus;* Vitt 1995), in trees, on rocks, and on ground *(T. montanus, T. itambere,* and others), and even on rock surfaces of Sugarloaf Mountain in Rio de Janeiro *(T. torquatus).* Species occur on isolated rock outcrops *(T. oreadicus, T. hispidus,* and several undescribed taxa; Vitt 1993; Vitt, Zani, and Caldwell 1996; Vitt, Caldwell et al. 1997) in the Amazon basin, but they are absent from the forest. Other tropidurines occur in the canopy of huge hardwood trees *(Uracentron flaviceps;* Vitt and Zani 1996a) or on their tree trunks *(Plica plica* and *P. umbra;* Vitt 1991; Vitt, Zani, and Avila-Pires 1997).

Ecological and morphological diversity is great within tropidurines. Two species, *Tropidurus semitaeniatus* and *Plica plica,* are flattened dorsoventrally, for very different reasons. *Tropidurus semitaeniatus* lives on expansive rock surfaces and enters crevices under caprocks when disturbed. Many individuals can occur on a single surface, and when disturbed they move across the rock surface like birds in a flocklike wave of activity. The entire group of lizards disappears into a crevice, where they can be seen tightly packed. Payoff for being flat must be considerable because females not only produce clutches of only two eggs (one per ovary), but eggs are overly elongate such that gravid females remain just as flat as nongravid females. *Plica plica* is the largest tropidurine, reaching 165 mm SVL and weighing 130 g. It too lives on a relatively flat surface: vertical trunks of the largest trees in Amazon rain forest. The body is not quite as flat as that of *T. semitaeniatus,* but by tropidurine standards it is flat. This lizard clings to bark of tree trunks using its large recurved claws

Aqua blue eyes distinguish *Anolis transversalis* from all other Amazonian lizards. (Laurie Vitt)

and maintains a center of gravity close to the trunk because of its flattened body. In seconds, one of these lizards can climb 45 m of trunk, running almost spiderlike, hugging the trunk. Even at its large size, this tropidurine eats mostly ants. Its close relative, *P. umbra,* lives on trunks of smaller trees and exclusively eats ants.

Horizontal branches and the canopy of large rain forest trees are home to *Uracentron flavipes.* Among tropidurines, only this species and its two close relatives *U. azureus* (a spectacular blue-green banded lizard) and *Strobilurus strobilurus* (formerly *Strobilurus torquatus;* Rodrigues et. al. 1989) have large, nonautotomous, spiny tails, the function of which remains unknown. One possibility is that architecture of spines and increased surface area of the flat tail capture heat, allowing the lizard to warm rapidly. Another is that they block the entrance to their refuge with their tail. *Uracentron flaviceps* live in groups of as many as twenty individuals in the same tree. A group consists of one sexually active male in breeding coloration (orange head and neck), one or more males of similar size but without breeding coloration and with reduced testes size, several adult females, plus numerous offspring of different ages. At night, they all seek refuge in hollow cavities in the tree. Egg deposition also occurs in these cavities. When sunlight floods surfaces of large limbs, the lizards come out nearly simultaneously. The dominant male often climbs high in the tree, basking on an arboreal termite nest and scouting for male intruders. When foraging in the canopy and along branches, these lizards often move together in what appears to be social foraging. Of course, limited foraging routes exist along branches in trees so "social" may not be an appropriate descriptor—nevertheless, these lizards do appear to move together, in a manner not unlike the wavelike movements of groups of *T. semitaeniatus* across rocks.

POLYCHROTINAE

Best known for the genus *Anolis,* Polychrotinae was until recently considered to contain 10 additional genera: *Anisolepis, Chamaeleolis, Chamaelinorops, Diplolaemus, Enyalius, Leiosaurus, Phenacosaurus, Polychrus, Pristidactylus,* and *Urostrophus* (Frost and Etheridge 1989). This subfamily is now restricted to two genera, *Anolis* and *Polychrus;* the others have either been synonymized or placed in other clades (Frost, Etheridge, et al. 2001). *Chamaeleolis, Chamaelinorops,* and *Phenacosaurus* are now considered subgroups of *Anolis.* Most striking is the remarkable radiation within the genus *Anolis;* more than 250 species are known from Central and South America, and more than 130 species from the West Indies. Diversity is so great that in the late 1980s *Anolis* was split into five genera, *Anolis, Ctenonotus, Dactyloa, Norops,* and *Semiurus* (Guyer and Savage 1986, 1992; Savage and Guyer 1989). However, because monophyly of all five of these groups has not been established, we continue to refer to all as *Anolis.* Because of their high diversity and abundance, especially in the West Indies, *Anolis* species have assumed a central role in testing ecological and evolutionary theory.

Most polychrotines are sexually dimorphic, with males having large, often brightly colored, extendable dewlaps that are used in social communication. Most species sleep on the end of branches or twigs, a behavior that protects them from many nocturnal predators, particularly those that approach from within the tree or bush. However, this behavior makes them vulnerable to capture by biologists at night: all one must do is approach silently with a headlamp and pick them off; they can then be measured, weighed, marked, and released as desired. Moreover, they are easily recaptured (though they may suffer from bad dreams).

Most polychrotines climb to at least some degree, and all have toe lamellae. *Anolis auratus,* in northern South America, spends considerable time on the ground darting among grass clumps but frequently climbs into vegetation to forage and bask. In central Brazilian cerrado, *A. meridionalis* is ecologically similar to *A. auratus.* Both are considered grass anoles. At the opposite extreme is *A. transversalis,* a spectacular lizard that lives in the canopy and on trunks of trees in Amazon rain forest. In this species, color patterns of males and females differ: whereas males have finely dotted transverse markings, females have large transverse bands. In between are species that live on tree trunks, so cryptically colored they are easily mistaken for a piece of bark (e.g., *A. capito* in Central America), and others associated with horizontal branches of rain forest trees (e.g., *A. ortonii*). A few, however, such as species in the *A. nitens* complex, are largely terrestrial, spending most of their time on leaf litter (Vitt and Zani 1996c; Vitt et al. 2001), though like most other polychrotines they sleep above ground.

Various species of *Polychrus* are almost chameleonlike in behavior. They move slowly through the canopy of vegetation using their tail both as a counterbalance and to hold on to limbs. Rather than quickly attacking insects like most anoles, *Polychrus* are slow and methodical (Vitt and Lacher 1981). Their sharp eyes detect even the slightest movement. As a result, they can locate even the most

Almost chameleonlike, Amazonian *Polychrus marmoratus* move slowly through vegetation in search of food and mates.

(William W. Lamar)

Clockwise from left: **Only rarely observed in the field, *Enyalioides palpe-bralis* is restricted to the far western Amazon rain forest.** (Laurie Vitt)

***Hoplocercus spinosus* is a poorly known lizard from the cerrado of Brazil.** (Dante Fenolio)

***Enyalioides laticeps* lives on the floor of Amazon rain forest, often sleeping in low vegetation.** (Laurie Vitt)

cryptic insects, which they slowly approach and pluck from the vegetation. When an observer is close by, *Polychrus* often stop midstride and remain perfectly still, blending into the background. Some species, such as *P. liogaster,* perform a striking open-mouth threat display when disturbed, with their body inflated.

LEIOSAURINAE

Seven genera, *Enyalius, Anisolepis, Aperopristis, Diplolaemus, Leiosaurus, Pristidactylus,* and *Urostrophis,* make up this South American subfamily (Frost, Etheridge, et al. 2001). The six species of *Enyalius* are moderate in size and exceptionally cryptic. All but one, *E. bilineatus,* are restricted to forest habitats (Jackson 1978; Vitt, Avila-Pires, and Zani 1996), though even *E. bilineatus* appears associated with bands of gallery forest that cut through Brazilian cerrado.

At the southern end of South America, species of *Diplolaemus, Leiosaurus,* and *Pristidactylus* are predominantly terrestrial (Cei 1986). They have shorter and slightly fatter bodies than those of polychrotines, similar to some terrestrial lizards in other iguanian families. All species are cryptically colored, matching backgrounds of microhabitats where they live. All are insectivorous. Two species, *Diplolaemus darwini* and *D. bibroni,* are active at very low temperatures, down to 5°C.

Common in vegetation along rivers in Central America, *Basiliscus plumifrons* drop into the water when approached. (Paul J. Gier)

AN OPENING IN THE FOREST

During the late dry season of 1993 I worked in Caribbean rain forest along the Río San Juan in Nicaragua. The first tier of vegetation was high enough that straight-line visibility was rather low, adding a structural complexity to the forest habitat such that many species of lizards were difficult to find. The top tier of vegetation was canopy formed by large trees, many with expansive buttresses. While walking through the forest one morning, conducting my typical searches for lizards, I looked up on a 10-cm-diameter tree trunk and noticed a large adult male *Corytophanes cristatus* about 4 m off the ground looking down at me. The lizard was oriented with

its tail down but had stopped still, apparently responding to my movements below. The crest on top of its head and its overall morphology were so bizarre that all I could do was stare at it. After gazing at the lizard for a few moments, I began climbing the tree to capture it. The lizard never moved, making it easy to catch. While in the tree, with my legs wrapped around it for support, I happened to glance out into the forest and was struck by what I saw. A window of visibility occurred at about 4 m off ground: I could see at least 50 m in every direction. I was entirely above the first tier of vegetation and well below the next tier. Sitting in that precarious po-

sition and putting the lizard into a cloth bag while waging a minor war with ants I had disturbed, I thought: "The large crest on these males' heads would make them visible across long distances. They could use this window for long-distance social interaction."

Whether social behavior in *C. cristatus* takes place at this level remains unknown. However, after my ah-ha thought, I changed my search image for these lizards and found several more at the same level, always on trunks with their heads oriented up. In all cases, the lizard's position was in the window of visibility I had observed. *(VITT)*

This subfamily contains only three genera, *Hoplocercus, Enyalioides,* and *Morunasaurus* (Frost and Etheridge 1989). *Morunasaurus* occurs from Panama through the Pacific lowlands of Ecuador. *Enyalioides* is restricted to the western Amazon region of Colombia, Ecuador, Peru, Brazil, and northwestern Bolivia. *Hoplocercus* occurs in the cerrado of Brazil. All these lizards are strange and very little is known about any of them. With the exception of *Hoplocercus,* all are rarely encountered.

Hoplocercus spinosus is most common in cerradão, a Brazilian cerrado habitat dominated by trees. This terrestrial lizard seeks refuge in hollows and crevices of live and dead tree trunks. It uses its enlarged, spiny tail to block entrances. When agitated, this lizard performs a spectacular defensive display in which it stands with its body expanded and elevated off the ground, feet extended, with the tip of its wide, spiny tail touching the ground, likely for stabilization. Turning its body sideways to an observer, its mouth is opened in a threatening fashion. Confronted at one end—the tail—with a bed of sharp spines, and at the other with a head ready to bite, a predator would surely think twice before launching an attack (Sick 1951).

Morunasaurus and *Enyalioides* are restricted to forest, where they apparently occur at low density. *Enyalioides laticeps* lives on forest floor and low vegetation, often running across the ground only to disappear into another animal's burrow. After about five minutes, the lizard peeks out of the burrow and, if the coast is clear, moves out onto leaf litter. *Enyalioides palpebralis,* one of the rarest species, looks just like a medieval knight in armor. The male's neck is ornamented with large dorsal spines and its body is constructed very much like that of Central American *Corytophanes.* Locomotion is slow and methodical; based on foot structure and the ease with which these lizards move into vegetation, we can surmise that they are likely arboreal. On land they are clumsy, moving across flat surfaces with difficulty.

Three genera make up this subfamily: *Corytophanes, Basiliscus,* and *Laemanctus* (Frost and Etheridge 1989). The distribution extends from northwestern Ecuador and Venezuela north through Central America to about 20° N latitude in Mexico on both coasts. All are characterized by large crests on the head (males only in *Basiliscus*). All are arboreal. One species of *Corytophanes, C. pericarinatus,* is viviparous; all others are oviparous.

Corytophanes are found on trunks and limbs of relatively small to medium-sized trees as well as on tangles of vines in rain forest. Three species are known, but only one, *C. cristatus,* an extreme sit-and-wait predator that feeds on relatively large insects, has been studied ecologically (Andrews 1979). Its bizarre cryptic morphology, color, and use of immobility as defense mechanisms make it extremely difficult to find. Algae and bryophytes grow on the skin of *C. cristatus,* likely enhancing crypsis (Gradstein and Equihua 1995). *Laemanctus* is very similar to *Corytophanes* in ecology and behavior.

Basilisks (genus *Basiliscus*) are often called Jesus Christ lizards because of their ability to run bipedally across water surfaces. In much of Central America, one has only to approach a small pond to observe *B. vittatus* run from the shore and across the water, disappearing in vegetation of the opposite shore. Their long, thin bodies, long tails, and long hind limbs facilitate high-speed locomotion, and fringed flaps on their toes provide a large surface area of contact with water. Along most rivers and large streams in Central America and northern South America, basilisks can be observed by the hundreds basking on shoreline vegetation. When approached, most drop into the water and disappear. Others swim back to shore, taking refuge in vegetation. Large crests on heads of males likely result from sexual selection, but tail crests may enhance crypsis and provide stability during high-speed running, as well as facilitating swimming.

CHAPTER 9

FROM GECKOS TO BLIND LIZARDS

Goniurosaurus hainanensis, a gecko from Diao Luo Shan, Hainan Island, China. (L. Lee Grismer)

Taken together, geckos, flap-footed lizards, worm lizards, and blind lizards are among the most remarkable of all lizards. Geckos stand out because of their often striking colors, soft skin, and adhesive toe pads, which allow them to scale smooth vertical surfaces or even walk upside down. Some Australian flap-footed lizards (Pygopodidae) are snakelike, with only tiny remnants of hind limbs (hence the common name), and one genus, *Lialis,* swallows its prey—other lizards—whole, similar to snakes. Some flap-foots are fossorial, living a subterranean existence. Worm lizards (sometimes called ringed lizards) are so strange and nonlizardlike that they must be seen to be believed. Their elongate bodies with scales arranged in juxtaposed rings and peculiar accordionlike locomotion conjure up visions of a subterranean world riddled with tubes in which life flourishes. Also evoking such a "tube" world are blind lizards (often called blindskinks), tiny, wormlike lizards either legless (females) or with tiny remnants of hind legs (males).

Relationships among these groups—especially where worm lizards and blind lizards fit in—are uncertain, but we do know that gecko families are closely related, and flap-footed lizards are closely related to geckos. Blind lizards are allied either with one of the worm lizard families or with skinks, while worm lizards may share a common ancestor with gekkotans (geckos and flap-footed lizards) or may be related to skinks or teiids.

GECKOS

When people think "lizard," they tend to envision a sun-loving, quick, hard-to-catch critter. Geckos defy this image in just about every way. Most exploit the night and its virtual cornucopia of nocturnal arthropods, including crickets, moths, termites, and spiders. Nocturnality allows avoidance of diurnal competitors and predators. Geckos have reevolved diurnality multiple times. They are among the most unusual of all lizards, with ancient roots that reflect much of Earth's history. A great deal more remains to be learned about these exquisite creatures.

The large monophyletic clade Gekkota contains three groups of geckos plus flap-footed lizards. We suggest that three families of geckos should be recognized: Eublepharidae, Diplodactylidae, and Gekkonidae. Some researchers consider these subfamilies (Kluge 1967), but we consider it convenient to elevate them to families so that the very distinctive and nongeckolike pygopodids can retain familial-level status (higher-level taxonomy being somewhat arbitrary).

Because Diplodactylidae are the sister group to Pygopodidae (Kluge 1976, 1987), "geckos" are paraphyletic: like "reptiles" and "lizards," they do not contain all descendents of a common ancestor (the definition of monophyly). All six genera of eublepharids (about 25 species) are terrestrial and retain movable eyelids (the ancestral state). Diplodactylids (14 genera, about 115 species), found only in the Australian biogeographic realm (including New Caledonia and New Zealand), include both terrestrial and arboreal forms, as do gekkonids (about 75 genera, more than 900 species), of which two very distinct subfamilies are recognized: Gekkoninae and Sphaero-

Few geckos are as brilliantly colored as the Philippine *Pseudogekko smaragdina.* (Rafe M. Brown)

dactylinae. Diplodactylids, gekkonids, and pygopodids share a derived state: an immovable spectacle covering the eye; using their tongue, these geckos lick their eye spectacle to clean it—it is replaced with each molt.

Eublepharids, diplodactylids, and gekkonids probably arose when Pangaea broke up 180 million years ago (Kluge 1987). Eublepharids are Laurasian, whereas diplodactylids and sphaerodactylines are likely Gondwanan. Eublepharids have an interesting relictual geographic distribution, being found in a dozen isolated areas scattered around the world, including North America, Africa, Asia, Borneo, and Malaysia (fig. 9.1; Grismer 1988). Gekkonines are cosmopolitan, found in both North and South America, Africa, many parts of Asia, central Asia, China, Vietnam, Borneo, Malaysia, and Australia. They have rafted to many remote oceanic islands as well. All sphaerodactylines are small, diurnal Neotropical lizards. Compared to the Old World, geckos in the New World are impoverished (Duellman and Pianka 1990).

A eublepharid gecko, _Aeluroscalabotes felinus,_ from Borneo presents an open-mouth threat display when disturbed.

(Louis Porras)

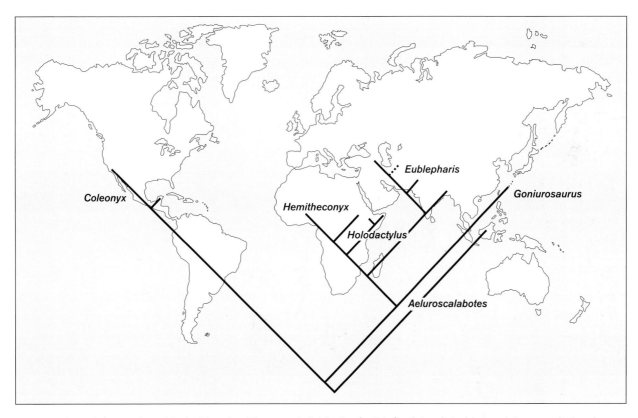

Figure 9.1 Area cladogram for eublepharid geckos. The current distribution for this family is relictual, but evolutionary radiations have occurred independently on different continents. (Redrawn from Grismer 1988 in Estes and Pregill, _Phylogenetic relationships of lizard families,_ by permission of Stanford University Press, © The Board of Trustees of the Leland Stanford Jr. University)

Geckos vary widely in size, ranging from tiny—*Sphaerodactylus parthenopion* and *Coleodactylus amazonicus* of the Neotropics, at 17 and 24 mm SVL, respectively—to gigantic—*Rhacodactylus leachianus* of New Caledonia, at 280 mm SVL. On New Caledonia, *Rhacodactylus* are fierce top predators that eat other lizards (Bauer 1995). The first gecko genus ever described, the Tokay *Gekko gecko,* is also extremely large, reaching a total length of 350 mm. Bites from both *Rhacodactylus* and *Gekko* can be quite painful (one of our colleagues sports a scar from a Tokay bite he received as a small boy in India).

Although most geckos are nocturnal (presumably the ancestral state), many are diurnal, including *Lygodactylus, Naultinus, Phelsuma, Pristurus, Quedenfeldtia,* and *Rhoptropus,* as well as all sphaerodactylines. Some high-latitude *Tenuidactylus* in Asia are diurnal during cooler months but crepuscular or nocturnal in warmer seasons (Szczerbak and Golubev 1996). Nocturnal geckos have elliptical pupils that open wide in the dark but narrow to a tight vertical slit in bright light; the slit can close completely to form tiny notches, giving the lizard a series of pinhole pupils.

Most geckos have small, smooth scales, but some African, Asian, and Australian species, as well as Neotropical *Phyllodactylus,* have lumpy, knobby, or tubercular scales, and a few, like *Geckolepis,* have fishlike overlapping scales. Geckos have smaller, thinner teeth, as well as more of them, than most other lizards (Greer 1989). Some geckos seem to have a "sweet tooth," being known to lick saps and nectar—the nectivore *Christinus guentheri* from Lord Howe and Norfolk Islands off Australia has even been observed breaking into sugar bags!

All geckos lay eggs, except for live-bearing diplodactylids from New Zealand *(Naultinus* and *Hoplodactylus)* and one from New Caledonia *(Rhacodactylus trachyrhynchus).* Clutch size is usually two, but sphaerodactylines and the gekkonines *Thecadactylus* and *Gehyra* lay only a single egg at a time. Many gecko species deposit more than one clutch per season. Communal nesting is widespread among geckos, as in *Gonatodes humeralis* of South America (Vitt, Zani, and Barros 1997) and *Oedura tryoni* of Australia (Milton 1980). Communal nesting could be a result of a shortage of suitable nesting sites but may also represent repeated use by one or more females of nest sites with a history of high hatching success. Temperature-dependent sex determination has been documented in *Eublepharis* and some *Phelsuma:* eggs incubated at lower temperatures hatch out as females; most eggs incubated at intermediate tempera-

tures become males; and at still higher incubation temperatures, females again become preponderant.

Eggs of sphaerodactyline and gekkonine geckos are nearly spherical. Like eggs of most other lizards, gekkonid eggs are soft when first laid. Upon exposure to air, however, they mineralize and harden (as do those of some diplodactylids), becoming desiccation resistant. Such eggs can also withstand exposure and even immersion in seawater (Brown and Alcala 1957), a major advantage for dispersal. Eggs of *Hemidactylus frenatus* are sticky when laid on vertical surfaces but not when laid on horizontal surfaces, suggesting that females can control egg stickiness (Greer 1989). Eggs of eublepharids and some but not all diplodactylids are elongate, remain soft, and are quite sensitive to changes in aridity and humidity. Although they must be incubated in a moist environment, a benefit of such eggs is that they can take up water from the surrounding environment.

Members of several gekkonine genera *(Lepidodactylus, Hemidactylus, Gehyra, Hemiphyllodactylus, Nactus,* and *Gehyra)* have dispersed to many remote South Pacific islands, very likely as hard-shelled eggs laid under the bark of a tree that subsequently was washed out to sea and drifted to another island. Some gecko females can store sperm for extended periods of time, which might also facilitate founding of new populations. In *Coleonyx variegatus* (Eublepharidae) and *Phyllodactylus homolepidurus* (Gekkonidae), sperm storage receptacles are found in a tube between the oviduct and infundibulum (Cuellar 1966b). The distribution of similar structures across gecko families suggests that sperm storage is widespread. Sperm storage also occurs in *Christinus marmoratus* (Gekkonidae), in which mating occurs after oviposition (Greer 1989; King 1977; King and Hayman 1978).

Some gekkonine species in the genera *Heteronotia, Hemidactylus, Hemiphyllodactylus, Lepidodactylus,* and *Nactus* are parthenogenetic; that is, females reproduce asexually without males, producing offspring genetically identical to themselves (see chapter 6). These clones arise via hybridization of two sexual parental species, and many are polyploids (they have more sets of chromosomes than either parental sexual species and these include chromosome sets from both parental species). A single parthenogenetic female can found a new population; moreover, in all-female "species," populations increase much faster than populations of sexual species because all individuals produce only female young, which in turn produce eggs.

Cyrtodactylus peguensis is a handsome gecko from Southeast Asia. (Louis Porras)

If the environment faced by offspring of parthenogenetic females is identical to that in which the parental female survived to reproduce, daughters would be expected to do well because they are genetically identical to their mother. However, environments are extremely complex and typically change through time. As a consequence, parthenogenetic individuals are at a selective disadvantage relative to sexually reproducing individuals because they don't have the genetic variability required for adaptation to changing environments.

The physical environment isn't the only thing that can affect parthenogenetic geckos. Their ability to respond evolutionarily to competitors and predators is also limited. A predator able to capture one member of a clone should be able to capture all members because genetic variation among clone members is near zero. Behavioral variation is restricted to phenotypic variation associated with that clone's genotype. The same goes for competition: a lizard vying for the same resources that could outcompete one member of a clone could outcompete all of them. Uncertainty of the biotic and abiotic environment therefore places unisexual populations at high risk. Nevertheless, some parthenoforms have been very successful in the short term, either because they live in relatively stable environments, such as *Hemidactylus garnotii,* which is associated with human structures, or because they can track habitat patches that remain stable through time, even though their spatial distribution might change.

An example of the latter type of habitat is an ecotone, a zone where two different habitat types meet. The exact location of an ecotone may shift through time, but because the habitat constitutes a mixture of vegetation (it is "weedy"), it often has high resource availability, which fosters parthenogenetic species that can respond reproductively.

Gecko feet have diversified greatly and often provide diagnostic features useful in identifying species. Indeed, well over a dozen gecko genera are named for characteristics of their digits: *Pachydactylus* means "thick toe," *Stenodactylus* means "narrow toe," *Gymnodactylus* means "naked toe," *Hemidactylus* means "half toe," *Phyllodactylus* means "leaf toe," *Lepidodactylus* means "scaly toe," *Carphodactylus* and *Lygodactylus* both mean "twig toe," *Chondrodactylus* means "grain toe," *Coleodactylus* means "sheathed toe," *Crenadactylus* means "notched toe," *Cyrtodactylus* means "curved toe," *Diplodactylus* means "double toe," *Ptyodactylus* means "fan toe," *Rhacodactylus* means "ragged toe," *Saurodactylus* means "lizard toe," *Sphaerodactylus* means "ball toe," *Tenuidactylus* means "thin toe," and *Thecadactylus* means "sac toe."

Lamellar toe pads adapted for clinging—specifically, for holding on to vertical surfaces—have arisen independently a number of times within geckos. Scanning electron micrographs show millions of elaborate, very fine hairlike setae, each bearing tiny hooks and typically assembled into hundreds (up to a billion altogether) of spatulae (Hiller 1976), which allow these lizards to gain purchase on almost any surface. Large arboreal geckos can climb vertically up panes of glass. Sometimes geckos even travel across ceilings hanging completely upside down. Several mechanisms of adhesion have been proposed—and eliminated—including suction (gecko feet still stick in a vacuum), glue (they lack glands), electrostatic attraction (disproved by experiments using x-rays to ionize air), and friction (a smooth pane of glass offers very little in the way of friction, although friction would certainly be important on rough surfaces) (Hiller 1976).

The most recent study of gecko toe pads demonstrates the remarkable power of natural selection (Autumn et al. 2000). Kellar Autumn and his colleagues removed a single seta from a Tokay gecko and, under a microscope, glued it with epoxy to an extremely fine wire. Each seta ends in hundreds of spatulae, which press up and conform to the substrate. Direct forces of setal attachment were measured with a micro-electromechanical sensor. Earlier

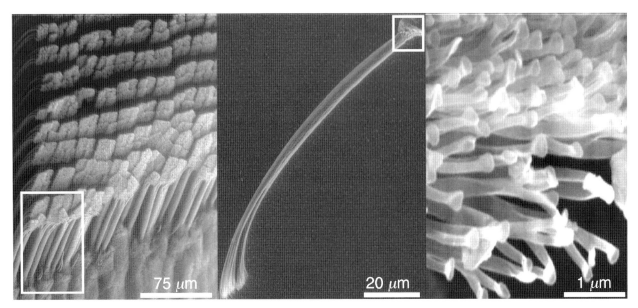

Scanning electron micrographs of a *Gekko gecko* foot pad. From left to right, rows of setae from a toe; an individual seta; and tiny terminal branches of a seta, the spatulae. (From Autumn et al. 2000)

work had demonstrated that intermolecular attractive forces, known as van der Waal's forces, provided adhesion (Hiller 1968). Autumn's study provides further indirect support for intermolecular forces, which require exceedingly intimate contact between the adhesive structure and the surface and are extremely weak at distances greater than atomic distance gaps. If the entire billion spatulae a gecko has were simultaneously engaged with substrate molecules, the force holding the gecko to the

Hind foot of a Tokay gecko *(Gekko gecko)* showing toe lamellae of the adhesive pad. (Chris Mattison)

substrate would be about 40 atmospheres (about 40 kg per square centimeter, or nearly 600 pounds per square inch)!

With such powerful forces, one might expect geckos to be plastered against their substrates unable to move, or at least to be picking up all sorts of debris. Indeed, gecko feet are "overdesigned" by at least an order of magnitude (Bauer and Good 1986). During a powerful cyclone, *Phelsuma,* "hanging on for dear life," were actually beaten to death by the furious flapping of the leaves they were on—yet still these dead geckos remained attached! How do geckos break such strong bonds? How do they control their powerful feet and toes? The mechanism of release was first studied by Tony Russell (1975). Whereas toe uncurling during attachment can be likened to blowing up an inflating party favor, toe peeling during detachment is analogous to removing a piece of tape from a surface (Autumn et al. 2000). When running, geckos peel the tips of their toes away from a smooth surface. Toe peeling may have two effects: first, to place an individual seta in an orientation or at a critical angle that aids its release; and second, to concentrate the detachment force on a small subset of those setae attached at any instant.

Although manufacture of small, closely packed arrays mimicking gecko setae is currently beyond the limits of human technology (Autumn et al. 2000), the natural technology of gecko foot-hairs might provide biological inspiration for future design of reusable dry adhesives. Perhaps one day, people will wear gecko skin gloves to climb cliffs and buildings. If so, natural selection will hold the patent. Scansorial toe pads have also evolved in diplodactylids as well as in iguanians (*Anolis;* Peterson and Williams 1981) and in some skinks (Williams and Peterson 1982).

Terrestrial species have considerably simpler pointed toes (but even these are highly variable in detail). One odd group of geckos, living on very soft sands in southern Africa *(Palmatogecko),* has webs of skin between its toes. Some tiny geckos, *Coleodactylus, Pseudogonatodes,* and *Lepidoblepharis,* have claws enclosed in a sheath of scales producing a slightly bulbous toe tip. Number of scales enclosing the claw varies among genera but in all cases results in the claw being totally hidden. These tiny lizards walk across unstable leaf litter, often leaping from leaf to leaf, their leaps exceeding their body length. Sheaths may keep them from getting snagged on leaves during such leaps, and increase surface area of the toe making contact with the substrate. In addition, sheaths may facilitate locomotion on water surfaces for these unwettable little geckos.

All geckos possess very fragile tails, which autotomize easily when attacked by a predator—though as always, exceptions exist: Australian *Nephrurus asper,* for example, appear not to lose their tails. Tail movements distract a predator away from the more vulnerable head or body of the gecko so that it strikes the tail (Congdon et al. 1974). Thrashing of an autotomized tail either while it is in the predator's mouth or in the immediate vicinity of the attack if a predator misses distracts a predator while the tailless lizard escapes (Vitt et al. 1977; Arnold 1984b). Costs associated with tail loss in geckos include loss of fat reserves and reduction in egg size (Dial and Fitzpatrick 1981), but the benefit is escape from predation. Survival clearly outweighs high costs. Broken tails usually regenerate quickly, though they are often distinguishable from original unbroken tails. In some species, such as *Hemidactylus agrius* of the semiarid caatinga of northeastern Brazil, regenerated tails are much larger than original tapering tails. One could speculate that natural selection has favored regeneration of a better decoy than the original tail in species involved in risky behaviors. In many geckos, caudal autotomy is strongly temperature dependent: tails break off more easily at very low and at high temperatures (Greer 1989). Thus a gecko can lose its tail when it is so cold that it can barely move. Geckos with actively functional tails are as likely to lose their tails as geckos with tails that do not serve an auxiliary function (Bauer and Russell 1994). Geckos have evolved a wide variety of caudal adaptations—indeed, some genera are named for aspects of their tails: *Oedura* means "swollen tail"; *Phyllurus,*

Stenodactylus sthenodactylus, **one of the terrestrial gekkonines, lacks enlarged adhesive toe pads.** (Louis Porras)

Clockwise from top left: **The Australian diplodactylid gecko *Diplodactylus elderi* is semiarboreal, foraging in spinifex grass.** (Eric Pianka)

Dunes of the Namib Desert are home to the gecko *Palmatogecko rangeri*. (Chris Mattison)

Australian leaf-tailed geckos, *Phyllurus ossa,* are difficult to detect against natural backgrounds. (Steve Wilson)

"leaf tail"; *Uroplatus,* "tail flat"; *Pristurus,* "sawed tail"; *Nephrurus,* "kidney tail"; and *Rhynchoedura,* "beaked swollen tail."

Many geckos store fat in their tails and can survive long periods without food provided they have access to water. A unique Australian species, *Diplodactylus conspicillatus,* has a short bony stub of a tail, used to close off diurnal retreats (abandoned trapdoor spider burrows). Prehensile tails are used as fifth legs for climbing by some arboreal geckos, such as *Diplodactylus elderi, Gehyra membranacruralis, Oedura filicipoda, Pseudothecadactylus,* and some *Rhacodactylus* (Pianka 1986; Greer 1989; Bauer 1998; King and Horner 1993). Some geckos have scansorial pads on the tail tip similar in structure to toe pads. The scansorial pad on the tail tip of the Brazilian gecko *Lygodactylus klugei* is used as a fifth landing point as these tiny geckos jump from one arboreal perch to another. Rapid regeneration following tail loss includes regeneration of the microstructure of the tail pad (Vitt and Ballinger 1982). Other geckos with scansorial pads on prehensile tail tips include members of the New Caledonian *Rhacodactylus* and the New Guinea *Gehyra membranacruralis, Pseudothecadactylus lindmeri,* and *Oedura filicipoda* (King and Horner 1993). Flattened leaflike tails, which presumably serve as camouflage by disrupting a gecko's shape, have evolved several times, in Australia *(Nephrurus* and *Phyllurus)* as well as in Madagascar *(Uroplatus).*

A New Caledonian diplodactylid gecko, *Eurydactylodes,* and some Australian *Diplodactylus* (subgenus *Strophurus*) have glandular tails, which exude a noxious smelly sticky fluid when these geckos are annoyed. These geckos have no fat glands in their tails, and their mucus glands could be modified versions of what were originally fat glands. Some of these geckos have bright mouth colors (dark blue or yellow), exhibited when they open their mouths in defensive displays. Some appear warningly colored with red and yellow splotches. The sticky mucus produced by some species is also colored black or yellow and can be squirted fairly long distances (up to half a meter) from the tail in thin streams. One night, Pianka pit-trapped a *Diplodactylus ciliaris* along with a small elapid snake and two other lizards in the same five-gallon bucket; in the trap the next morning, all four were glued together by the sticky mucus.

Both sexes of Australian knob-tailed geckos (genus *Nephrurus,* 7+ species) have a curious round knob at the tip of their tails, the function of which remains somewhat

The Australian knob-tailed gecko *Nephrurus vertebralis* has a sensory structure at the end of its tail. (Brad Maryan)

obscure (broken regenerated tails lack knobs). These knobs are rigid, highly vascularized, and possess numerous sensilla as well as a muscled slim peduncle, permitting free movement of the knob. Knobs may serve as mechanoreceptors but might also be used for heat exchange in thermoregulation (Russell and Bauer 1987). Another theory is that the knob is vibrated in the substrate to create a buzzing sound that distracts predators (Greer 1989). Perhaps these knob-tailed geckos press their knobs against one another in social encounters? For obvious reasons, observations on the behavior of nocturnal geckos in their natural microhabitats remain distressingly scant.

Twelve species of Australian desert geckos vary greatly in their degree of dietary specialization (Pianka and Pianka 1976). Some species are relatively generalized, eating a wide variety of arthropod prey *(Heteronotia binoeii, Diplodactylus ciliaris, D. damaeus, D. strophurus, Nephrurus laevissimus),* but three sympatric species *(Diplodactylus conspicillatus, D. pulcher,* and *Rhynchoedura ornata)* consume virtually nothing but termites, sitting and waiting along active termite trails at night catching the insects as they pass by. Termite specialists are much more likely to have empty stomachs than are geckos with broader, more generalized diets, suggesting that dietary specialization is accompanied by feast-or-famine benefits versus costs (Pianka and Pianka 1976; Huey et al. 2001).

Some geckos have weakly attached skin that tears away when they are attacked by predators, allowing the gecko to escape (e.g., *Ailuronyx* from the Seychelles, African *Afroedura,* Asian *Teratoscincus, Geckolepis* in Madagascar, and Australian *Gehyra mutilata*). Like many other lizards, geckos often eat their own shed skin, presumably recycling a significant amount of protein in the process (Szczerbak and Golubev 1986).

An interesting Southeast Asian gecko, *Ptychozoon* (= fold animal), has wide flaps of skin along its sides between its front and hind legs, enabling it to parachute. *Ptychozoon* also has flaps of skin on the sides of its head as well as along the hind margins of its limbs. Its tail is flattened with wide scalloped margins and its toes are webbed. When at rest on tree trunks, these geckos hug the tree with their flaps and margins closely pressed against the tree, effectively smoothing out their outline and making them very difficult to see. Members of *Luperosaurus,* the sister genus to *Ptychozoon,* also have flaps and folds and also parachute to safety. These are not, however, the only "flying geckos." During field work near the Rio Curuá-Una in the central Amazon, one of Vitt's graduate students, Pete Zani, was searching through a bromeliad about 36 m above the forest floor in a large tree when an adult *Thecadactylus rapicauda* jumped into the air, stretched out its legs, and glided about two-thirds of the way to the ground. The enormous scansorial pad on one of its front legs caught a leaf on an adjacent tree, which

was enough of a purchase to allow the gecko to grab hold and scurry to safety (Vitt and Zani 1997). Australian *Pseudothecadactylus lindmeri* have been observed to escape by leaping as far as 2 m between branches and dropping 6 m to the ground (Greer 1989).

Ground-dwelling geckos tend to have larger eyes than climbing species (Werner 1969), perhaps increasing their field of vision. Many terrestrial geckos *(Agamura, Alsophylax, Chondrodactylus, Coleonyx, Eublepharis, Teratoscincus)* walk with stiff legs holding their bodies well up off the ground, often with their tails held erect or arched over their back (perhaps they are mimicking scorpions as a predator defense tactic). When threatened, some geckos (e.g., *Nephrurus*) adopt stereotyped defensive postures, inflating their bodies and lunging and hissing open-mouthed at any enemy. In contrast, arboreal geckos typically assume low-profile clinging positions, holding their bodies close to the substrate with their legs extended away from their bodies.

Unlike most other lizards, geckos at night are generally thermoconformers with body temperatures very close to ambient thermal conditions. Environmental temperatures are usually quite low at night and change little from

Australian *Diplodactylus pulcher* specialize on termites. (Brad Maryan)

After parachuting from the treetops and closing its body flaps and pouches, a *Ptychozoon kulhli* gecko stalls and veers sharply upward as it comes in for a landing. (Time-lapse photo of a single lizard by Tim Laman)

place to place, offering little or no opportunity for active thermoregulation. Some nocturnal geckos *(Hoplodactylus, Nephrurus, Palmatogecko, Teratoscincus)* are active at very cool temperatures (11–15°C). Interestingly, when offered a wide range of temperatures in a laboratory thermal gradient, most geckos select high-temperature microenvironments and maintain high body temperatures, sometimes as high as those of diurnal lizards (Huey et al. 1989). To be active at night, geckos must compromise their optimal body temperature and sacrifice some degree of performance. Many, perhaps most, nocturnal geckos may select high-temperature diurnal retreats, enabling them to digest food captured during cold nights. Some species achieve high body temperatures during the day simply by locating themselves under the bark on the sunny side of a tree, while others *(Diplodactylus conspicillatus, Rhynchoedura ornata)* move up and down vertical trapdoor spider tube burrows in the ground.

A small gekkonine gecko, *Gymnodactylus,* occurs across Brazilian cerrado and caatinga habitats, entering Atlantic rain forest in the southern part of Brazil (Vanzolini 1968a,b). These active diurnal geckos are most frequently found in pairs under a surface rock or in crevices and

Occurring from Pakistan to Assam, the ground-dwelling gecko *Teratolepis fasciatus* is the only species in its genus. (R.D. Bartlett)

spaces within piles of medium-sized rocks, suggesting that long-term pair bonds may exist. Scales on the dorsal surface are unequal in size, making it quite easy to distinguish these from some of the South American sphaerodactyline geckos that they superficially resemble. Although active in dark places during the day, *Gymnodactylus* frequently bask in sun during late afternoon, rapidly retreating into

An Australian gecko, *Rhynchoedura ornata*, emerges from its burrow to begin a night of activity. (Eric Pianka)

rocks when approached. As in the sphaerodactyline *Gonatodes,* males are more brightly colored than females, a characteristic that undoubtedly serves important social functions.

In cerrado north of Brasília, *Gymnodactylus geckoides darwinii* lay clutches of two eggs. In caatinga, however, a different subspecies, *Gymnodactylus geckoides geckoides,* deposits only a single egg, with ovulation alternating between the left and right ovary (known as allochronic ovulation). Given that most other gekkonines produce two eggs (synchronic ovulation), this subspecies has likely undergone a relatively recent evolutionary reversal of a supposedly fixed trait. Alternatively, what has always been considered an all-or-none process (allochronic ovulation) might be nothing more than lack of synchrony in ovulation by the two oviducts. If so, the expectation would be high variance among species in the interval between clutch production, with some species having long intervals, others having short intervals, and some, such as *G. geckoides darwinii,* having no interclutch interval. (The same argument can be applied to anoles.)

Phyllopezus pollicaris, a large-bodied gekkonine with a flattened body, inhabits rock crevices in broad rock faces *(lajeiros)* in the caatinga of northeastern Brazil. It forages at night on dark exposed rock surfaces, which, while heat-

ing to lethal temperatures of more than 60°C during the day, retain enough heat at night to allow these geckos to maintain body temperatures higher than ambient air temperatures. During the day, these geckos remain hidden in crevices but position themselves such that their body temperatures while inactive are higher than body temperatures at night when they forage (Vitt 1995).

Nocturnal geckos sometimes bask in the sun during the day, staying very close to cover. Some members of one Australian group of diplodactylids (subgenus *Strophurus*) have black pigmented peritoneum like many diurnal lizards, which is thought to screen solar radiation (Greer 1989).

Certain gekkonines are established almost worldwide at every portside settlement with a warm climate. Sometimes these invaders spread inland, but they usually do not stray far from human habitations. The Mediterranean gecko *Hemidactylus turcicus,* for example, arrived in Baltimore before 1990. It reached the port of Houston over a decade ago and has moved northward along Interstate 35 hundreds of kilometers, all the way inland to Dallas. This species reached Tempe, Arizona, more than thirty years ago, catching a ride in vans loaded with ichthyological field equipment returning from Cuatro Cienegas, Mexico. A congener, the Southeast Asian gecko *Hemidactylus frenatus,* has spread to Africa, Guam, western Mexico, Madagascar, and Australia.

On many islands in the Pacific, the common house gecko *Hemidactylus frenatus* is currently invading and displacing other species of geckos. One such is the mourning gecko, *Lepidodactylus,* an all-female parthenoform that invaded remote Pacific islands hundreds to thousands of years ago, often facilitated by early human travelers (Petren and Case 1996). *Hemidactylus* are larger and faster than *Lepidodactylus* and invariably win in aggressive encounters with the smaller asexuals. Moreover, *Hemidactylus* are more pursuit-oriented foragers and are more successful at long-distance strikes than *Lepidodactylus.* Humans build flat walls and light these up at night, which attracts insects and creates a relatively simple microhabitat with concentrated food resources. As a result, interspecific competition is intense and *Hemidactylus* outcompetes *Lepidodactylus,* especially around human habitations.

Keith Petren and Ted Case (1998) studied these two competing species of geckos using manipulated physical environments. Ten individuals of each species were housed in enclosures inside World War II airplane hangars on Oahu, Hawaii. The control treatment, intended to represent a typical urban environment, consisted of a single light in the center of an enclosure, which resulted in a concentrated source of insect foods. One experimental treatment involved four evenly spaced lights, resulting in a more dispersed distribution of insect foods. Another experimental treatment modified the topographic structure of microhabitats by placing a complex array of baffles around the lights. Effects of competition were assessed by weighing geckos at regular intervals. Lizards were also observed from behind a blind. Dispersed lights attracted from one-third to one-fifth fewer insects than single lights in control treatments. Increased food dispersion reduced competition between the two species. Adding topographic structure reduced spacing patterns (geckos were closer together) and lessened the advantages enjoyed by the larger, faster invasive species. In turn, competition increased between territorial individuals within the smaller resident species *Lepidodactylus* (more fighting was observed).

Geckos are the most vocal of all lizards, with some calling much like frogs. Two common local names for *Gekko gecko* are Tokay and Tucktoo, derived from its loud calls "tow-kay" and "tuck too." Similarly, *Hemidactylus frenatus* is called "cheechak." The word "gecko" itself is said to have originated from the sound of a gecko's call. The barking gecko *(Ptenopus garrulus)* of the Kalahari chirps at dusk, chorusing like frogs. These lizards are difficult to collect because they seldom stray very far from their burrows, but a colleague of ours, Ray Huey, perfected a "shovel trick" to catch *Ptenopus.* To attract females, males sing from the entrance to their burrow with just their head and foreparts out. If approached slowly and silently, they will often hold their ground at this relatively safe spot. Holding a shovel out at arm's length angled about 45 degrees downward, Ray crept slowly toward a calling male until within a couple of meters, then suddenly lunged forward, inserting the shovel below the lizard and cutting off its retreat.

One night when termites were swarming, sending out their winged reproductive alates, Pianka's field crew found dozens of *Ptenopus* out well away from their burrows feeding on the juicy soft-bodied insects. Some of the geckos were so full that they could scarcely move, their throats and mouths stuffed with termites. Arron Bauer and his colleagues (1990) found *Ptenopus* individuals with 60 percent of their body weight constituted in termites.

Such gluttony has its risks. One unusual overcast day when winged termites were swarming on mating flights during daylight hours, *Ptenopus,* normally strictly nocturnal, were active during daytime feeding on termites. Several species of shrikes (small predatory birds) were capturing these geckos, saving them for future meals by pinning them up on thorny *Acacia* bushes. Dozens of crucified *Ptenopus,* many still alive, festooned nearly every thorny bush like Christmas tree ornaments that day, ironically very near "red Christmas" in the hot Southern Hemisphere summer (Pianka 1994c).

Finding geckos at night by the dim glow of their eyeshine is tricky. You have to be looking from just the right distance away: too close or too far, and you simply cannot see them. Spiders and certain insects have bright sparkling eyeshine, and since these are plentiful, one must learn to ignore them but perceive and distinguish the much fainter glow of gecko eyes. Several of Pianka's assistants proved to be excellent geckoers; night after night they would return with two to three times as many as he could catch. In contrast to diurnal lizards, many geckos are relatively easy to capture once they have been located. Pianka's ex-wife Helen likened geckoing to an Easter egg hunt, basing this analogy on both their beauty and their approachability.

Thecadactylus can often be found on tree trunks in the dense Amazon rain forest by eyeshine reflected from the collector's headlights. *Thecadactylus* eyeshine is faint orange, but with practice can be distinguished from that of spiders (usually green or blue and "sparkling") and most frogs (usually much brighter). Another of Pianka's clever field assistants, Bill Giles, invented a useful method to catch speedy, wary nocturnal arboreal Australian geckos *(Gehyra)* that stay on the side of marble gum tree trunks opposite the collector. This "shovel trick" is simple but efficient: wearing a headlight, one simply holds a shovel in one hand, and with the other hand free, poised, and ready, sticks the broad side of the shovel around the backside of the tree and moves it up or down; very frequently, a gecko darts around into the light, where it can be grabbed.

Using eyeshine to collect geckos in the Kalahari, we found a beautiful semiarboreal gecko, *Pachydactylus rugosus,* that had previously been thought to be rare. This species' favored microhabitat is about half a meter to a meter above the ground on branches of the "wait-a-while" thorny bush *Acacia mellifera* (protected by vast numbers of recurved, cat's claw–like thorns that snag your

Sometimes called "wonder geckos" or "frog-eyed geckos," Asian *Teratoscincus scincus* arch their back and rub scales together to produce a rattling sound when disturbed. (John Murphy)

Madagascan day geckos, like this *Phelsuma grandis,* are brilliantly colored; they have undergone convergent evolution with diurnal lizards in other families. (John Murphy)

Two of my Kalahari study sites were within the Kalahari Gemsbok National Park (now known as the Kalagadi Transborder Park) in South Africa. This large nature reserve supports herds of half a dozen species of antelope plus the full spectrum of large predators, including both leopard and lion. Hearing a wild lion roar makes the hackles on the back of your neck stand on end; it certainly made me appreciate how our African ancestors must have felt. Tourists are allowed within the park only during daylight hours. As biologists needing to collect geckos at night, however, we were given permission to camp out.

In benign Australia, the only real hazard of geckoing was getting lost. In the African bush, however, hunting for geckos at night was dangerous, as lions and leopards were prowling about too. Every individual big cat is known to park authorities, and it is strictly illegal to shoot one for any reason. Even so, we applied for, and were issued, a permit to carry a big .45-caliber handgun for use in self-defense. Although this probably wouldn't stop a lion, we fully intended to go down fighting if we were attacked.

Having heard repeated warnings never, ever to turn our backs on, or to run from, a big cat, we ventured—with considerable trepidation—out into the dark African night to look for geckos by eyeshine, staying as close to our vehicle as possible. (Here, getting lost was not a problem!) However, our efforts proved extremely ineffective because we couldn't help but scan the black distance in a 360-degree sweep every few minutes with our headlamps, checking for the bright (relative to geckos) eyeshine of predators. Springhares were quite common, and these large rodents also had large bright eyeshine. Whenever we saw one, we had to wait until it bounced away to reassure ourselves that it was indeed harmless.

Once we had collected all the species of geckos we expected to find in a particular area, we ceased geckoing because of the danger. However, at one study site deep within the park, we had been unable to collect one highly expected but also highly elusive uncommon species of gecko (*Pachydactylus capensis*). On our last night out at this site we were joined by South African herpetologist Wulf Haacke. That day at a waterhole only about 4–5 km from our camp, Wulf had seen fresh tracks of a large lone male lion—the most dangerous, because without females to hunt for them, males tend to be hungry and ill tempered. So on that last night out, we were all quite keyed up; those who smoked were chain-smoking away. All of our mouths were dry and our hearts were beating faster than usual. At five- to ten-minute intervals, Wulf would climb on top of his vehicle and scan the horizon with a portable spotlight. We all stayed close to the vehicles and to each other. I was geckoing with Larry Coons; we were walking silently along about 10–15 m apart, each of us scanning our 180 degrees of horizon. Suddenly Larry whispered loudly: "Eric, right between us!" We had walked up on a serval, a long-legged cat about the size of a bobcat, without even seeing it! Soon thereafter, we abandoned our quest for that elusive species of gecko, leaving it as "highly expected." *(PIANKA)* clothing and skin). Extracting these geckos from wait-a-while bushes proved as difficult as locating them in the first place!

A very unusual genus of Asian geckos, *Teratoscincus,* has evolved large, overlapping rhomboid scales like those of skinks. Tail scales are especially large and fingernail-like, and produce a hissing sound when the muscular tail is moved rapidly. A thrashing autotomized tail also makes this noise (Szczerbak and Golubev 1986), which may attract a predator's attention. *Teratoscincus* occur in sandy deserts as well as on hard clay and saline flats, where they dig their burrows.

A beautiful group of arboreal diurnal geckos known as day geckos (*Phelsuma,* about 40 species), varying in size and color, occurs in Madagascar and on many islands in the Indian Ocean, including the Seychelles. Most are bright green, often marked with spots of vivid red or blue. Because day geckos also breed and do well in captivity, they are favorites among herpetoculturalists. *Phelsuma* feed on nectar and are important pollinators of many insular plants. Curiously, however, the vast majority of gekkonines—of which *Phelsuma* is one—are nocturnal and relatively drab colored. Why did the ancestor to *Phelsuma* become diurnal, allowing this adaptive radiation? And why did this group of geckos evolve such striking colors? The green could be a background color match, since these geckos spend most of their time among green leaves. Some diurnal diplodactylid geckos in New Zealand are also green. Bright red and blue markings are probably used as signals in social encounters. The *Phelsuma* clade of geckos appears to have undergone convergent evolution with Neotropical sphaerodactyline geckos.

SPHAERODACTYLINE GECKOS

Throughout the Caribbean, Central America, and much of South America, geckos in the subfamily Sphaerodactylinae have radiated and in many places occur at high densities. As a group, they are all small-bodied, predominantly diurnal (*Sphaerodactylus* are crepuscular), and tend to be arboreal or scansorial, although members of one clade containing three genera, *Coleodactylus* (5 species), *Pseudogonatodes* (5 species), and *Lepidoblepharis* (16 species), live in leaf litter. The largest genus, *Sphaerodactylus,* with approximately 90 species, occurs from northern South America through Central America and southern Mexico and across the Caribbean islands to south Florida.

Many *Sphaerodactylus* are very small, with one species, *S. parthenopion,* likely being the smallest lizard in the world at 17 mm SVL and weighing only 0.12 g! The greatest diversity of *Sphaerodactylus* occurs across Caribbean islands, where they could be considered evolutionary counterparts of the highly diverse *Anolis* lizards. The genus *Gonatodes* contains 17 species, most of which are about 40 mm SVL. Greatest diversity occurs in the Dutch West Indies and Venezuela (Kluge 1995).

Sexual dimorphism and dichromatism are common in sphaerodactyline geckos. Male *Gonatodes albogularis,* for example, are brilliant orange on their neck and head, whereas male *G. humeralis* are spectacularly colored with red and yellow speckling on neck and head. In most species, males are slightly larger than females.

A ***Phelsuma grandis*** hatchling eats its own shedding skin.
(Bill Love)

Males of the sphaerodactyline gecko species *Gonatodes albogularis* have brilliant colors used in social displays. (Laurie Vitt)

***Gonatodes humeralis* shed their skin in a single piece and then eat the skin.** (Dante Fenolio)

Leaf litter sphaerodactylines are ecologically quite similar. *Coleodactylus* and *Pseudogonatodes* tend to occur on leaf litter in terra firma forest, above the flood zone, but *Coleodactylus* is often found in elevated patches of flooded forest as well. *Lepidoblepharis* tend to be associated with low, wet areas and often jump into water, scurrying across the surface using serpentine movements of the body to "swim" while suspended on the surface film. In the leaf litter of the Brazilian Amazon rain forest, *Coleodactylus amazonicus,* which reaches about 24 mm SVL maximum size, leaps from leaf to leaf in search of tiny insects. To these geckos, the world is a mat of leaf litter—there are no planets nor stars—and light is little more than transient patches that meet the forest floor, filtered through the canopy of giant trees.

When not moving, *Coleodactylus* is nearly invisible against leaf litter. Remaining perfectly still on a leaf five times its own length, it carefully scans its surroundings for the slightest indication of movement. Preferred prey of these minuscule lizards are tiny insects called springtails. Terminal segments of a springtail's body are tucked underneath its soft abdomen like a loaded spring. When disturbed, a flick of the "spring" sends the insect flying into the air and presumably out of reach of potential predators. Movement of a leg or repositioning of a body is all that is necessary for a gecko to detect a springtail's presence. From then on it is a careful game of approaching the insect at a slow enough rate to avoid detection but attacking rapidly enough to capture the springtail before it reacts. A single bite and the exoskeleton of the insect is broken, signaling to the lizard that the insect can be swallowed.

The genus *Gonatodes* is represented by species of intermediate size, usually about 35–40 mm SVL. Most are arboreal, living on trunks of forest trees or in catacombs formed by roots of stilt palms and buttresses of large rain forest trees. One species, *G. humeralis,* has an enormous distribution, stretching from gallery forests of rivers embedded in Brazilian cerrado across the entire Amazon and Orinoco basins, east to foothills of the Andes and north to Panama. These geckos are active on vertical trunks of forest trees and retreat into spaces between the trunk and the ground or into crevices formed by large, intricate buttresses. Females deposit a single egg, and the shell hardens shortly after deposition. Often many eggs and eggshells are found at a single oviposition site, suggesting either that individual females return to the same site repeatedly or that multiple females use a single site.

Populations of *Gonatodes humeralis* appear ecologically similar across Amazonian localities (Vitt, Zani, and Barros 1997), where they appear to be territorial, behaving in much the same manner as territorial iguanid lizards. More often than not, one or more females live within a male's territory. During morning and late afternoon, they often bask in direct sunlight, but during much of the rest of the day they remain in shade. Whenever males interact, their body coloration spontaneously changes from a dull gray with intricate patterns of darker colors to brilliant red and yellow. Males frequently chase each other around buttresses or on tree trunks. Females are drab-colored and appear incapable of color change.

In the same forest at some localities, but on the underside of large decaying logs, *G. hasemani* live, also usually in pairs, clinging to logs and other surface objects with their tiny claws. Males have a dark orange neck and head; the remainder of their body is black. Females are drab and do not change color. Social interactions are restricted to the home log, and these geckos appear to do nearly everything upside down—making even the swallowing of an insect a major operation! A combination of dark coloration and activity restricted to dark places renders these lizards difficult to locate and capture. Even though *G. humeralis* and *G. hasemani* often live in close proximity, social interactions between the two species do not occur.

PYGOPODIDS

Pygopodids (= rump-footed) are either the sister group to Australian diplodactylid geckos (Kluge 1974) or embedded within the Diplodactylidae (Kluge 1987). Commonly known as flap-footed lizards, pygopodids are essentially legless: only tiny vestigial remnants of hind legs are present in all species. They arose and underwent an adaptive radiation in Australia, where all 8 genera (about 40 species) are found (one species *Lialis jicari* is endemic to New Guinea, which also has been reached by the Australian *L. burtonis*). Some pygopodids are burrowers *(Aprasia* and *Ophidiocephalus),* others are terrestrial *(Lialis, Paradelma, Pygopus,* and some *Delma),* while still others are "grass swimmers" (many *Delma* and *Pletholax). Aclys* and some *Delma* are actually fully arboreal, crawling around in shrubs (Bryan Jennings, pers. comm.). Fossorial species have reduced eyes, countersunk lower jaws, and short tails. Terrestrial species are relatively heavy-bodied with

This Australian flap-foot, *Aprasia smithi*, is a burrower. (Brad Maryan)

An Australian pygopodid, *Paradelma orientalis*, exposes its gekkotan affinities by licking its face. (Steve Wilson)

moderately long tails. Arboreal species tend to be elongate and thin with very long tails. The so-called javelin lizard *(Aclys)* is the most arboreal, spending much of its time a meter or more above the ground in tall shrubs (Bryan Jennings, pers. comm.) but diving rapidly to the ground when threatened by birds. *Delma* can lift themselves completely off the ground, jumping by using their powerful muscular tails (Bauer 1986). Pygopodid tails are exceedingly fragile and easily autotomized. If a pygopodid sheds most of its tail, the lizard is essentially helpless; however, while the highly muscularized tail thrashes around wildly, attracting a predator's attention, the py-

gopodid can slip away unnoticed. Like their gekkotan ancestors, all pygopodids lay two eggs. And like geckos, they clean their eye spectacle by licking it.

Most pygopodids are diurnal (*Pygopus nigriceps* is nocturnal, and *Lialis* is both crepuscular and often active at night), but their eyes are those of nocturnal lizards with no fovea and elliptical vertical pupils (Greer 1989). Rick Shine (1986) suggests that *Pygopus lepidopodus* may have evolved to become diurnal because that is when its prey, spiders, are most likely to be "at home" and vulnerable in their dead-end burrows. Its close nocturnal relative *Pygopus nigriceps,* in contrast, captures spiders and scorpions at night by sitting and waiting for them to move past.

Most pygopodids are insectivorous, but *Lialis* has specialized on lizard prey, especially skinks. Skinks have osteoderms, plates of interconnected bones underneath each scale, which presumably confer some degree of armored protection from predators. *Lialis* (Patchell and Shine 1986) and several species of skink-eating snakes (Savitsky 1981) have evolved hinged teeth—when a tooth hits an osteoderm, it folds, allowing other teeth to penetrate between scales and obtain a firm grip on skink prey. Moreover, if the skink squirms and tries to struggle backward, the ratchetlike teeth erect, locking the skink in the pygopodid's mouth. Harry Greene (1997) evokes an image of a skink ratcheting itself down a predator's gullet! They do go down rather fast. Skulls of *Lialis* are very flexible, exhibiting cranial kinesis, which presumably facilitates capture of lizard prey.

This Australian pygopodid, *Ophidiocephalus taeniatus,* is fossorial. (Dr. Hal Cogger)

The distinctive *Ophidiocephalus taeniatus* was first described in 1897 from a single specimen collected in the southern Northern Territory of Australia (Lucas and Frost 1897). It was rediscovered eighty years later, in northern South Australia (Ehmann and Metcalfe 1978; Ehmann 1979). Indeed, some feared that this monotypic genus might have gone extinct. With a very restricted geographic range (probably relictual) in central Australia, it burrows in litter under trees and shrubs, in microhabitats reminiscent of those used by some fossorial skinks *(Lerista* and *Typhlosaurus)*.

Another interesting feature of the evolution of pygopodids was the acquisition of enlarged head scales, a feature often associated with burrowing. (These might have been present in common ancestors of geckos and pygopodids, but lost only among geckos. If so, pygopodids may have branched off from geckos earlier than currently thought—and that would make geckos monophyletic rather than paraphyletic.) Some pygopodids (certain *Delma, Paradelma,* and *Pygopus*) mimic snakes, holding their dark heads up high on arched and flattened necks, flicking their black tongues, and lunging (with mouth closed) as if they are dangerous. Enlarged head scales, which appear on numerous burrowing lizards, resemble those of snakes, of which most Australian species are venomous.

There are two reasons a gecko ancestor might have lost its legs and evolved into an ancestral pygopodid: (1) to increase rate of movement through vegetation such as in grass swimming, leading ultimately to arboreality; or (2) to gain access to large prey such as spiders living in crevices. Subsequently, natural selection would have favored further elongation and limb reduction due to advantages of lateral undulatory locomotion in escape from predators (Gans 1975; Shine 1986b; Greer 1989).

AMPHISBAENIANS

Amphisbaenia means "double moving," an allusion to the short, blunt tails of these peculiar squamates. Commonly called worm lizards, they are dubbed *cobras com duas cabeças* (two-headed snakes) by rural Brazilians, since they appear to have two heads. Four families, 19 genera, and about 150 species are recognized within this distinct evolutionary group of subterranean lizards. Their origin and affinities remain obscure, causing the group to be labeled *incertae sedis:* of uncertain taxonomic position.

An ancient group, amphisbaenians are widely distributed around the world, being found in tropical South America, Mexico, Florida, the Caribbean, eastern and sub-Saharan Africa, Morocco and Algeria, Turkey, as well as the Arabian and Iberian peninsulas (Gans 1990). Two of the four families have restricted geographic ranges and contain only a single genus: the last surviving member of the Rhineuridae, *Rhineura floridana,* is found only in central Florida (this family has an extensive fossil record in North America); and the family Bipedidae—comprising *Bipes,* with 3 species—occurs in Baja California and mainland Mexico. Trogonophidae contains 4 genera (6 species); these sand specialists are found in the Middle East and North Africa. The largest and most diverse family, Amphisbaenidae (about 19 genera, 140 species), has an extensive geographic distribution, including the West Indies, South America, and sub-Saharan Africa; one very primitive genus in this family, *Blanus* (6 species), has a disjunct distribution in the Mediterranean region (Spain, Morocco to Algeria, Syria to Turkey and Iraq). Indeed, although disjunct geographic distributions of amphisbaenian genera are not uncommon, their relationship is nonetheless supported by numerous shared derived characteristics (Estes et al. 1988). However, its members also share some unique derived characters with several other distinct clades, including gekkotans, scincomorphans, anguimorphans, dibamids, and snakes, making the position of this family within Scleroglossa uncertain. Because amphisbaenians have diverged from all other squamates in such a variety of ways, several paragraphs are required to list their many exceptional traits.

All so-called worm lizards are elongate, and all but members of one genus are completely limbless. The exception is the bizarre Mexican *Bipes,* which has no hind legs but does have well-developed hypertrophied front legs. Amphisbaenian bodies are cylindrical, covered with smooth square scales arranged in rings (annuli). Two body rings are present for each vertebra in all amphisbaenians except the Mediterranean *Blanus,* in which only one annulus occurs per vertebra. Many amphisbaenians are pink and, in fact, very wormlike in general appearance. Others, like *Amphisbaena fuliginosa,* are brightly colored, often with relatively intricate patterns. Amphisbaenians all have reduced or vestigial eyes, no external ears, and backward-pointing nostrils. The skin moves independently of the trunk, facilitating rectilinear locomotion—though one's first impression when observing one in the

Amphisbaenia fuliginosa **is the most colorful South American worm lizard.** (Laurie Vitt)

field is that locomotion is accordionlike. Many squamates with long bodies (pygopodids, anguids, skinks, snakes) have lost one lung—always the left—but amphisbaenians have lost the right lung. When Pianka rolled a rock over in Puerto Rico and picked up his first amphisbaenian, he was surprised when it bit him: these squamates have sharp teeth and quite a powerful bite.

Head shape varies considerably among amphisbaenians (Gans 1971, 1978; Vanzolini 1999). Some heads are blunt and nearly rounded off, such as in the wormlike *Amphisbaena cunhai.* Others, like *Leposternon,* have spatulate-shaped heads. Heads of still others, such as *Anops bilabiatus,* are laterally compressed. Most species construct burrows as they move, using their heads as a digging tool (head-first burrowing). Species with spade-shaped heads move their heads up and down, forcing the soil to pack above and below their head. Those with rounded and laterally compressed heads push forward, rolling their head from side to side, packing the walls of the burrow. One group of desert dwellers, the trogonophids, rotate their heads from side to side, appearing to screw their way through surface sand. Neck and body muscles are well developed to accommodate head-first burrowing (see also chapter 2), with forward thrust generally resulting from rectilinear locomotion.

When threatened, amphisbaenians move their short tails around in a very menacing, headlike manner. If a potential predator is fooled by such head mimicry and grabs the reptile by its tail (tails of many are scarred), its head remains free to bite back. If a predator grabs an amphisbaenian by its head, the lizard presses its blunt tail hard against the attacker, perhaps making it think the tail is the lizard's head and causing the predator to release it, thus facilitating escape. Tails of many amphisbaenians autotomize, but they do not regenerate.

Most amphisbaenians lay eggs, but a North African species of *Trogonophis* retains eggs internally until they hatch and gives birth to living young, as do some Tanzanian forms *(Chirindia).* Amphisbaenians eat invertebrates, primarily arthropods, but may also take small vertebrates. Evidence from studies of *Blanus cinereus* indicates that males can distinguish between the sexes by chemical cues, and can tell their own chemical cues from those of other males (Cooper, López, and Salvador 1994; López et al. 1997).

Only a single detailed field study has ever been made on an amphisbaenian. After examining a large number of *Amphisbaena alba* from a site in the cerrado of central Brazil, Guarino Colli and his colleagues from the Universidade de Brasília concluded that this species, the largest

Many South American worm lizards, like this *Amphisbaenia cunhai,* are tiny, secretive, and poorly known ecologically.

(Laurie Vitt)

***Amphisbaenia alba,* the largest South American worm lizard, raises its tail as a head mimic when disturbed.** (Laurie Vitt)

in the Neotropics, eats a diversity of prey, including some items that might not be expected for such a large species (Colli and Zamboni 1999). *Amphisbaena alba* reaches 735 mm SVL, yet ants, beetles, insect larvae, and termites dominate its diet volumetrically. Sometimes found in ant nests—usually those of leaf-cutter ants, *Atta* (Riley et al. 1986)—they undoubtedly feed on ants and termites in microhabitats where prey density is high. Unlike most amphisbaenians that have small clutches, *A. alba* produces clutches of eight to sixteen eggs, which are often deposited in ant or termite nests. Not surprisingly, no detectable sexual dimorphism exists. Even though *A. alba* performs an impressive display with its tail when disturbed, tails do not autotomize as they do in many other amphisbaenians. Effectiveness of the tail display may have offset the advantages of tail loss in this species.

DIBAMIDS

Little is known about these small wormlike burrowing scleroglossans (which, like amphisbaenians, remain *incertae sedis*). Males have small flaplike hind limbs, somewhat like those of some pygopodids; females are entirely limbless. Eyes are vestigial, and dibamids have no external ear opening. Their clutch consists of a single egg.

Only two genera are known. The single species of *Anelytropsis* is found in arid habitats in eastern Mexico, including deciduous scrub and pine-oak forests. *Dibamus* includes 9 species in secondary forests and rain forests in southeastern Asia. Although geographic distributions of the two genera of dibamids are quite disjunct, their relationship to each other is supported by about forty shared derived characteristics, some also shared with members of various other lizard families (Estes et al. 1988).

CHAPTER 10

FROM RACERUNNERS TO NIGHT LIZARDS

This tropical African lacertid,
***Holaspis guentheri,* often sleeps with its tail coiled,**
resembling an armored millipede. (Louis Porras)

In this chapter we introduce the highly active teiids and lacertids, the diminutive gymnophthalmids, and the secretive, long-lived xantusiids. Teiids, lacertids, and gymnophthalmids are in constant motion, maintaining a distance between themselves and other creatures—including humans—and frequently looking back to keep tabs on what could, after all, be a potential predator. Any sudden move by an observer, and the lizard darts off; although it immediately initiates foraging behavior again, its vigilance never ceases.

These are the lizards that dominate terrestrial habitats in the New World and much of the Old World (excluding Australia). Most of these lizards live in open habitats—even Amazonian forest teiids, such as *Dracaena* and *Crocodilurus*, favor large swamps or edges of rivers with sun exposure. Tiny gymnophthalmids appear at first glance to be exceptions because many are found in undisturbed rain forest. However, when scale is considered, most live in the open: a patch of leaf litter in western Amazonia is as much an open habitat to *Prionodactylus eigenmanni* as the bank of a small stream flowing through the same forest is to *Neusticurus ecpleopus*.

Teiids, gymnophthalmids, and lacertids (all of which form the clade Lacertiformes) are tied together evolutionarily because they share a common ancestor. Teiids are most closely related to gymnophthalmids, and both are presently restricted to the New World. Lacertids, the sister taxon to the teiid-gymnophthalmid clade, are an Old World group. Ecological counterparts in their respective habitats, lacertiforms share a number of identifying features: they have elongate, streamlined bodies compared with many iguanians; alert and often fast moving, they forage actively, primarily on the surface of the ground

(with notable exceptions), discriminating between prey and nonprey using chemical cues; and many, but not all, are active at high body temperatures.

Xantusiids, or night lizards, differ considerably from teiids, gymnophthalmids, and lacertids, and their phylogenetic position is less certain. They may be the sister taxon to Lacertiformes, but they could be more closely allied to the Annulata, the group containing, among other

Pantodactylus schreibersii is a common gymnophthalmid in Brazilian cerrado. (Janalee Caldwell)

The gymnophthalmid *Micrablepharus atticolus* lives in tailings of leafcutter ant nests. (Laurie Vitt)

things, wormlike amphisbaenians (see chapter 9). Another possibility is that they are related to Gekkota. We include xantusiids here recognizing that future studies may place them elsewhere within the evolutionary tree of lizard families. Xantusiids live in enclosed spaces. Some occupy crevices in rocks *(Xantusia henshawi)*, others *(X. vigilis)* live under decaying remains of Joshua trees or clumps of large beargrass, *Nolina bigelovi.* Xantusiid eyes are capped over like those of most geckos; they have elliptical pupils and are active in dark places. Their body temperatures while active are considerably lower than those of Lacertiformes.

TEIIDAE

The teiids we know today are a New World family, ecological counterparts of Old World lacertids. Historically, teiids had a wider distribution, at least in the Northern Hemisphere. Their northern distribution is restricted to deserts, the southern half of the Great Plains, and the coastal plain of the southeastern United States. A now extinct group, polyglyphanodontines, was diverse in the late Cretaceous of North America, and they also occurred in the Upper Cretaceous of Mongolia. The fossil record indicates that New and Old World polyglyphanodontines had diverged considerably from each other by mid to late Cretaceous. A faunal exchange during early Cretaceous sent some from east to west; others, however, went from west to east. The fossil record does not clearly indicate whether teiids originated in the Old or New World, though ancient teiids apparently colonized much of the New World. By the end of the Cretaceous, all Northern Hemisphere teiids had gone extinct (Nydam 2000). Modern teiids, therefore, diversified from ancestors remaining in subtropical and tropical areas of the New World, where they achieved the remarkable evolutionary success we see today. A single genus, *Cnemidophorus,* reinvaded and diversified in the northern half of the New World.

Several characteristics distinguish teiids from other lacertiforms. Teiid head scales are separate from skull bones; those of lacertids are fused to the skull. Teiid teeth have solid bases and are "glued" to jaw bones with cementum; those of lacertids are hollow, and cementum is absent. Indeed, cementum is such a prominent teiid character that it can be used to distinguish fossil teiid jawbones from those of all other fossil lizards. Additional characteristics include generally small, granular scales on the dorsal surface, with large, rectangular scales forming distinct transverse rows ventrally. All teiids have fully formed legs and a fairly distinctive overall morphology consisting of streamlined bodies, long tails, relatively pointed snout, eyelids, and long hind limbs. They are active foragers and lay eggs.

Teiids occur throughout the southern and western continental United States, across Mexico and Central America, on many Caribbean islands, and into much of South America (excluding the high Andes and extreme southern parts of South America). Nine genera are recognized: *Ameiva, Callopistes, Cnemidophorus, Crocodilurus, Dicrodon, Dracaena, Kentropyx, Teius,* and *Tupinambis* (Presch 1974). One genus, *Cnemidophorus,* with 56 named species (Wright 1993), has undergone an extensive adaptive radiation. *Ameiva* and *Kentropyx* are well represented also, but *Tupinambis, Callopistes, Teius, Dracaena,* and *Dicrodon* contain only a few species. *Crocodilurus* is monotypic, that is, represented by a single species, *C. lacertinus.* Teiids are divided into two subfamilies, Teiinae *(Ameiva, Cnemidophorus, Dicrodon, Kentropyx,* and *Teius)* and Tupinambinae (the remaining four genera). Teiids vary in body size, ranging from the small whiptail *Cnemidophorus inornatus* (55 mm SVL) to large tegus, *Tupinambis* (500 mm SVL), and caiman lizards, *Dracaena* (approx. 300–450 mm SVL).

The 56-plus species of *Cnemidophorus* (= carrying leg armor), commonly known as racerunners or whiptails, range from southern Idaho through Central America and the Caribbean all the way to Argentina. They reach their largest body sizes in northern South America and on some Caribbean islands, with size generally diminishing as latitude increases (although exceptions exist: *C. tigris,* for example, is smaller in southern parts of its geographic range). Several species of *Cnemidophorus* remain unnamed, including one found in Manaus, Brazil. Many "species" of whiptails are unisexual; in these, males do not exist and females reproduce by parthenogenesis. As soon as a female reaches sexual maturity, she begins producing daughters that are genetically identical to herself (see chapter 6). Parthenogenesis occurs in some other teiids (e.g., *Kentropyx*) as well, and in some other squamate families.

Herpetologists have had difficulty identifying whiptails and working out their relationships. As Charles Lowe points out in his insightful introduction to *Biology of Whiptail Lizards (Genus* Cnemidophorus) (Wright and

Vitt 1993), the famous herpetologist Edward Drinker Cope considered this genus the most difficult in all of herpetology. Yet Cope had seen only the tip of the iceberg: parthenogenetic *Cnemidophorus* were not discovered until 1958 (not published until 1962), adding a new element of confusion to the genus.

Discovery of parthenogenesis in East Asian *Lacerta* by Ilya Darevsky (1958) sent the American herpetologists Richard Zweifel and Charles Lowe deep into their collections searching for the answer to a question implicit in an observation first made by Sherman Minton (1958); why were there no males in *Cnemidophorus tesselatus*? Indeed, not only did *C. tesselatus* prove to be parthenogenetic, but a swarm of parthenogenetic *Cnemidophorus* were found to occur all across the southwestern United States and northern Mexico, all very difficult to sort out. Zweifel and Lowe (1966) discovered three characters—number of scales around midbody, number of scales between paravertebral stripes, and number of scales on toes—that were sufficient to distinguish one species of *Cnemidophorus* from another. Of course, in those days we didn't have the fancy molecular techniques that now dominate systematic herpetology, but at least Cope's most difficult genus could finally be sorted out into species.

Among teiids, whiptails are most widespread geographically and appear tied to open habitats to a much greater extent than *Ameiva, Kentropyx,* or most larger-bodied genera. These lizards are most frequently encountered on beaches and desert flats, in tropical dry forest, and along edges of relatively closed habitats such as forests. They often use roads and trails to get to open patches within forest in tropical regions. As a group, they are also among the most terrestrial of teiids, rarely entering water or climbing into vegetation.

In addition, whiptails—which appear quite nervous while foraging, often darting off at the slightest provocation—are among the most active of Lacertiformes. Their high activity levels are supported in part by their high body temperatures. *Cnemidophorus deppii,* for example, which live on open beaches in western Central America, average nearly 40°C while active.

In temperate North and South America, whiptails are active during summer, with activity falling off rapidly as fall approaches. In North America, adults disappear underground in August (though juveniles remain active until September or October), and reproduction is highly seasonal, occurring in spring and early summer (as, for example, in *Cnemidophorus inornatus* and *C. neomexicanus;* Christiansen 1971). In tropical environments where annual temperatures are relatively constant, activity and reproduction can occur nearly year round (as in *C. ocellifer* in northeastern Brazil; Vitt 1983).

A North American species, *Cnemidophorus tigris,* ranges from southern Idaho through Sonora and Baja California. Because of its wide distribution, it exemplifies geographic variation in ecological traits. In the north, *C. tigris* are active at both lower body temperatures and lower ambient environmental temperatures than in the south. Their seasonal period of activity is also shorter in the north. Frequencies of broken regenerated tails are higher in the south (Pianka 1970a). Most whiptails eat a variety of insects (particularly termites) and spiders, but some, such as *C. lemniscatus,* add fruits to their diet, and a few, such as *C. murinus* and *C. arubensis,* are herbivorous.

Ameiva, which occur through southern Mexico, Central America, and much of South America, with numerous island species in the Caribbean, are very much like *Cnemidophorus* in general morphology and behavior. As in *Cnemidophorus, Ameiva* are terrestrial. Diets are varied but include a diversity of invertebrates, small vertebrates, and fruits (Vitt and Colli 1994; Censky 1996). *A. ameiva* in South America has been best studied. This alert, fast-moving, large (190 mm SVL) lizard is common in virtually every open habitat within its range, even entering cities to forage alongside dogs in garbage heaps! It is among the most conspicuous of lizards in Venezuelan llanos, cerrado, caatinga, and lowland rain forest habitats. *Ameiva ameiva* maintain body temperatures of 37°C and higher, regardless of habitat, and can be seen in strikingly large numbers in lowland forest along roads, trails, and river edges where direct sunlight hits the ground (Sartorius et al. 1999).

Like many other teiids, *Ameiva ameiva* digs a burrow where it remains while inactive. Burrows generally have a single entrance and are left open while the lizard is inside. When on flat ground, burrows are shallow, with the terminal chamber just under the surface. If an intruder digs into one of these burrows, the lizard typically emerges through the roof of the terminal chamber and dashes off. On steep hillsides, burrows often go straight into the bank and may likewise be rather shallow. Although lizards have no escape routes from these shelters, their location—sometimes as much as 4.5 m up the bank—likely discourages many predators.

The large teiid *Ameiva ameiva* occurs in nearly all tropical habitats of South America. (Laurie Vitt)

In Central America, several species of *Ameiva* occur along beaches and in dry forest of the west coast, including *A. festiva* and *A. quadrilineata*. *Ameiva festiva* also occurs in rain forest of eastern Central America, along the Caribbean. Just north of Río San Juan in Nicaragua, *A. festiva* is a common forest species near treefalls. In many respects, including coloration and color pattern, it appears nearly identical ecologically to *Kentropyx pelviceps* of lowland Amazonian forest, foraging and basking in treefalls but also entering forest to forage for brief periods before its body temperature falls (Vitt and Zani 1996d). However, it doesn't climb like *Kentropyx*.

At least 12 species of *Ameiva* occur on the Lesser Antilles. Each island has a single species, with one exception: the Anguilla Bank has two, *A. corax* and *A. pleii* (Censky and Paulson 1992). Unlike *Anolis*, which colonized the island arc from the west (Puerto Rico), *Ameiva* colonized from the mainland to the south. When sea levels were much lower during the Pleistocene, islands were larger and separated by less water. The fact that most islands have but a single species suggests that *Ameiva* are their own strongest competitors. On the Anguilla Bank, the two species are different in body size, with adult male *A. pleii* reaching 181 mm SVL and *A. corax* reaching only 132 mm SVL. Those 50 mm may represent the minimum body size difference allowing coexistence. Of course, juveniles are much more similar in size and probably face intense interspecific competition as a result.

This divergence in body size, both within and among genera, is in fact fairly common among teiids (see below). Unlike iguanian species, which vary considerably in overall morphology and often segregate by microhabitat, teiid species are strikingly similar morphologically (with a few exceptions). Yet because they cover large areas while foraging, and forage at about the same time of day, encounters among individuals and between species occur frequently. Hence, divergence in body size appears to allow coexistence. Examination of body sizes and DNA-based phylogeny for populations of *C. tigris* and *C. hyperythrus* from Baja California, Mexico, and associated islands, for example, reveals that body size has evolved in both species and that when the two species occur together, one *(C. hyperythrus)* is small and the other *(C. tigris)* is large (Case 1979, 1983; Radtkey et al. 1997). When these two species occur alone, however, they are intermediate in size, indicating that optimal body size in a single-species guild differs from that in a two-species guild. This phenomenon, in which a character shifts away from the norm in response to interaction with another species, is referred to as "character displacement." One consequence of divergence in body size in teiids is that prey can be partitioned on the basis of size, offsetting competition for food.

Kentropyx (= spur or point rump, referring to spikelike enlarged scales on either side of the male's cloaca) is easily distinguished from all other teiids by a single feature: keeled ventral scales. These scales may help hold lizards against branches and leaves while climbing in vegetation. The eight species of this genus are restricted to mainland South America (Gallagher and Dixon 1991). Some (e.g., *K. pelviceps, K. altamazonica,* and *K. calcarata*) occur only in Amazon forest habitats or gallery forests along rivers that extend into savannalike grasslands, or cerrados, of central Brazil. One, *K. altamazonica,* though generally associated with waterways in lowland rain forest, has also been found far from water in patches of cerrado isolated in tropical rain forest of Rondônia. Others, such as *K. vanzoi* and *K. striata,* live in grasslands, the former in portions of cerrados and the latter in patchily distributed Amazon savannas. Most *Kentropyx* are excellent climbers, often ascending forest-edge trees where they lie atop leafy vegetation or on branches basking in the morning sun to gain heat. Body temperatures of *Kentropyx* are slightly lower than those of whiptail lizards but still higher than those of lizards in most other families. Some forest species, such as *K. calcarata,* can often be found in large numbers by searching exposed surfaces along forest edge. Juvenile *K. calcarata* climb into shrubs and work them much as birds do, gleaning insects and spiders from tops and bottoms of leaves as well as from twigs, limbs, and trunks. *Kentropyx* forage continuously, ceasing only when body temperatures drop below their preferred range or when they are copulating.

Two genera of large teiids, *Crocodilurus* and *Dracaena* (maximum SVL being about 220 mm and 450 mm, respectively), are semiaquatic. The single species of *Crocodilurus, C. lacertinus,* is widespread in the Amazon and upper Orinoco regions, *Dracaena guianensis* is widespread in Amazonia, and *D. paraguayensis* occurs in the Pantanal region of southwestern Brazil and extreme northwestern Paraguay. Members of both genera climb into trees along watercourses to bask, often lying on branches with their

Dracaena guianensis **is an aquatic Amazonian teiid, a snail specialist during wet season.** (William W. Lamar)

limbs and tail hanging free, appearing to be precariously balanced on the limb (Vanzolini 1961). Both are also graceful swimmers, folding their legs back against their body and swimming in a serpentine manner with head up. Tails are laterally compressed in both genera and have a double dorsal crest, providing forward thrust as it is moved back and forth. While foraging in shallow water, these lizards frequently walk on the bottom in search of prey. Not only is the water relatively warm, but the lizards' large size probably offsets some heat loss as well. When disturbed in water, they typically swim away from an intruder; they can also move quite rapidly on the water's surface, using a combination of serpentine locomotion powered by tail thrusts and rapid movement of front and hind limbs. Locomotion across water does not appear to be bipedal, and their body never raises off the surface. They can switch from a methodical swimming locomotion to rapid surface escape nearly instantaneously, demonstrating that tail thrust assisted by leg action is powerful.

Crocodilurus is a drab brown to greenish above, though juveniles are mottled with orange, particularly on their limbs, which renders them nearly invisible against water and leaf litter. Most striking are the large, rectangular scales of the ventral body surface, which are brilliant enamel-white in juveniles and enamel yellow in adults. No ecological studies have been conducted on this species, so most natural history observations are anecdotal. The few individuals that have been examined fed on juvenile toads and insects (Martins 1991), suggesting that their diet probably includes most catchable invertebrates and small vertebrates.

Dracaena, often found in the same swamps and waterways as *Crocodilurus,* is primarily a snail specialist, crushing mollusk shells with its powerful jaws and large molariform rear teeth and expelling pieces of crushed shell with the tongue. *Dracaena* enters flooded forests *(igapo)* during the wet season to forage. During dry season, when flooded forests dry out, *D. guianensis* forages in trees, apparently searching for insects and possibly bird eggs (Goulding 1989).

Tupinambis are by far the most impressive terrestrial teiid lizards, with two species, *T. merianae* and *T. rufescens,* reaching a snout-vent length exceeding 500 mm; their tail is nearly twice SVL, bringing maximum total length up to about 1300 mm (Fitzgerald et al. 1991). Six species are currently recognized, two of which have been described only in the last decade: *T. longilineus,* from the western Amazon (Avila-Pires 1995), and *T. quadrilineatus,* from central Brazil (Manzani and Abe 1997; Colli et al. 1998). These impressive animals march around during the heat of midday in search of prey, covering large areas while foraging. One species, *T. teguixin,* is most common along watercourses, where it feeds at the edge, often walking through shallow water. *Tupinambis merianae,* occurs in relatively open areas of northeastern and central Brazil and on into southern South America. The largest *Tupinambis, T. rufescens,* the red tegu, occurs in rather arid habitats of the Chacoan biotic province (Fitzgerald et al. 1991). The somewhat diminutive and slightly more streamlined *T. longilineus* occurs along rivers and streams in the western Amazon of Brazil. The first individual was found in central Rondônia in 1986 and described nine years later (Avila-Pires 1995); since then, three others have been collected and several more seen near Rio Ituxi, a tributary of Rio Purus in western Amazonas, Brazil.

All species of *Tupinambis* are dietary generalists. Arthropods, mollusks, and small vertebrates, particularly frogs, are common in their diets, as is carrion. They also feed seasonally on plant materials, especially fruits. The large *T. teguixin* raid nests of turtle eggs *(Podocnemis unifilis)* on breeding beaches and most likely prey on eggs of other oviparous reptiles as well (Avila-Pires 1995). All species can swim; some, such as *T. teguixin,* enter large rivers when disturbed and swim to the other side.

Tupinambis have larger clutch sizes than most teiids, partly a consequence of their large body size. For example, *T. rufescens* in Argentina average twenty-one eggs per clutch, with larger females producing larger clutches

Crocodilurus lacertinus is one of two large teiids that spend much of their time swimming in search of prey. (Laurie Vitt)

Tupinambis longilineus is known from only four specimens.

(Janalee Caldwell and Laurie Vitt)

(Fitzgerald et al. 1993). Nesting habits of some species have caught the attention of naturalists throughout history. William Beebe, a well-known naturalist working with the New York Zoological Society in the first half of the twentieth century, stated, for example, that "direct or indirect evidence of six separate nestings of tegus in the nests of termites were found at Karatabo and Caripitu. These varied from two feet above the ground to as high as twelve. Five were three to four feet up in low growths" (Beebe 1945). Female *Tupinambis,* when ready to deposit eggs, climb tree trunks and vegetation seeking out nests of arboreal termites in the genus *Nasutitermes.* Although these primarily terrestrial lizards have no obvious specializations for climbing (e.g., prehensile tails, recurved claws, toe fringes), nevertheless they do climb trees to dig large holes in the massive, dark-colored, and bulbous-shaped termite nests—ideal substrates for egg development— and deposit their clutches of ten or more eggs. Termites in the meantime swarm over the nest, but their chemical defenses and minuscule bites have little apparent effect on these big lizards. Once a lizard finishes depositing its clutch, termites repair the nest, sealing the lizard eggs inside. Nests remain relatively warm and humid due to the termites' own metabolism. When eggs hatch, one or more juveniles dig out and the remainder follow (Dixon and Soini 1986). Several large varanids use termitaria for nesting chambers as well (see chapter 13).

One should never underestimate the ingenuity of young children when money can be made. In 1977, while I was a postdoctoral research fellow of the Museu de Zoologia da Universidade de São Paulo working in semiarid caatinga in interior northeastern Brazil, I marshaled the help of about a dozen very enthusiastic local youths (all boys— young girls were not allowed to go out alone) to help me capture lizards. On any given day, more than a hundred lizards of sixteen species might be brought in. I paid the children 1 cruzeiro (about $0.05) per lizard, but since *Tupinambis merianae* were quite large and difficult to capture alive, I increased the price to 20 cruzeiros for adults of this species. One small kid, measuring about 1.5 m, seemed especially good at collecting them. I recall one day when he and his dogs had dug out two large adults. Arriving at my field lab at about 8:30 A.M., he held the lizards by their heads, with each lizard's tail dragging on the ground behind him. Later, at the end of the dry season, he brought me the first juvenile *Tupinambis* I had seen. He wanted 20 cruzeiros for it, but I told him that it was a small lizard and not worth as much as an adult. We finally settled on a compromise price of 10 cruzeiros. After we exchanged money and lizard, he sort of sheepishly asked me if I wanted any more "small" ones, to which I replied, "Of course, I'll take all you can catch." He went running off, with a sly grin on his face; it never occurred to me that he might have found a nest. About half an hour later he returned with thirteen juvenile *T. merianae* for which I was of course compelled to pay 10 cruzeiros each, for a total of 130 cruzeiros! I got him to tell me that he had found the nest in the process of hatching, but he refused to show it to me. The most I coerced out of him was that it was in a tree inside a "casa de cupim" (termite nest). He brought me more adults over the course of the following six months but never found another nest. *(VITT)*

Tupinambis teguixin is widespread in South America and likely will prove to be several species. (Laurie Vitt)

Some smaller teiids include fruit and flowers in their diets. For example, in the Amazon savanna region of Roraima in northern Brazil, species in three genera, *Cnemidophorus, Kentropyx,* and *Ameiva,* eat yellow fruits of a locally common plant, the Muruci (Vitt and Carvalho 1995). Three teiids, *Cnemidophorus arubensis, C. murinus,* and *Dicrodon guttulatum,* are herbivores. The first two apparently can detect alkaloids, such as quinine, in plants, and thereafter can presumably selectively avoid alkaloid-containing plants or at least balance their intake of potentially toxic plants against their recent dietary history (Schall 1990). These lizards' ability to detect potentially toxic chemicals in plants is not surprising, considering that most scleroglossan lizards use chemical signals to discriminate among prey types. Apparent teiid avoidance of insects such as ants, hemipterans, and particularly noxious beetles likely reflects this ability as well, since many defensive chemicals produced by insects are alkaloids.

GYMNOPHTHALMIDAE

Gymnophthalmids (= naked eye), also known as microteiids, are the sister taxon to teiids. All species are small,

most are terrestrial, some are semiaquatic, and a few are partly arboreal. About 36 genera and 160 species are known. Most species are elongate and thin bodied when compared to other lizard taxa, have relatively short limbs, reduced to varying degrees in some species and nearly absent in others. Most have transparent windows in their lower eyelids, allowing them to see when their eyes are closed. These are strictly New World lizards (Presch 1980). As a group, they are pretty much limited to tropical latitudes, but gymnophthalmid diversity is high in both lowland Amazonian forest and foothills, valleys, and hillsides of the Andes. Some species, such as *Pholidobolus macbrydei,* even reach high elevations in the Andes (Montanucci 1973; Hillis 1985). *Proctoperus bolivianus* occurs as high as 4000 m in the Peruvian Andes. Aside from a few tiny geckos *(Sphaerodactylus, Coleodactylus, Pseudogonatodes,* and *Lepidoblepharis),* these are the smallest species in New World tropical lizard assemblages.

As many as ten species of gymnophthalmids can occur together. For example, in the Juruá River system of western Brazil in the state of Acré, the following species have been found within one square kilometer: *Neusticurus ecpleopus, N. juruazensis, Alopoglossus atriventris, A. angu-*

This Amazonian rain forest gymnophthalmid, *Alopoglossus atriventris*, lives in leaf litter. (Laurie Vitt)

***Neusticurus ecpleopus* is a semiaquatic gymnophthalmid that lives in small Amazonian streams.** (Janalee Caldwell)

latus, Iphisa elegans, Prionodactylus argulus, P. oshaughnessyii, Ptychoglossus brevifrontalis, Cercosaura ocellata, and an as yet undetermined species of *Bachia*. These species account for about one-third of the lizard species in this particular assemblage.

Unlike teiids, which often vary by size when occurring together, all gymnophthalmids are small, with many about the same size. Also unlike teiids, in which the maximum number of species occurring together seems to be about four, gymnophthalmids can coexist in larger numbers of species. Reasons for this are complex, but their body size relative to the structural diversity of habitats occupied may provide a partial explanation. Large animals must range over larger areas than small ones and so tend to use a greater range of habitats. Small species, in contrast, perceive smaller environmental elements than do larger species. During a typical foraging bout, for example, a large-bodied lizard such as the teiid *Ameiva ameiva* might traverse leaf litter, bare ground, stream banks, and clusters of debris associated with treefalls—all relatively small areas compared to the extent of land an individual *Ameiva* covers during a normal day. In an absolute sense, then, the structurally diverse forest habitat is a mosaic of accessible patches to an *Ameiva*. Yet what appears as a mere patch within an *Ameiva*'s foraging area would be an entire landscape to a small lizard whose foraging bouts cover just a few square meters on any given day.

Gymnophthalmids divide their space equitably. *Neusticurus ecpleopus,* to cite one representative gymnophthalmid, lives at the forest-stream interface and is nearly always associated with stream banks. The closely related *N. juruazensis,* in contrast, lives one step removed: in low, damp areas associated with forest streams but not directly on banks with *N. ecpleopus.* Where low vines and shrubs occur, the long, streamlined lizard *Prionodactylus argulus* lives above the ground on shrubs and vines. *P. oshaughnessyii* forages at the bases of trees and on and around fallen logs, often seeking patches of sunlight breaking through the forest. In leaf litter, where layers of leaves produce a microhabitat with an enormous surface area available for exploitation, tiny *Iphisa elegans* moves in and among leaves, often seeking those same migrating patches of sunlight. In piles of debris associated with treefalls, *Cercosaura ocellata* negotiates twigs, branches, and vines, occasionally climbing off the ground to forage. In the structurally diverse microhabitat created by packs of leaf litter along streams and in low areas, *Alopoglossus angulatus* moves in and out of crevices, while under leaf litter and rotted logs the nearly limbless *Bachia* moves about like a miniature animated sausage in search of termites. Each of these small gymnophthalmids, along with their many cousins, is associated with specific habitat patches.

Species of *Bachia* are undoubtedly among the strangest gymnophthalmids. Their bodies are sausagelike: front limbs, back limbs, or sometimes all limbs are reduced to paddlelike structures that appear to be used for locomotion associated with foraging. With SVL just under 30 mm, *Bachia* pull and push themselves by their tiny limbs as they move about under leaf litter and along edges of decaying logs. Although the Guiana earless microteiid *Bachia cophias* uses its tiny three-toed appendages when moving slowly, when in a hurry it resorts to serpentine locomotion, undulating body and tail laterally from side to side

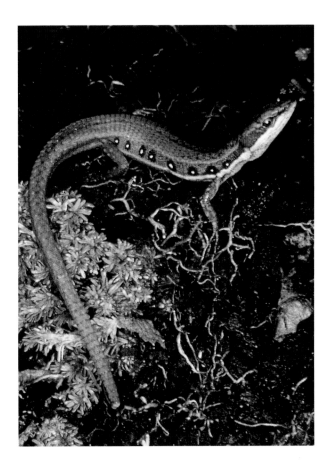

Clockwise from left: **Male *Prionodactylus oshaughnessyi* are brick red on the sides during the breeding season.** (Laurie Vitt)

***Cercosaura ocellata* is often found in leaf litter within rain forest treefalls.** (Laurie Vitt)

***Iphisa elegans,* with its two dorsal and two ventral scale rows, is among the easiest of Amazonian gymnophthalmids to identify.** (Laurie Vitt)

while holding its diminutive legs against its sides, dangling uselessly. In extreme emergencies, this 100–120-mm-long lizard flicks its tail like a spring, lifting itself completely off the ground about 25 cm and a full 30 cm forward (some Australian pygopodids in the genus *Delma* also leap like this) (Schmidt and Inger 1957). *Echinosaura* of northwestern South America and southern Central America is equally bizarre. In litter on forest floors in Panamanian and Colombian rain forests, little rough dark brown *Echinosaura* move slowly, resembling twigs. If touched, these lizards become rigid, looking even more strongly like a fallen twig.

Although most gymnophthalmids are associated with wet tropics, an impressively wide array of species occurs throughout relatively open areas of South America, particularly northeastern and central Brazil. Because so many areas remain to be explored in central Brazil, their diversity could be considerable. A species-rich gymnophthalmid fauna—comparable to or even more diverse than the lizard fauna found at most localities within Amazon rain forest—occurs in open habitats along sand dunes associated with the Rio São Francisco in northeastern Brazil. Miguel Rodrigues, currently director of the Museu de Zoologia da Universidade de São Paulo, discovered an entire reptile fauna in this area composed primarily of fossorial species, nine of which are gymnophthalmids (Rodrigues 1996). Of these nine, seven (representing four genera: *Calyptommatus, Procellosaurinus, Psilophthalmus,* and *Notobachia*) are endemic to the area; all are fossorial. The remaining two species are terrestrial gymnophthalmids, *Vanzosaura rubricauda* and *Colobosaura mentalis,* widespread in open habitats of South America. Of course, not all these lizards live at the same exact spot. One hypothesis is that the dunes date back to the Quaternary and were separated into two isolated sets as the Rio São Francisco made its way to the Atlantic Ocean sometime after the last glaciation. As a result, "sister species" occur on each side of this river. The river probably changed course, cutting off populations from one another (perhaps multiple times), generating high diversity through a process known as allopatric speciation. The wormlike *Calyptommatus sinebrachiatus,* for example, occurs in dunes on the river's north side, whereas *C. leiolepis* and *C. nicteris* are found on dunes on the south side. Interestingly, the latter two are isolated from each other as well, each occurring in its own set of isolated sand dunes.

Among dune gymnophthalmids, *Vanzosaura rubricauda*

is the most widespread, occurring from caatinga of northeast Brazil to deserts of Argentina, where it can be locally abundant. Two distinct color morphs are known for this species: one is a black and white lined version with an orange or red tail; the other is a nearly unmarked gray lizard with a red or orange tail.

The Andes appear to be a center of diversity for some gymnophthalmid genera. In Ecuador, *Pholidobolus* has diversified into a nearly linearly arranged set of five species occurring both in valleys and on Andean slopes to elevations exceeding 3000 m (Montanucci 1973). Distributional ranges of two species, *P. montium* (north) and *P. affinis* (south), appear to be dynamic. After 1973, when Richard Montanucci first described these species' distributions, a considerable amount of human activity caused drastic habitat change at the zone of contact between these two species. As a result, the contact zone shifted at least 30 km south in only twelve years (Hillis and Simmons 1986). At slightly lower elevations in Ecuador and extending northeast through Colombia as far as Panama and Costa Rica, the genus *Ptychoglossus* has also diversified, into 15 known species (Harris 1994). The distribution of this genus includes tropical lowlands of the Brazilian Amazon, a difficult habitat in which to collect, as the paucity of records from tropical forest suggests.

At elevations of 2000–3000 m in the Peruvian Andes, the gymnophthalmid genus *Macropholidus* can be found in a variety of habitats, including montane cloud forest, disturbed forest, brushy hillside, and even some open areas (Cadle and Chuna M. 1995). Like other gymnophthalmids, *Macropholidus* is diurnal and active in leaf litter, seeking refuge under rocks. Like low-elevation gymnophthalmids, *Macropholidus* reproduces over an extended time period (January–June) corresponding to the wet season and early to mid dry season. Nesting is communal, with many eggs being laid in a single location. At one site, Cadle and Chuna M. found remains of 220 eggs in addition to 8 unhatched ones! In this case, nesting sites for small reptiles would appear to be limited because snake eggs were found in lizard nests as well.

Parthenogenesis occurs within a number of gymnophthalmid genera and may ultimately prove to be widespread. The best-known parthenogenetic gymnophthalmid is *Gymnophthalmus underwoodi,* a small, slender, smooth-scaled lizard with a bluish tail that occupies disturbed areas in lowland forest of northeastern South America. In the Brazilian state of Roraima, three species

The fast-moving gymnophthalmid *Arthrosaura reticulata* lives near streams at the interface of downed logs and the forest floor.
(Laurie Vitt)

of *Gymnophthalmus* are known. One is *G. underwoodi.* The second is *G. leucomystax,* a similar lizard but with a brown tail. Endemic to Roraima, it is associated with termite nests in the Lavrado area, a savanna grassland that is part of a system of Amazon savannas extending from the north Amazon across the central Amazon, with patches elsewhere. It is a sexually reproducing species with typical males and females. A third *Gymnophthalmus,* a small, red-tailed lizard, is found along gallery forests associated with major waterways such as the Rio Branco in Roraima. This undescribed species often occurs at high density in leaf litter underlying gallery forest. When approached, it dives into the leaf litter, leaving its tail exposed and moving about, presumably to distract a predator's attention from its body. Like *G. underwoodi,* it is parthenogenetic. Chromosomal data reveal that karyotypes of the three species are distinct, even though the parthenogenetic ones were once both classified as *G. underwoodi* (Yonenaga-Yassuda et al. 1995). Moreover, *G. underwoodi* from Roraima has a different karyotype than *G. underwoodi* from other localities, indicating that all three parthenogenetic species (the two *G. underwoodi* and the undescribed red-tailed species) arose independently from other *G. underwoodi.*

Sexual dimorphism in size, relative head size, and coloration exists in some gymnophthalmids but is absent or reduced in subterranean species. Because this large family contains so many completely different genera, no single description suffices. In *Neusticurus, Cercosaura,* and some species of *Prionodactylus,* males have slightly larger heads than females and sport conspicuous, circular, target-shaped markings on the sides of their bodies. Male *Alopoglossus* have black lateral stripes and black markings on the face. In the widespread open-habitat species *Vanzosaura rubricauda,* males are smaller than females but have relatively larger heads. In addition, while reproductively active, males have red or orange on their jaws, throat, and front legs.

Unlike members of their sister taxon the teiids, which generally have high body temperatures while active, many gymnophthalmids are active at lower body temperatures. For example, *Arthrosaura reticulata,* common along streams, averages 26.6°C; *Neusticurus ecpleopus,* which lives on stream banks and frequently enters water, averages 23.8°C; semiarboreal *Prionodactylus oshaughnessyii,* which is often found on tree trunks or logs, averages 26.2°C; and the tiny *Leposoma percarinatum,* also associated with damp areas, averages 28°C. All these very active lizards live in Amazon rain forest and can often be found in patches of filtered sunlight. Body temperatures of gymnophthalmids in open habitats remain unknown, but considering the secretive habits of most, they are probably low. Some terrestrially active species in open areas, like *Vanzosaura rubricauda,* may have high body temperatures.

LACERTIDAE

Lacertids are found across Eurasia and Africa and into Southeast Asia but do not reach Australia. About 27 genera and 220 species are recognized. Appropriately named (*lacerta* is Latin for "lizard"), lacertids are small to medium-sized diurnal lizards with movable eyelids, long cylindrical bodies, and well-developed legs. Most species have small smooth body scales, but a few have enlarged or spiny scales. Head scales are large. Lacertid teeth are hollow at the base (teiid teeth are solid). Virtually all lacertids are terrestrial or rock-dwelling lizards, although a few species, including *Holaspis* and *Takydromus,* climb in

vegetation, and at least one appears to live high in trees. Most are insectivorous, and all but one species lay eggs. Some populations of European *Lacerta vivipara* bear living young, whereas others lay eggs. Sometimes quite colorful (as, for example, the large—250 mm SVL—*Lacerta lepida* from Spain), lacertids cannot change color rapidly, though some do change color over the course of their lives. Most lacertids are sexual, but five unisexual (all-female) species exist, all of which occur in the genus *Lacerta* and are found in southwestern Asia (Darevsky 1958).

Some species are highly adapted to loose sands. In northern Africa, sandy desert regions support lacertids (*Acanthodactylus*) with fringed toes, shovel noses, and countersunk lower jaws. Far away in the Southern Hemisphere, on windblown sand dunes of the Namib Desert of southwestern Africa, an independent clade of lacertids has evolved similar life forms, containing the species *Meroles ctenodactylus, M. cuneirostris,* and *M. anchietae.* The impact that wind-blown sand habitats can have on evolution of lizard morphology is striking. The North American iguanid genus *Uma,* the Saharan skink genus *Scincus,* and the Namibian desert gerrhosaurid *Angolosaurus*

***Lacerta schreiberi* is often referred to as the Iberian emerald lizard.** (Chris Mattison)

skoogi have all converged on body plans very similar to those of sand-dwelling lacertids, including countersunk lower jaws, shovel noses, and fringed toes.

A beautiful Kalahari lacertid, *Nucras tessellata,* consumes considerably more scorpions than most other lizard species. *Nucras* are active during the heat of midday at body temperatures of 39.3°C (probably a predator avoidance tactic), foraging widely to find and dig up large scorpions in their diurnal retreats (by day, scorpions are exceedingly patchily distributed, nonmobile prey items) (Pianka et al. 1979). Scorpions are extremely large and nutritious, a benefit that presumably facilitated evolution of dietary specialization in *Nucras*.

Holaspis guentheri lives in forests in central Africa. Scales on each side of its unusual tail project out, down, and backward. These lizards press their tails against trees while climbing, using projecting scales to catch the bark. They also have rows of enlarged scales under their feet and other projecting scales along the base of their toes that help them cling to trees with smooth bark. *Holaspis guentheri* are boldly striped with yellow and black on their backs and display orange ventral surfaces (Schmidt and Inger 1957). Their tails are also brightly banded, and these lizards coil their tails, particularly while inactive, perhaps mimicking centipedes or millipedes.

Gastropholis echinata is bright green and lives in trees in African rain forests. These moderately large lizards (about 100 mm SVL) have a broad ring of spiny scales around the base of the tail that facilitates climbing (Schmidt and Inger 1957). In some respects, *Holaspis guentheri* and *G. echinata* appear convergent with *Kentropyx* (Teiidae) in that they are climbing members of a generally terrestrial group of lizards.

Still another genus of climbing "grass-swimming" lacertids is *Takydromus* (= speed runner), a group of slender, long-tailed, green lacertids from Southeast Asia (some are found in Taiwan and Japan as well). These lizards look like grass blades and are at home climbing in the top of tall grasses. A lizard with a snout-vent length of 60 mm may have a tail 300–350 mm long, which allows it to distribute its weight over many grass stems. *Takydromus* jump from stem to stem, often turning sharply, making it difficult to follow their movements through the grass.

The unusual *Ophisops* (= snake eyed) is found in northern Africa and southwestern Asia. While some lacertids have a large transparent disk in the center of the lower eyelid through which they can see even when their eyes

This Kalahari lacertid, *Nucras tessellata,* is a scorpion specialist.
(Eric Pianka)

Lizarding in the Kalahari with Ray Huey and Larry Coons was usually a lot of fun, but at times it got pretty scary. Our biggest concern was lions and leopards, but cobras and puff adders were also hazards. Horned sand vipers, for example, bury themselves in the sand waiting for an unsuspecting lizard to move past; many widely foraging lacertids are ambushed by these snakes. We disturbed many a buried viper while probing around grass tussocks looking for legless burrowing skinks. When one of these little red snakes comes leaping out of the sand, it lifts its entire body completely off the ground as it hisses and lunges. We aren't afraid of snakes, but we were exceedingly surprised every time one of these vipers came thrashing at us with such menace. Puff adders, large vipers the size of big rattlesnakes, though much more dangerous, were less common than horned sand vipers. A large one can have fangs 3 cm long, and they are highly venomous. These big snakes are camouflaged with blotches that break up their outline. Usually you don't see a puff adder until it hisses loudly when you're right on top of it: needless to say, this can be quite startling! *(PIANKA)* are closed, in *Ophisops* these disks are so large that the eyelid is no longer movable.

The Kalahari lacertids *Meroles suborbitalis* and *Pedioplanis lineo-ocellata* sit and wait for prey, whereas two other species found in the same microhabitats, *Heliobolus lugubris* and *Pedioplanis namaquensis,* forage widely for their food. Judging from their relative stomach volumes, those species that engage in this active mode of food gathering capture more prey per unit time than do sit-and-wait species—as would be expected, given that active foraging is energetically expensive, requiring a greater overall energy budget (Huey and Pianka 1981; Anderson and Karasov 1981). Compared with sit-and-wait species, widely foraging lacertid species eat more termites, which are sedentary, spatially and temporally unpredictable, but clumped prey.

Another ramification of foraging mode in these Kalahari lizards concerns exposure to predators. Because of their more or less continual movements, widely foraging species tend to be more visible and, as a result, suffer higher predation rates by some kinds of predators. Active foragers fall prey to predators that hunt by ambush such as the horned sand viper, *Bitis caudalis,* whereas sit-and-wait species tend to be eaten by predators that forage widely, such as secretary birds. "Crossovers" in foraging mode thus occur between trophic levels.

Among the most spectacular lacertids is the giant (by lacertid standards) *Gallotia simonyi* (260 mm SVL), an endangered species living on the island of El Hierro in the Canary Islands. A smaller species, *Gallotia caesaris,* also occurs on El Hierro, and several other *Gallotia* occur on surrounding islands. These lizards are apparently omnivores, eating both animal and plant material (primarily fruits).

Some *Lacerta* and *Podarcis,* especially those on islands in the Mediterranean, eat large numbers of ants. Whether the ants eaten by these scleroglossans contain defensive chemicals remains unknown.

The most striking evolutionary divergence within lacertids is dietary specialization on ants in various species of *Acanthodactylus* (Arnold 1984). A detailed analysis of the diet of one species, *A. erythrurus,* combined with sampling of insects in the lizard's habitat over a six-month period, for example, indicates that although beetles and Heteroptera (true bugs) dominate the diet during four of six months, and beetles are more frequently eaten in five, during the sixth month ants are more frequently eaten; indeed, no heteropterans were captured in the lizard's habitat during that month, whereas 68 percent of insects trapped were ants.

XANTUSIIDAE

Xantusiids are commonly referred to as night lizards, though that name may be something of a misnomer (see below). Affinities of these distinctive lizards with other lizard groups remain uncertain. They could be the sister group to Lacertiformes, or they might even belong with Gekkota, as suggested by the fact that they, too, lick the spectacles over their eyes (Greer 1985a). Scales on their backs are small, granular, and, with the exception of *Lepidophyma,* nearly uniform in size, while those on their bellies are large and rectangular shaped. Eyelids are absent; instead these lizards possess a fixed spectacle like those seen in many geckos, pygopodids, some skinks, and snakes. The lower eyelid has become permanently fused in the "up" position, fitted with a clear scale through which the lizard can see. Pupils are elliptical, suggesting activity when light levels are low. Their catlike pupils no doubt led to the notion that they are strictly nocturnal, but in fact some night lizards are as much diurnal as nocturnal; their daytime activity is simply restricted to dark, protected places where they are difficult to observe. All xantusiids are viviparous and have evolved a mammal-like

The tropical xantusiid *Lepidophyma mayae* is found in Central America and southern Mexico. (J.A. Campbell)

Long-lived yucca night lizards, *Xantusia vigilis*, live in rubble under Joshua trees. (Jim Rorabaugh)

arrangement whereby the mother nourishes her litter "in utero." Placentation in xantusiids is not as extensive as in certain skinks, but at least a portion of nutrients required for development pass from mother to offspring via a placenta.

Found only in North America, Central America, and Cuba, these generally small lizards (less than 50 mm SVL) are sedentary, patchily distributed, and long lived. Species occurring on islands off the coast of southern California are an exception in terms of size, being by xantusiid standards quite large: 100 mm SVL. The three genera—*Xantusia, Lepidophyma,* and *Cricosaura*, comprising some 21 known species and subspecies (Bezy 1989; Papenfuss et al. 2001)—are similar in appearance, differing primarily in minor details of scalation and color pattern. Most species in the genus *Xantusia* live under exfoliating granite, other rocks, or in Joshua tree rubble in deserts of

SECRETS IN THE ROCKS

As a graduate student, I was in awe of several herpetologists. One was Donald W. Tinkle of the University of Michigan, who came to my school, Arizona State University, as a Maytag visiting professor. (For me, interested as I was in lizard life histories, this was something like having John Lennon drop in on a rehearsal of my rock band.)

At the time of his arrival, a fellow graduate student, Justin Congdon, and I had been collecting around the state; one favorite site was the boulder piles near the one store (and only building) in the "town" of Sunflower, about 50 km north of Phoenix. We collected good numbers of *Xantusia* there, always under exfoliating granite or in deep crevices. We were struck by the fact that these *Xantusia* were larger than the *X. vigilis* we could collect a mere 18 km away under plant debris on the desert floor. In addition, they were flattened and had large spots on their bodies, similar to *X. henshawi*.

When we showed these lizards to Tinkle, he was convinced they represented a new taxon and proceeded to contact one of his colleagues, Robert Webb, with the intention of describing it as a new species. After examining the lizards closely and assembling data on scale counts used in xantusiid taxonomy, however, they could find no characters that separated it from *Xantusia vigilis* except that it was larger, flattened, and colored differently.

As it turned out, Robert Bezy had explored the problem a few years earlier while a graduate student at the University of Arizona in Tucson (Bezy 1967). A flattened *Xantusia* from rock crevices in Arizona had been described decades earlier by Lawrence Klauber and named *X. arizonae* (Klauber 1931). When Bezy examined all collections and added new material, he concluded that *X. arizonae* was simply an ecomorph of *X. vigilis:* at a number of places, when *X. vigilis* colonized rock crevices, its morphology had changed. Bezy concluded that these flatter *Xantusia* were a separate sub-

species, recognizing that origins of crevice dwelling must have occurred more than once within *X. vigilis.*

Bezy had not examined lizards from the Sunflower locality, and Don Tinkle thought our lizards differed enough to warrant attention. Observing his thought process as he progressed from absolute conviction that our *Xantusia* was an undescribed species to finally determining that it was exactly what it should be, and already described, impacted my approach to saurology in an important way. We are often convinced that we understand something based on minimal observations. When hard data are brought to bear and answers differ from what we first thought, the data must prevail or we aren't doing science. Ironically, in this case, Tinkle had been right from the start. The Sunflower lizards have just recently been described as a new species based on molecular data (Papenfuss et al. 2001). *(VITT)* the southwestern United States and Mexico (Bezy 1988). Species of *Xantusia* that live in crevices are dorsoventrally flattened. Species in the genus *Lepidophyma* live primarily under and inside rotted logs in tropical habitats of Central America. Most are rounded in cross section, although the species that inhabit crevices, such as *L. gaigeae,* are dorsoventrally flattened like crevice-dwelling *Xantusia*. Some populations of the Central American species *Lepidophyma flavimaculatum* are unisexual. Both the use of crevices and the flattened morphology have arisen several times independently within xantusiids (Bezy 1989). This has occurred even within the single species of *X. vigilis:* populations living under Joshua tree logs and debris are cylindrical in morphology, whereas those living in crevices under exfoliating granite are flat. *Cricosaura* has smaller limbs than other xantusiids and typically moves in a serpentine manner. Xantusiids have relatively low body temperatures (about 29°C) and rapidly die when exposed to temperatures above 35°C for any length of time.

Xantusia vigilis, a small lizard (only about 40 mm SVL as an adult), lives in the Mojave Desert in rubble under Joshua trees and introduced saltcedars, where it can avoid the temperature extremes of the surrounding desert (Zweifel and Lowe 1966). Growth rates are surprisingly slow for such a small lizard, and sexual maturity is reached only after three years. During September, females produce from one to three live young. Survivorship is unusually high, with 80 percent of a cohort born in any given year living to reach sexual maturity. Some individuals live as long as nine years. *Xantusia henshawi,* which frequents granitic rock outcrops in southern California deserts, is larger, with adult males averaging 56 mm SVL and females averaging 62 mm SVL (Lee 1975). By living within crevices in exfoliating granite it, too, avoids high daytime temperatures. As in *X. vigilis,* live young are born in fall,

and neonate number varies from one to two. Survivorship is also quite high. Diets of *Xantusia* species include beetles, ants, and spiders (Brattstrom 1952; Pianka 1986). Substantial numbers of seeds and other plant parts are eaten by the large-bodied *X. riversiana,* and all *Xantusia* eat their own shed skins, similar to geckos.

In a combination field and laboratory study, Julian Lee (1974) demonstrated that xantusiids move about actively during the day, a time period when these lizards are difficult to observe in nature (see also Mautz and Case 1974). These lizards are also active on exposed rocks shortly after sunset and may remain out until midnight, demonstrating nocturnal activity. Because lizards out at night tend to be adults, reproductive-related social behavior is likely, but it generally involves little movement; metabolic rates are slow as well, consistent with low activity levels (Mautz 1979). These lizards can therefore be considered both diurnal, in that substantial activity occurs within crevices during the day, and nocturnal, in that lizards position themselves outside crevices at night, likely in a social context.

Other xantusiids, it seems, are similar ecologically to *Xantusia vigilis* and *X. henshawi.* The giant *X. riversiana* and Central American *Lepidophyma* are secretive, living in holes, crevices, or under decaying logs. *Cricosaura typica* lives under rocks, logs, and other surface features in dry forest. *Cricosaura* differs from other xantusiids in that its limbs are reduced, body elongate, and its locomotion snakelike (Schwartz and Henderson 1991). Nevertheless, it could not be mistaken for anything other than a xantusiid: it looks like a stretched-out *Xantusia vigilis!* Even though the distribution is limited to a few areas in southern Cuba, these lizards are apparently fairly common. They feed on a variety of insects, including house crickets, insect larvae, and moths.

SKINKS

The casque-headed skink, _Tribolonotus gracilis._ (Chris Austin)

Skinks are the largest lizard family, with some 120 genera and about 1,400 species described so far (and many more remaining to be described). Probably monophyletic, they exemplify diversity in all aspects of their biology. Terrestrial, arboreal, fossorial, and semiaquatic species are known, and they have diversified in all types of environments, from Australian and African deserts to Amazonian lowlands, from temperate forests and cool montane habitats to African savannas and Brazilian cerrado. Size varies from tiny to large, and morphology varies from short and robust with strong, well-developed limbs to elongate and fragile with tiny limbs or none at all. In some arid lands, notably Australia, they dominate the lizard fauna; in others, such as the Sonoran and Great Basin Deserts, they are essentially absent. Like certain geckos, some have rafted across oceans, colonizing other continents and remote islands.

Skinks are characterized by large, symmetric, shield-like scales on their heads (though of course many other lizards have head shield scales also). Typically, skinks have smooth, glossy, cycloid scales, although a few do have sharp keeled scales. Another extremely distinctive feature of skinks is the bony secondary palate in the roof of the mouth, separating the respiratory passage from the digestive one. (In most other lizards, these two passages are confluent.) Whereas almost all other lizards pant when thermally stressed (thus cooling the roof of the mouth, into which large blood sinuses above can dissipate heat), skinks do not pant, perhaps because their secondary palate reduces heat exchange (Greer 1989).

Four major evolutionary groups (subfamilies) of skinks are formally recognized (Greer 1970): Acontinae, Feylininae, Lygosominae, and Scincinae. Of these, Scincinae is probably paraphyletic (that is, it does not contain all descendants of a common ancestor). Scincines are primitive skinks with smooth cylindrical bodies and small legs. Acontines and feylinines appear to be derived from as yet unknown scincine ancestors (Greer 1985b). Acontines *(Acontias, Acontophiops, Typhlosaurus),* found in Africa, are specialized legless burrowing skinks found only in leaf litter, loose sandy soil, or underneath logs. Feylinines, comprising a single genus *(Feylinia)* with six species, are found in central tropical Africa.

Eumeces (probably a paraphyletic genus) in the Scincinae displays several important ancestral character states (referred to as pleisiomorphic by systematists) that have sometimes been thought to closely resemble ancestral traits for all scincids (Greer 1974, 1979; Hutchinson 1993). However, recent work by Tod Reeder of San Diego State University suggests that New World *Eumeces* are embedded well up in the skink phylogeny; moreover, Reeder considers certain African and Asian scincines *(Brachymeles, Chalcides)* to be some of the most basal scincid taxa (T. Reeder, pers. comm.).

Within the more advanced lygosomine skinks, three

Occurring in western Africa, *Riopa fernandi* lives in rain forest habitats. (R.D. Bartlett)

Emoia longicaudata, of the Torres Strait and Cape York Peninsula of Australia, is arboreal. (Christopher Austin)

distinct and presumably monophyletic lineages occur: the *Egernia* group, the *Sphenomorphus* group, and the *Eugongylus* group, all of which occur in Australia (a major center for skink diversity). Some members of each group also occur outside Australia, especially in New Guinea and New Caledonia. The *Egernia* group contains six genera: *Corucia, Cyclodomorphus, Egernia, Hemisphaeriodon, Tribolonotus,* and *Tiliqua* (including the formerly recognized *Trachydosaurus*). The large *Sphenomorphus* group is probably monophyletic, with 29 genera: *Ablepharus, Anomalopus, Ateuchosaurus, Calyptotis, Coeranoscincus, Coggeria, Ctenotus, Eremiascincus, Eulamprus, Fojia, Glaphyromorphus, Gnypetoscincus, Hemiergis, Isopachys, Laurutia, Leptoseps, Lerista, Lipinia, Lobulia, Ophioscincus, Notoscincus, Papuascincus, Paralipinia, Parvoscincus, Prasinohaema, Saiphos, Scincellla,* and *Sphenomorphus* (this genus is probably paraphyletic), and *Tropidophorus.*

The *Sphenomorphus* group of skinks is cosmopolitan, reaching its highest diversity in Australia (23 genera and 230-plus species); 8 of these 23 genera occur outside Australia, primarily in Southeast Asia (Greer 1974, 1979, 1989; Greer and Parker 1974). Five species of *Scincella* occur in the New World, while other species of *Scincella* are found far away in Taiwan and Korea. Three species of *Sphenomorphus* occur in Mexico and Central America (Greer 1974). Monophyly of the Australian *Sphenomorphus* group is strongly supported by immunological and DNA sequence data, although the genus *Sphenomorphus* is probably paraphyletic (T. Reeder, pers. comm.).

The *Eugongylus* group is even larger, with about 41 named genera: *Acritoscincus, Afroablepharus, Bartleia,*

An arboreal skink, *Prasinohaema prehensicauda* of New Guinea has green blood. (Christopher Austin)

Sometimes referred to as the "bush crocodile," *Tribolonotus novaeguineae* is found in forests of New Guinea near water.

(Louis Porras)

Carlia jarnoldi **forages in leaf litter of dry sclerophyll forests of northeastern Queensland.** (Dr. Hal Cogger)

Bassiana, Caledoniscincus, Carlia, Cautula, Cophoscincopus, Cryptoblepharus, Cyclodina, Emoia, Eroticoscincus, Eugongylus, Geomyersia, Geoscincus, Graciliscincus, Lacertaspis, Lacertoides, Lampropholis, Lankascincus, Leiolopisma, Leptosiaphos, Lioscincus, Lygisaurus, Marmorosphax, Menetia, Morethia, Nangura, Nannoscincus, Niveoscincus, Oligosoma, Panaspis, Phoboscincus, Proablepharus, Pseudemoia, Ristella, Saproscincus, Sigaloseps, Simiscincus, Tachygyia, and *Tropidoscincus* (Storr et al. 1999; Bauer and Sadlier 2000). Various attempts to break this very large group into monophyletic subgroups have been made but are still disputed (Greer 1979, 1989; Hutchinson et al. 1990; Bauer and Sadlier 2000). Exactly where *Mabuya* (probably a paraphyletic genus) and several other Old World lygosomines should be placed within Lygosominae remains uncertain. Hutchinson and Donnellan (1993) recognize a *Mabuya* group, which includes *Apterygodon, Dasia, Eumecia, Mabuya,* and other Asian lygosomines. Tod Reeder (pers. comm.) recognizes a *Lygosoma* group consisting of *Haackgreerius, Lamprolepis, Lygosoma, Vietnascincus,* and other Asian lygosomines. New skink genera are still being described.

Skinks are cosmopolitan, occurring on all continents except Antarctica as well as on many oceanic islands. (They appear to have rafted almost as often as geckos.) Skinks defy stereotyping, having adopted a wide variety of habits: Most are diurnal, but a few, such as some species of *Egernia* and *Eremiascincus,* are nocturnal. Many are terrestrial, though some are fossorial, still others are arboreal, and a few—*Amphiglossus astrolabi, Eulamprus quoyii,* and *Tropidophorus grayi*—are even semiaquatic. Skinks are particularly diverse in tropical areas of the Old World,

and include some unusual species. *Riopa haroldyoungi* from Thailand, for example, is snakelike, with tiny front and hind limbs, brownish yellow on the body with a series of black reticulated markings on the back, and a long, thick tail with a striking mixture of black and blue bands. Skinks have colonized the New World only a handful of times; as a result, only five genera are represented—*Eumeces, Mabuya, Neoseps, Scincella,* and *Sphenomorphus*—though of these, two have been particularly successful: the scincine *Eumeces* in North America and the lygosomine *Mabuya* in South America. *Neoseps* is a derivative of *Eumeces egregius,* while *Scincella* and *Sphenomorphus* are lygosomines (T. Reeder, pers. comm.), indicating that several lineages reached North America.

Many North American skinks are secretive, spending much of their time under fallen logs or rocks, and thus they are not very conspicuous. Central and South American skinks *(Mabuya* and *Sphenomorphus),* in contrast, are active during the day and frequently observed basking or foraging. They are even more conspicuous and diverse in some parts of Africa and Australia, being active during the heat of midday, often in very arid areas.

Skinks have repeatedly evolved reduced appendages, including complete limblessness in *Acontias, Anomalopus, Barkudia, Brachymeles, Coeranoscincus, Lerista, Ophiomorus, Ophioscincus, Scelotes, Sepsophis, Typhlacontias,* and *Typhlo-*

Eumeces schneideri, **which occurs in arid lands of northwestern Africa and western Asia, is popular in the pet trade.** (R.D. Bartlett)

Many skinks, like this *Brachymeles bicolor* of the Philippines, have reduced limbs associated with subterranean habits. (Rafe M. Brown)

saurus, found in Asia, Africa, and Australia. A single North American genus, *Neoseps,* approaches limblessness, with both front and hind limbs reduced to tiny appendages, as do many in Southeast Asia (species of *Larutia* and *Riopa haroldyoungi,* for example). Legless skinks, as well as those with reduced limbs, are typically burrowers and usually have small eyes and consolidated head shields.

All degrees of limb reduction can occur even within a single genus. *Lerista,* an Australian scincid (about 75 described species), constitutes the best example. Several species of *Lerista* retain the ancestral state, with all five front toes and all five rear ones (digital formula 5/5). These species are active surface dwellers. Other species, such as *Lerista ameles* and *Lerista apoda* (meaning "without feet"), have not only lost all their digits but also become completely legless. These are burrowing subterranean lizards. Digital formulae span the entire gamut of possibilities: 4/5, 4/4, 3/4, 3/3, 2/3, 2/2, 1/3, 1/2, 0/2 (*Lerista bipes*), 0/1, and 0/0. Digit loss follows an orderly pattern, with those in the midline of the foot (digits 3 and 4) being the last to be lost (Greer 1989). Among *Lerista,* toes retain the relative proportions of a normal five-toed skink even when reduced greatly in absolute size (in other words, digits 3 and 4 re-

Australian *Lerista gascoynensis,* which occur in the Gascoyne River basin, have no front limbs and reduced hind limbs, with only two digits. (Brad Maryan)

main much longer than digits 1, 2, or 5), suggesting that phalangeal reduction is "held off" until the digit is on the verge of being lost (Hutchinson 1993).

Viviparity has also arisen multiple times among skinks, although many species retain the ancestral condition and lay eggs. Separate populations of two Australian species,

Lerista bougainvillii and *Saiphos equalis* (Greer 1989), actually exhibit both egg laying and live bearing. Brood sizes vary greatly among skinks, from one to two in some species (such as *Lobulia* and *Prasinohaema,* which appear to have a fixed clutch size as in anoles and geckos) to as many as sixty-seven in Australian *Tiliqua gerrardii* (Greer 1989). Some viviparous skinks give birth to a single extremely large neonate *(Corucia zebrata, Tiliqua rugosa,* and *Typhlosaurus gariepensis).* Within viviparous skinks, the entire range of fetal nutritional modes occurs. Many species ovulate large eggs containing all nutrients necessary for development, such that developing offspring feed on their own yolk (lecithotrophy) before being born alive. Various degrees of placental development occur in which nutrients are passed from mother to offspring (matrotrophy) during development. For example, *Chalcides* have complex placental connections between mother and progeny through which pass substantial amounts of nutrients required for development (Weekes 1935). Still others, like *Mabuya heathi* in Brazil, have such an advanced placental arrangement that more than 99 percent of nutrients necessary for neonatal development are passed from the mother through a chorio-allantoic placenta (Blackburn et al. 1984), with gestation taking more than nine months.

The placenta of *M. heathi* has morphological attributes unknown in any other tetrapod (Blackburn and Vitt 2002). This placental type with extended gestation appears to occur in all New World *Mabuya* but in no Old World *Mabuya* (Blackburn and Vitt 1992), suggesting a single colonization event by an unknown ancestor. Colonization was probably trans-Atlantic, as there are no north temperate *Mabuya* (Greer 1970).

In most lizards, including some skinks, inguinal fat bodies protrude into the abdomen from the pelvic area, storing valuable energy reserves used in reproduction. Curiously, however, most members of the large *Sphenomorphus* group of skinks have lost these fat bodies (Greer 1989) and rely instead on their tails to store most fat reserves. Tail loss can thus be costly.

Nevertheless, almost all skinks exploit tail autotomy as a means of escape from predators. In some skinks, tails of juveniles are markedly brighter than those of adults, in reds, blues, and yellows, likely to attract attention away from the body (Cooper and Vitt 1985). Australian *Ctenotus* and North American *Scincella* (referred to as *Lygosoma* by Clark 1971) sometimes return to the site where their tail was lost and swallow the remains. Despite the prevalence of tail autotomy, a few species, such as some *Egernia,*

Often associated with rocky outcrops in arid central Australia, *Egernia depressa* has highly keeled scales on its body and tail. (Brad Maryan)

While a graduate student I was primarily interested in life history evolution and as a result had collected considerable data on reproduction in a variety of lizard species. I had also studied nearly every relevant bit of information available in the literature: I knew that clutch size varied among lizard species, and that in species with variable clutch sizes, number of eggs increased with female body size. I also knew that some lizards were oviparous whereas others were viviparous.

At the time, very few lizards were known to have complex placentae. However, one skink genus, *Chalcides*, was known to augment nutrients contained in yolk of developing embryos with nutrients passed directly to embryos from the mother's circulatory system. Likewise, night lizards, *Xantusia*, were thought to have some sort of placenta. I also believed, based on existing literature, that Temperate Zone lizards were seasonal in reproduction and that tropical lizards were aseasonal in humid environments but seasonal in tropical areas with distinct wet-dry seasonality. My perceptions of lizard reproduction, however, were about to change.

Following graduate school, I spent a year in Exu, Pernambuco, in the semi-arid northeast of Brazil. Every month I sampled a set of thirteen lizard species and worked out their reproductive biology in detail. What I found caused me to seriously question much existing dogma in the reproductive biology literature, but two things struck me particularly hard: (1) the local environment alone *did not* determine how and when lizard species reproduce (still a common misconception); and (2) ovulation of eggs with most developmental energy packaged in yolk was not a reliable characteristic separating squamates (and reptiles in general) from mammals.

When I put together all the reproductive data on my thirteen lizard species, which included both long-lived and short-lived species, I discovered that some reproduced year round, some in the wet season, and some in the dry season. I also found that egg size and number varied greatly among species. Even though most species were oviparous, one was viviparous. No environmental effect could explain the variation found at this single site. I submitted a manuscript pointing out that what we believed about tropical lizard reproductive seasonality was falsified by the simultaneous occurrence of all of these reproductive strategies in one place (i.e., time and place were controlled variables). The paper was rejected! I have since made the case through a wide variety of other published papers, however, as have others.

While collecting female reproductive data, I became interested in what appeared to be encysted parasites in oviducts of one lizard, *Mabuya heathi*. I had never seen anything other than eggs or embryos in lizard oviducts, so it was of enough interest that I dissected out the oviducts and saved them in

continued on next page ▶

Corucia, and *Tiliqua,* do not shed their tails. Some *Egernia* species, for example, have tough flattened spiny tails that they use to block off entrances to their crevice retreats (a convergent tactic with the Asian-African agamid *Uromastyx,* some African cordylids, the Madagascar oplurine *Oplurus,* and the Mexican iguanine *Ctenosaura clarki*).

Like many lizards, all skinks possess bony plates within their scales, or osteoderms, conferring protection from predators. In other lizards these are typically a single bone, but in skinks they are made up of compound plates of several interconnected bones underneath each scale. Several species of snakes (Savitsky 1981) and one pygopodid lizard, *Lialis* (Patchell and Shine 1986), prey on skinks; these predators have hinged teeth to facilitate obtaining a firm grip on skinks (teeth fold when they hit an osteoderm). If a skink struggles backward during ingestion, the teeth lock into place; the fact that skinks are swallowed rapidly suggests that they may actually facilitate their own demise, crawling away from the ratchetlike teeth down a predator's gullet (Greene 1997). Perhaps they feel as if they've escaped into a tight dark burrow—until, that is, they suffocate or digestive enzymes begin to act. Some skinks (such as the Australian *Ctenotus*), like many geckos, avoid such a fate by having very loose skin and scales that tear away when they are attacked, facilitating escape.

Skinks display a wide variety of eye types (Greer 1989). As in other lizards, only the lower eyelid moves, being lowered to open the eye and raised to close it. The ancestral condition is a freely movable scaly opaque eyelid, as found in tuatara *Sphenodon* and most lizards. Several more derived states exist among skinks. Some species have a freely movable eyelid with a clear disklike central scale, or window, through which the lizard can see even when its eye is closed, similar to gymnophthalmids. Other species have an eyelid fused immovably in the raised position with an expanded clear area through which they can formalin. Over several months I became aware that every female contained these growths. During the fourth month of sampling, as they enlarged, I realized what I was seeing: nearly microscopic implanted embryos! I went back to all the oviducts I had saved and reexamined them. The "encysted parasites" were evenly spaced and occurred in both oviducts. Of course, the first question that popped into my mind was "How could I possibly have missed this?" The reason was simple: this was not known to occur in lizards, so it simply didn't register in my "lizard things" set of memory banks.

After returning to the United States, I went to Ann Arbor, Michigan, as a postdoctoral fellow to work with Don Tinkle. I explained to him what I had found: essentially, a lizard that ovulated eggs the size of those ovulated by mammals. Still impressed with my discovery, I patiently awaited his response. After a few seconds of silence he said simply, "No you didn't," and went on to explain that no such thing occurred in reptiles.

I was so taken aback that I came close to doubting my own observations. Then I got the oviducts so that he could see for himself. When he examined them, he was blown away. For both of us, the experience was the same: we were so confident of our understanding of reptilian reproduction that we could not even accept the possibility of this degree of placentation in a lizard.

I ended up with an entire year's sample of embryos and eventually, with the help of Daniel Blackburn, then a graduate student at Cornell University and interested in reptile viviparity, worked out Mabuya heathi's reproductive cycle. Together with another Cornell graduate student, Carol Beuchat, we worked out details of placentation in this lizard.

When Dan Blackburn, Carol Beuchat, and I completed the placentation description, we fired off a short paper to Science, a prestigious scientific journal. Both reviews of the manuscript were excellent, with one stating emphatically that the paper should be published in Science. The editor, however, rejected

it. I called her and asked why the paper had been rejected. She first told me that the reviews were negative, which I refuted simply by reading them to her. She then said that Science gave highest priority to areas of immediate popularity, and so most of what they were then publishing in biology was molecular. In other words, our paper wasn't trendy enough!

I had frequently discussed the study with Everett Olson, a National Academy member and vertebrate paleontologist who considered it one of the major discoveries in vertebrate biology. When he learned that Science had rejected our paper, Everett offered to have it reviewed for an even more prestigious journal, Proceedings of the National Academy of Sciences, which is where we published it. The Science reviews were the best I have ever received for a manuscript, and it is the only time in my career that a journal turned down a manuscript with positive reviews. (VITT)

see. Still others exhibit a clear eyelid fused all around, forming a spectacle similar to those in geckos, pygopodids, xantusiids, and snakes. Larger skinks tend to display the ancestral condition, with movable opaque eyelids, but most smaller species have more derived eyelid conditions. Permanently capped eyes in small skinks reduce evaporative water loss and protect the eye. Most skinks, being diurnal, have round pupils, but one Australian nocturnal species, *Egernia striata,* has evolved an elliptical pupil.

Some Australian skinks, particularly *Egernia* and *Tiliqua,* are semisocial. Bobtail skinks, also called sleepy lizards *(Tiliqua rugosa),* form long-term monogamous pair bonds (Bull 1994). *Egernia* are often found in small groups, presumably family groups. This occurs in *Egernia* species that are arboreal *(E. depressa, E. stokesi)* and saxicolous *(E. cunninghami),* as well as in terrestrial species that dig extensive burrow systems *(E. kintorei, E. striata). Egernia*

The widespread nocturnal Australian skink *Egernia striata* lives in spinifex sandplains. It is unusual among skinks in having an elliptical pupil. (Dr. Hal Cogger)

Lamprolepis smaragdina **is one of a handful of brilliant green skinks.** (Dr. Hal Cogger)

striata, a large nocturnal skink, digs elaborate tunnel systems that are used as retreats by many other species of reptiles, both diurnal and nocturnal (Pianka 1986). These complex burrows, vaguely reminiscent of a tiny rabbit warren, with several interconnected openings often as far as a meter apart and up to half a meter deep, are an important feature of Australian sandy deserts. Most sand removed from an *E. striata* burrow is piled up in a large mound outside one "main" entrance.

Another smaller, more crepuscular species, *Egernia inornata,* digs a simpler shallow burrow, consisting of a U-shaped tube with but one arm of the U open (this, the sole entrance and only open exit to the burrow, is usually aesthetically located underneath a small woody shrub). The other arm of the U, which typically stops just below the surface, can be used as an escape hatch in an emergency simply by breaking through the sand's crust. Individual *E. inornata* often have two such burrows 10 to 20 m apart (Pianka and Giles 1982).

Effectiveness of such escape hatches is seen in interactions with a particularly effective lizard predator: the monitor lizard. In Australia, monitor lizards are called goannas, probably a corruption of "iguana." The sand goanna, *Varanus gouldii flavirufus,* a major predator on other Australian desert lizards, forages widely, relying heavily on chemosensory cues to locate prey. Holding its body off the ground and flexing it laterally while swinging its head from side to side on its long neck, this big varanid sweeps its long forked tongue across the sand in wide arcs. The tongue is then retracted and tucked up into the vomerolfactory sensory system. Upon detecting a scent signal, this monitor follows the trail to the source, usually a burrow, and digs up the intended prey. Digging is methodical, accomplished by the forearms and sharp claws of the forefeet, with the pointed snout, mouth, and sharp teeth right in between, ever ready to snatch up prey as it dashes to escape. Sand monitors consume many geckos captured in their dead-end diurnal retreats—but very rarely are *Egernia striata* and *E. inornata* captured, thanks to their burrows' escape hatches.

An interesting very large arboreal skink, *Corucia zebrata,* is found in the Solomon Islands. These herbivorous green or brownish green skinks, popular in the pet trade, are relatives of Australian *Tiliqua. Corucia* possess strongly prehensile tails and are excellent climbers. They are live-bearers, giving birth to one or two very large young. Juveniles eat the feces of adults, thus acquiring intestinal endo-symbionts, which aid in digestion. Another, much smaller, arboreal green skink from the Solomon Islands is the emerald tree skink, *Lamprolepis smaragdina.* These agile climbers, which eat both arthropods and flowers (petals, nectar, and fruits), are ecological counterparts of day geckos *(Phelsuma)* and anoles. Males are highly territorial (McCoy 2000).

Within the past four to ten million years the large scincid genus *Ctenotus* (meaning "comb ear") has undergone an extensive adaptive radiation within Australia. *Ctenotus* is that continent's most speciose lizard genus, with more than 100 species described (Wilson and Knowles 1988). The genus was named for the presence of scales entering the front side of the external ear opening (Storr 1964)—a trait that separates *Ctenotus* from all other *Sphenomorphus* group members, suggesting that this large genus is monophyletic. Most *Ctenotus* are active diurnal heliothermic lizards, and they have radiated into a wide variety of habitats, including tropical and temperate forests, savannas, mulga scrub, shrubland, dry lakebeds, mallee, rocky areas, coastal sands, desert sandridges, and desert sandplains. Except for Tasmania and high mountaintops, *Ctenotus* occur virtually anywhere one goes in Australia, as well as on many offshore islands (fig. 11.1). One species, *Ctenotus spaldingi,* is also found outside mainland Australia on islands of the Torres Strait and in southern New Guinea.

In the Great Victoria Desert of Western Australia, as many as 11 species of *Ctenotus* can be found living together at the same place (Storr 1968; Pianka 1969). These include

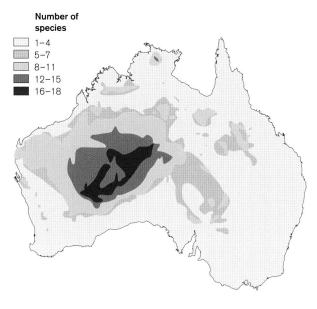

Figure 11.1 Species density of *Ctenotus* skinks across Australia.

several very small species, others of intermediate size, and several large species. Some of these sympatric lizards actively thermoregulate, while others are passive thermoconformers. Similarly, whereas some species stay very close to cover, seldom leaving tussocks of porcupine grass, others forage alertly in open spaces, retreating to grass tussocks only when threatened. Some are dietary generalists, eating a wide variety of insects, but others are dietary specialists, eating virtually nothing but termites. *Ctenotus* often exhibit strong habitat and microhabitat specificity as well, living only on sandridges or on flat sandplains, for example. Some species are associated with old, long-unburned spinifex, others require spinifex and leaf litter underneath marble gum trees.

What factors have allowed *Ctenotus* to undergo such a spectacular adaptive radiation? The absence of lacertids in Australia left open many diurnal active, widely foraging niches. A phylogenetic analysis of active body temperatures indicates that *Ctenotus* display appreciably higher body temperatures than related skinks (Garland et al. 1991). Acquisition of this "key innovation" may well have been a factor allowing *Ctenotus* to become active diurnal lizards and hence to radiate to fill many new niches. Aridification may also have facilitated the adaptive radiation of *Ctenotus*.

Among the smallest terrestrial vertebrates are tiny Australian skinks in the genus *Menetia* (7 species). Maximum snout-vent length of adults of the largest species is a mere 38 mm. Only geckos in the genera *Coleodactylus, Lepidoblepharus, Pseudogonatodes,* and *Sphaerodactylus* are smaller. *Menetia* belong to the *Eugonglyus* group and have a spectacle covering each eye and just four toes on their front feet (the innermost having been lost). Two species have lost the external ear opening and a third is variable in this regard (Greer 1989). All are diurnal, terrestrial, and oviparous, with clutches of one to three eggs. Neonate *M. greyi* measure only 20 mm and weigh around 0.1 g. Males of this species develop an orange throat and yellow venter during the breeding season. Both sexes mature during their first year, and some live to breed again the second year. Females can produce two clutches per season (Smyth and Smith 1974). In addition to coping with all the other hazards facing any small lizard, in hot open arid areas during summer these tiny diurnal lizards must often be precariously close to overheating (Greer 1989). Small size could, however, confer a significant advantage for such tiny ectotherms: rapid heat exchange allows opportunistic responses to brief periods of heat (Fraser and Grigg 1984).

The interesting skink *Gnypetoscincus queenslandiae* occurs in wet tropical habitats in Queensland. These cryptic lizards have a coarse skin covered with bizarre granular scales, which help keep the skin evenly moist through capillary spread along their edges. The aquatic Philippine skink *Tropidophorus grayi* has similar scales, as does the earless monitor *Lanthanotus borneensis* of Borneo and the gymnophthalmid *Neusticurus* of Amazonia.

Ristella is an intriguing genus of small lygosomine skinks found in southern India. Little is known about these presumably arboreal creatures, known as "cat skinks" because of their retractile claws.

In the windblown shifting sands of the Sahara, a skink known locally as a "sandfish," *Scincus philbyi*, dives into and swims through sand. Like members of several other lizard lineages (gerrhosaurids, lacertids, phrynosomatines), *Scincus* have enlarged lamellar fringes on their toes that enhance traction on loose sand. They also have countersunk lower jaws and shovel-nosed snouts.

The Kalahari Desert of southern Africa is home to two species of legless burrowing skinks. The larger species, *Typhlosaurus lineatus,* is widespread, occurring throughout the Kalahari, but the smaller species, *T. gariepensis,* has a more restricted geographic distribution, being found only in the sandveld of the interior part of the desert (Huey et al. 1974; Huey and Pianka 1974). *Typhlosaurus gariepensis* is more specialized than *T. lineatus,* requiring softer sands found on sandridges. Both species are termite specialists. In the zone of sympatry with *T. gariepensis, T. lineatus* are larger than they are in allopatry and have proportionately larger heads (they also eat larger termites). Such enhanced differences in sympatry constitute ecological character displacement, presumably reducing interspecific competition.

Compared to the Old World, skink diversity in the New World is extremely low, with, as we have seen, only five genera represented. The most abundant of these are *Eumeces,* with about 31 species, and *Mabuya,* with about 15 species. Both genera also have numerous Old World representatives. Members of the third New World genus, *Sphenomorphus,* also have many Old World relatives, particularly in Australia, and the 5 New World *Scincella* species have many eastern Asian relatives. The fifth genus, *Neoseps,* is a monotypic derivative of *Eumeces egregius* (Tod Reeder, pers. comm.). Excluding *Neoseps,* only *Eumeces* has higher diversity in the New World than in the Old.

In North America, *Eumeces* reaches its greatest diversity along a midlatitude band stretching from the central part of the United States to northern Mexico, with number of species appearing to drop off with both increasing and decreasing latitude. Southern Mexican and Central American species have relatively small distributions, often limited to mountain ranges.

Eumeces fasciatus is the only lizard in many parts of the northeastern United States. Across the central and southeastern United States, *Eumeces* species account for most of the lizard fauna—the one possible exception being north Florida and southern Georgia, where four species of glass lizards (*Ophisaurus,* Anguidae) are found. Three skink species (*E. fasciatus, E. laticeps,* and *E. inexpectatus*) occur together over a fairly large area of the Southeast, extending from Virginia to Louisiana. In long-leaf pine sandridges, where *E. fasciatus* drops out, the semifossorial *E. egregius* occurs, living under sand in pine-sawgrass associations (Mount 1963). In the plains of the central United States, *E. fasciatus, E. laticeps, E. anthracinus, E. obsoletus,* and *E. septentrionalis* have overlapping distributions, although seldom are they all found together. In forested regions, *E. fasciatus, E. laticeps,* and *E. anthracinus* occur together, and in more open habitats *E. obsoletus* and *E. septentrionalis* coexist, sometimes along with *E. fasciatus.* On the West Coast, the complex *E. skiltonianus* group extends from Baja California through British Columbia. Although currently recognized as two species *E. gilberti* and *E. skiltonianus*—with many subspecies, relationships could change drastically as molecular evidence becomes available (Tod Reeder, pers. comm.). In southern Texas, New Mexico, and Arizona and both eastern and western Mexico, *E. tetragrammus* is widespread (Lieb 1985).

Through Mexico and Central America, many *Eumeces* appear restricted to montane environments and as a consequence have very limited distributions (Taylor 1935). Others, however, like *E. tetragrammus* and *E. brevirostris* and its relatives, have rather extensive distributions (Lieb 1985; Robinson 1979).

Species of *Eumeces* have become model organisms for studies of chemoreception. Visual cues continue to play a role in intraspecific recognition, particularly mate choice, and in prey discrimination (Cooper and Vitt 1993; Hasegawa and Taniguchi 1994), but skinks also have exceedingly well developed olfactory abilities and recognize and determine species, sex, and sexual receptivity of other individuals by scent (Cooper and Vitt 1984a,b, 1986a,b,c,d).

Restricted to rain forest of northeastern Queensland, *Gnypetoscincus queenslandiae* is extremely secretive, usually found in damp microhabitats under logs while inactive. (Steve Wilson)

The semiaquatic Philippine skink *Tropidophorus grayi* has ridged dorsal scales similar to those of some semiaquatic gymnophthalmids. (Rafe M. Brown)

They also detect predators and discriminate prey using chemical cues (Cooper 1990; Cooper and Vitt 1989).

In South America, *Eumeces* drops out completely but is replaced by *Mabuya*. Some species of *Mabuya* enter Central America *(M. unimarginata)* and Caribbean islands *(M. mabouya)*, but the greatest diversity is in central South America. Five species occur within the Amazon Basin *(M. bistriata, M. carvalhoi, M. guaporicola, M. nigropalmata,* and *M. nigropunctata);* four species occur in the vast cerrado *(M. dorsivittata, M. frenata, M. guaporicola,* and *M. nigropunctata);* a single species, *M. heathi,* occurs in semiarid caatinga; and two species, *M. macrorhyncha* and *M. caissara,* occur in Atlantic forest and on coastal islands. Most lizard assemblages in mainland South America east of the Andes contain at least one species of *Mabuya.*

Conspicuous in the New World distribution of *Eumeces* is their absence in deserts. Although in the southwestern United States *Eumeces obsoletus* does occur on hillsides bordering deserts, while *E. gilberti* can be found in some desert riparian areas, and *E. tetragrammus* enters deserts in some mesic areas, these skinks have not been successful in the arid lands of North and South America, as other skinks have been in Australia and Africa.

This lack of success in New World deserts is in fact one of the most striking differences between Old World and New World skinks. Reasons for this failure are unclear, but two observations suggest a partial explanation. First, it seems significant that the Old World counterparts of the most successful New World skinks, *Eumeces* and *Mabuya,* have not been very successful in deserts either (though three or four species of *Mabuya* do occur in the Kalahari). Most likely, therefore, skinks in these two genera had life histories preadapting them to New World colonization, with the key lying in their reproductive strategies. All New World *Eumeces,* namely, either brood eggs or are viviparous, while all New World *Mabuya* are viviparous, with a unique placental type. This likely enabled *Eumeces* to migrate across the Bering Strait over a land bridge, and *Mabuya* to raft across the Atlantic. Because *Eumeces* ancestors were Temperate Zone lizards to begin with (none are found in the Southern Hemisphere), the very set of traits that allowed them to migrate through cold environments may have restricted their ability to successfully diversify in arid deserts. This hypothesis is supported by the fact that skinks in open habitats of the New World (e.g., *E. septenrionalis, E. obsoletus, E. egregius, E. gilberti*) spend much of their time

underground, where conditions are more mesic than at the surface.

The second part of the explanation centers on lizards that *have* diversified in arid regions of the New World, the teiids (see chapter 10). At one time, teiids were found in both the New and Old Worlds, but during the Cretaceous all teiids in the Northern Hemisphere disappeared. Those remaining in South America thus gave rise to the high diversity seen today, ultimately repopulating North America. Teiids are especially successful in arid and tropical environments, where direct sun for heat gain is available. Because teiids, especially *Cnemidophorus* and *Ameiva,* have high body temperatures and high rates of activity, they likely were formidable competitors of New World skinks. Within the New World, *Eumeces* have done well mostly in places where *Cnemidophorus* do poorly—at high latitudes and high elevations and within forest— partly because all *Eumeces* have some degree of extended parental care (Noble and Mason 1933; Vitt and Cooper 1989). Egg brooding not only provides physiological support for developing embryos, especially maintenance of appropriate humidity within the nest (Somma and Fawcett 1989), but it also likely affords protection from sources of mortality such as fungus and small predators. It also allows transportability: if conditions deteriorate, brooding females can simply move their clutches. Thus, some *Eumeces* can use habitats that are too cool for *Cnemidophorus.*

Competition may help explain a similar lack of diversity in small-bodied skinks of the New World tropics. Small skinks have done quite well in the Old World and on oceanic islands—in forested environments with warm climates, where they typically live in leaf litter or in damp areas. Although such habitats are certainly found in the New World, not a single small-bodied skink occurs in South America. Instead, gymnophthalmids are diverse and often abundant in these same environments. Like skinks, gymnophthalmids are active foraging lizards that use chemical cues for prey detection. Moreover, gymnophthalmids, like geckos and *Anolis* lizards, have small, fixed clutch sizes, which might provide additional advantages over New World skinks because they can deposit eggs in rapid succession and at different sites. The only places where small-bodied skinks occur in the New World are the southeastern United States *(Scincella lateralis)* and portions of Central America and Mexico *(Scincella* and *Sphenomorphus)*—where few or no gymnophthalmids occur. In Malaysia and Thailand, interestingly, many forest skinks—such as the tiny, very elongate, reduced-limbed *Larutia*—are quite similar in body form to tiny leaf litter gymnophthalmids of the Amazon such as *Bachia*.

The reason *Mabuya* has done relatively well in South America even in the presence of teiids remains obscure. Unlike teiids, which lay eggs and provide no parental care, all New World *Mabuya* are viviparous. Because teiids eat reptile eggs, which they locate by using their keen vomer-olfaction system, viviparity eliminates one source of mortality. (A similar dynamic obtains between some *Mabuya* skinks and lacertids, which are likewise sympatric in much of the Old World.) Perhaps more importantly, viviparity allows females to exercise total control over development of their offspring up until parturition. One advantage of this is that females can live in relatively shaded environments (places teiids avoid) yet still find patches of sun to bask and gain heat (e.g., off the ground) to aid in offspring development. In Brazilian cerrado, where *Cnemidophorus ocellifer* occurs alongside several species of *Mabuya*, *Mabuya* either are arboreal in the small, stunted cerrado trees or partly subterranean. In Amazon rain forest, one skink, *Mabuya nigropunctata*, has been very successful even in a setting in which as many as four teiids occur together. Because *M. nigropunctata* is arboreal, living on trunks of relatively large forest trees, interactions with teiids are reduced. (One sympatric teiid genus, *Kentropyx*, though also arboreal, uses small shrubs on the forest floor rather than vertical tree trunks.) Moreover, *M. nigropunctata* is smaller than most Amazonian teiids. Thus, whereas *M. nigropunctata* can live within forest, where it finds patches of sun to bask and gain heat, *Cnemidophorus* are limited to habitat edges and roads, with access to patches of open ground for basking. Although teiids are probably superior competitors in open habitats, the complex mosaic of habitat patches in New World tropics provides microhabitats for *Mabuya* that teiids cannot easily access.

Two species of climbing skinks are found on trees in the Kalahari Desert: *Mabuya striata* and *M. spilogaster*. Both are live-bearers with broad diets composed of a variety of insects. Both species maintain higher body temperatures in the summer, when they bask early and late in the day, than in the winter, when their basking occurs at midday. These skinks, seldom found together, seem to replace one another along a fairly sharp zone of contact (Huey and Pianka 1977b). Their diet and microhabitat utilization patterns suggest that their ecological requirements may be too similar for coexistence.

In contrast, two egg-laying, terrestrial species of *Mabuya* are found in broad sympatry throughout the Kalahari. *Mabuya variegata* is a small skink associated with grass tussocks, whereas *M. occidentalis* is larger and has more generalized microhabitat requirements.

Like many lizards, the North American *Eumeces laticeps*, locally known as broad-headed skinks or "scorpions," feed on a variety of insects and other arthropods, and occasionally on fruits. Although most prey are relatively innocuous, these skinks do consume a variety of stinging insects, including honeybees, bumblebees, and large paper wasps *(Polistes)*. Typically, skinks bite an insect, break its exoskeleton with a single crunch, and swallow it whole, head first (Cooper 1981), apparently not getting stung in the process. These active foragers often cover rather large areas in search of a meal. They go in and out of holes in live oak trees using chemosensory cues to locate prey that cannot be found visually.

On the coastal islands of the southeastern United States, these skinks climb up to paper wasp nests, peel the caps off the larval chambers, and wait for larvae to drop to the ground (McIlhenny 1937). The lizard then descends to the ground, eats the larva, and returns to the nest to uncap more chambers. The nervous system integration necessary for such a feat is impressive. First, the skink must locate a nest—probably by observing wasps landing on and moving about on its surface. Once a nest is located by sight, the skink must climb up to it and explore it, using its tongue to transmit particles from the nest to its vomeronasal organs, located in the roof of the mouth (functionally similar to olfactory organs). Although adult wasps occasionally attack the lizard, its flat, glasslike scales protected by osteoderms make it nearly impossible for a wasp to penetrate with its stinger. The skink determines that tasty morsels lie under the papery caps of each cell in the nest and must recognize that pulling a cap off will expose a larva. Whether the lizard anticipates that a larva will fall to the ground or just responds to the falling insect is not known. Once an insect falls, however, the lizard tracks it visually, then descends to search the area where the insect dropped; by flicking its tongue across the ground surface, it locates the prey and eats it.

Some prey, like the large velvet ant *Dasymutilla occidentalis*—commonly known as the cow killer!—have evolved defenses that are effective against even broad-headed skinks (Vitt and Cooper 1988). Velvet ants, which are the size of paper wasps but look like fuzzy ants, are in their own family, though they are more closely related to wasps than to ants. Wingless females march around on the ground in search of prey, usually when lizards are active. Cow killers are a brilliant red or orange: warning colors. Their body is covered with tiny hairs that have microscopic ends resembling tiny spears, and they have large biting jaws and an exoskeleton as hard as a rock. One can jump up and down on most species of velvet ants and they will walk away unharmed. To complete their defense, cow killers have an extremely long stinger coiled in their abdomen and a potent venom that causes severe pain even in humans (and likely cows!).

A naive broad-headed skink may attack a velvet ant once—but only once. When a skink bites down and the exoskeleton does not crunch, the lizard will reposition the velvet ant in its mouth and try again, and again. Eventually the velvet ant bites the lizard's tongue with its sharp and powerful jaws and extrudes its stinger, sinking it deep into the lizard's tongue; it injects venom with a pumping movement while continuing to hang on by its jaws. The lizard responds by spitting out the insect and rubbing its face against the ground. The velvet ant walks away as though nothing happened. Once a lizard has experienced the velvet ant's sting, it avoids these insects like the plague.

Broad-headed skinks exhibit striking sexual dimorphism, generally with bright coloration on the head or neck region, larger body size, and relatively larger heads in males. The reason for sexual dimorphism in *Eumeces* has to do with sexual selection (Vitt and Cooper 1985b): large males, especially those with larger and redder heads, outcompete smaller males in aggressive interactions, winning access to more females and passing on their genes — genes related to body and head size, aggressiveness, and extreme head coloration. Until reaching sexual maturity, males are similar in both head size and color to like-sized females. In their third year, however, shortly after emerging from overwintering, they experience their first major release of testosterone, the hormone (an androgen) that will drive most of their sexual behavior. This androgen causes their heads to develop red or orange coloration and increases head muscle mass and size.

Still small (a newly sexually mature male is only about 80 mm SVL, compared to more than 140 mm SVL for a large male) and having little chance yet of obtaining mates, males enter the breeding population inexperienced in the weaponry that will determine their success in the long run. Larger males aggressively attack young males, driving them away from females, and females avoid mating with small males. Although young sexually mature males do attempt to court and mate with females, occasionally with success, they rarely escalate their encounters with large males to the point of actual fighting or even physical contact. Fighting is dangerous: if a large size difference exists, the smaller lizard may well suffer life-threatening wounds. A much better strategy is to wait until they reach a size at which they might, if not win, at least avoid injury. Large enough body size is reached at an age of three to four years.

In addition to experiencing behavioral and morphological changes associated with male-male interactions, males become particularly sensitized to social signals picked up via the vomeronasal sensory system. During breeding season, males identify chemicals produced in the female's cloacal glands (Trauth et al. 1987); these contain a lipid-based chemical (pheromone) that allows a male to determine whether he is interacting with a female and, if so, whether she is sexually receptive. These chemicals are so effective that sexually active male *Eumeces laticeps* can follow a female's chemical trail over and under logs, across ground, and up and down tree trunks.

Not all is lost in the likely event that a young male does not breed during his first summer as an adult. Experience gained then will make him a better competitor in the future (and increased size will help greatly as well). From a female's perspective, a male becomes more attractive as he grows in size, not just because he might win in aggressive encounters with other males but also because the fact that he has lived that long enhances his quality (individual fitness). Whatever a male did right that resulted in survival for an additional year must be better than whatever nonsurvivors did.

As individual females become sexually receptive and emit pheromones signaling their receptivity, males attempt to court them. Not all females are receptive synchronously (some might become sexually receptive later than others, while those that have mated drop out of the breeding population); however, all males are always in reproductive condition. Consequently, the operational sex ratio—the ratio of fertilizable females to males—is male-biased. Competition among males for females is therefore intense: not only are there not enough receptive females to go around at any one time, but males that have targeted specific females will do all they can to keep other interested males away. In this game, large, aggressive males have a clear competitive advantage, and so tend to enjoy higher reproductive success than smaller males.

Once a male reaches a competitive size, the odds in this behavioral game change. Exactly what that size threshold is depends on other males in the immediate area. In one place, a 110-mm-SVL male might be the largest and thus most successful breeder. If that male entered an area with a 140-mm-SVL male, however, it might not breed. Males usually interact with encounters that escalate with similarity in size and experience. A male courting a female typically runs out to meet any other male that approaches. The two then may face off, moving in a circle with one male's head facing the other's in what appears to be a race to catch up with the other as they go around and around. When one male does catch up, he often bites at the head and neck of the other male; sometimes males open their mouths wide, measuring the head width of the other. This information allows a male to assess his probability of winning should the encounter escalate. If the combatants are equal in size, interactions can become full-blown battles as they attack each other, biting at the head, body, and sometimes tail. Such encounters result in tail loss and wounds on the head and neck. Depending on the level of battle, the loser might retreat with wounds severe enough that it drops out of the breeding population to recover. A lost tail, for example, might result in a severe depletion in energy reserves, making it difficult to engage in social encounters or even survive through the winter. A male's best strategy then is to forgo reproduction and invest time and energy feeding to regenerate his tail.

Given that head and body size in *Eumeces laticeps* provide a significant advantage for males in terms of breeding success, why haven't these lizards evolved even larger heads and bodies? One reason is that natural selection sets limits on what is practical: a skink with a giant head might not be able to escape predators as easily as one with a smaller head, and a huge red head would likely attract more predators. Another is that alternative strategies may exist for sexually mature males. For example, a male large enough not to be rejected by a female but still too small to compete with other males might follow a courting pair using the female's chemical scent trail, but at a discrete distance so as not to elicit an attack. If the other male, who, having already courted the female and so primed her for mating, sees a second large male, he might initiate a prolonged aggressive interaction, providing an opportunity for the "satellite" male to sneak in and mate. The wide variation in head size among males of equal body size certainly suggests this possibility. However, alternative mating strategies have not been examined in this otherwise well studied skink.

CHAPTER 12

FROM GIRDLED LIZARDS
TO KNOB-SCALED LIZARDS

Elgaria multicarinata basks in sun while overlooking Temecula Canyon in southern California. (L. Lee Grismer)

Like skinks, girdled lizards (Cordylidae), African plated lizards (Gerrhosauridae), alligator lizards (Anguidae), and knob-scaled lizards (Xenosauridae) have some of their body scales underlain by osteoderms, and their foraging behavior is similar. Cordylids that have been studied *(Platysaurus, Cordylus,* and *Pseudocordylus)* are sit-and-wait foragers; most maintain territories (Cooper et al. 1997; Whiting 1999), an exception being the snakelike cordylid *Chaemasaura.* Gerrhosaurids and anguids move about while foraging but at such a slow rate that their behavior is often referred to as "cruising" rather than "wide or active foraging." Girdled lizards and African plated lizards as a group are quite distant evolutionarily from alligator lizards and their allies, and the secretive knob-scaled lizards differ from all of these. Cordylids (with the exception of *Platysaurus*), gerrhosaurids, and anguids are heavily armored thanks to their large scales and osteoderms.

Cordylids and gerrhosaurids—which, except for some Madagascar gerrhosaurids, are currently restricted to Africa south of the Sahara (a fossil is known from France)—are skinklike in form (scincomorphans); moreover, they are each other's sister group, and skinks (chapter 11) are considered the sister group to both (Lang 1991). The common ancestor to cordylids and gerrhosaurids was skinklike. Gerrhosaurids must have evolved before Madagascar separated from mainland Africa, about 80–100 million years ago, because they occur on both the African mainland and Madagascar. Cordylids presumably arose later, as they are found only in southern and central Africa. All gerrhosaurids lay eggs, but most cordylids are live-bearers. Both families gave rise to elongate forms with reduced appendages. Gerrhosaurids remained somewhat skinklike, but many cordylids have converged on a bauplan and ecology similar to those of iguanians. Many cordylids are spiny and colorful, territorial, sit-and-wait ambush foragers. Because of a large number of similarities, these two families are sometimes considered subfamilies (Cordylinae and Gerrhosaurinae) rather than families (Zug et al. 2001). Nevertheless, they are easily distinguished: cordylids have four parietal scales, versus only two in gerrhosaurids; and cycloid scales are found on the throat of cordylids but not on gerrhosaurids.

Anguids and xenosaurids are anguimorphans and therefore are allied to varanoids (see chapter 13). Many anguids (especially *Gerrhonotus* and *Diploglossus*) and all xenosaurids can inflict painful bites with their vice grip–like jaws. Like Gila monsters, they often hold on, repeatedly clamping down, reminding herpetologists of their own ancient hunter-gatherer origins.

CORDYLIDAE

Cordylids, or "girdled lizards," have body and tail scales arranged in regular rings or whorls, giving them a very distinctive look. Four genera—*Cordylus, Pseudocordylus, Chamaesaura,* and *Platysaurus*—and about 40 species are recognized. All except *Chamaesaura* are flattened dorso-

ventrally, with a ventrolateral fold typically separating dorsal plates from ventral ones. In *Platysaurus* this fold is not evident, but a marked size difference between ventral and lateral scales gives the appearance of a lateral fold. Some species have strongly keeled scales and are very spiny. Except for *Platysaurus,* all cordylids are viviparous. Most live in rocks, retreating to crevices when disturbed (Broadley 1978), though snakelike *Chamaesaura* are terrestrial and live in grassy habitats. The largest cordylid, *Cordylus giganteus,* is also terrestrial and digs extensive burrows, which are defended against conspecifics. Endemic to the highveld grasslands of the northeastern Free State of South Africa, it is highly threatened due to habitat loss (van Wyk 2000).

Some cordylids use their spiny tails in defense, rather like certain iguanids (*Ctenosaura* and oplurines) and agamids *(Uromastyx).* Like New World chuckwallas (*Sauromalus:* Iguaninae), too, many cordylids avoid predators by wedging themselves into crevices and inflating their bodies (Cooper et al. 2000). A variety of postures are used to maintain their hold as they press their head, back, and tail against upper surfaces of the rock and block access with the tail. One remarkable species, *Cordylus cataphractus,* known as the armadillo lizard, defends itself by flexing forward ventrally and grasping its tail in its mouth, forming a ring with the belly inside. In this posture, they are virtually impossible to swallow (Schmidt and Inger 1957). This species lives in groups of up to thirty individuals and, like most cordylids, is a sit-and-wait forager (Mouton et al. 2000).

Two particularly interesting genera, *Platysaurus* and *Pseudocordylus,* are the sister clade to *Cordylus* (Lang 1991). Both are associated with rocky habitats in South Africa, but in mutually exclusive distributions: whereas *Platysaurus* occurs in Tropical Transitional Zone of southeastern Africa, with two isolated species (*P. capensis* and *P. broadleyi)* in the northern Cape region, *Pseudocordylus* occurs in Cape/Temperate Transitional Zone of the northern Cape (Branch and Whiting 1997). The genus *Platysaurus* (= flat lizard) has undergone an adaptive radiation in and around Zimbabwe, with very flattened bodies that allow them to squeeze into tight crevices under slabs of rock. Restricted to certain types of rock such as granite and sandstone, these lizards are often found in isolated populations, sometimes in dense colonies. Diversification and speciation likely occurred as the Mozambique Plain underwent alluvial filling and as Kalahari sands shifted eastward during the Pliocene and Pleistocene, covering extensive portions of rock habitats and isolating populations of these cordylids.

Male *Platysaurus* defend territories during the breeding season. During territorial encounters, males raise their head and forebody displaying their colorful throat and chest. Unlike other cordylids, *Platysaurus* are oviparous, laying two large eggs. Females often nest communally. Most *Platysaurus* are insectivorous, but some species also eat flower petals, leaves, and seeds. *Platysaurus broadleyi,* for example, though typically insectivorous, can shift its diet to take advantage of rich resources that are available only fleetingly. When Namaqua figs are available in South Africa's Augrabies Falls National Park it will feed on them, discriminating between ripe and unripe figs and opting preferentially for ripe ones, usually on the ground. If ripe figs are not available on the ground but are in trees, these lizards sometimes climb to feed on the fruit in situ

(Whiting and Greeff 1997). Even more interesting, they use birds as cues to find fruiting fig trees (Whiting and Greeff 1999).

The crag lizards, *Pseudocordylus,* occur on rocks and rocky cliffs of the Natal Drakensberg, a huge lava flow forming a high-elevation escarpment. Unlike *Platysaurus, Pseudocordylus* possess highly keeled dorsal and lateral body scales. Diversification in this genus may have resulted from climatic change isolating populations, both in elevation and across the landscape, in suitable areas on massive crags (Broadley 1964). These viviparous crevice dwellers produce from one to five young and eat a variety of arthropods. One species, *P. langi,* also consumes some vegetation (leaves and flowers).

Three species in the cordylid genus *Chamaesaura* (= ground lizard) have evolved greatly reduced limbs and very long tails (three to four times body length). These southern African lizards are vaguely reminiscent of leg-

FLAT IS FLAT IS FLAT!

In 1979, while a postdoc at the University of Michigan with Don Tinkle, I was putting together data I had collected the preceding year in northeastern Brazil. I had brought a small collection of lizards with me to be deposited in the collections at the Museum of Zoology, and was spending a lot of time looking at the specimens already on the museum's shelves. One day as I was working my way along, seeing species I had never before laid eyes on, I found a jar of lizards that I thought were the *Tropidurus semitaeniatus* I had brought back from Brazil. I was impressed that the curators had already incorporated my material into the collection. Yet when I read the label on the jar, I saw that the genus was *Platysaurus,* one I had never heard of. A quick check revealed that these lizards were from South Africa, and examination of the specimens themselves told me they had to be cordylids or something related to them. Rather than having tiny ventral scales like *Tropidurus,* they had enlarged rec-

tangular ventral scales. Other differences were apparent as well.

Two lizards so similar morphologically, I thought, had to be ecologically similar. I dug into the literature and learned that *Platysaurus,* like *Tropidurus semitaeniatus,* live on rock surfaces and retreat into narrow crevices—just on a different continent. At the time, I was interested in lizard life histories and I had already determined that *T. semitaeniatus* had clutch sizes of two eggs, occasionally one or three (all other species of *Tropidurus* had clutch sizes of from four to thirteen), and that its eggs were much more elongate than eggs of other lizards. The basic explanation was that use of narrow crevices required a flat morphology, and the flattest females enjoyed a selective advantage: they could enter narrower crevices and go deeper in them than less flat females. Females carrying fewer eggs are presumably flatter. Natural selection had reduced clutch size and caused eggs to elongate so that gravid females were as

flat as nongravid females. A potential predation cost of reproduction had been circumvented.

But what about *Platysaurus,* this remarkable look-alike from southeastern Africa? A small paper by Donald Broadley in 1974 contained sufficient information to convince me that not only had morphology evolved similarly in *T. semitaeniatus* and *Platysaurus,* but clutch size and number of eggs had as well. *Platysaurus* deposit clutches of two elongate eggs, whereas all of their close relatives are viviparous and produce more young. The observation that some life history traits had evolved independently in lizards from different families living on different continents in response to identical morphological responses to crevice dwelling suggested that morphological constraints could influence life history traits predictably (Vitt 1981). Such convergent pairs of species are known as "ecological equivalents" (Grinnell 1924). *(VITT)* less anguids *(Ophisaurus)* and Australian pygopodids except that their body scales are sharply keeled (FitzSimons 1943; Branch 1988), although the pygopodid *Pletholax* also has strongly keeled scales. Commonly called grass lizards, they take advantage of their elongate shape to move freely in tall grass, through which they "swim" rapidly like snakes. (The agile lacertid *Takydromus* also moves gracefully through grass.) On smooth or sandy surfaces, *Chamaesaura* are not nearly as agile. On losing their tail, too, these lizards are slow and clumsy, though tails

Some cordylids, like these *Platysaurus intermedius,* are highly dimorphic in color, with the male brightest. (Louis Porras)

A kaleidoscope of colors reflect off the head of a *Zonosaurus haroldschultzei.* (R.D. Bartlett)

regenerate rapidly. In dry grass, *Chamaesaura* are camouflaged and difficult to find. *Chamaesaura* bear litters of five to nine living young in late summer.

GERRHOSAURIDAE

Four genera of Gerrhosauridae, or African plated lizards—*Angolosaurus, Cordylosaurus, Gerrhosaurus,* and *Tetradactylus*—occur south of the Sahara; two genera—*Tracheloptychus* and *Zonosaurus*—occur in Madagascar. These slender, long-tailed lizards, vaguely reminiscent of skinks, are all diurnal and terrestrial, and all lay eggs. Most gerrhosaurids are terrestrial, but the bright green *Zonosaurus boettgeri* is arboreal, while *Z. maximus* is semi-aquatic, diving and hiding underwater for extensive periods (Pough et al. 1998).

Species of *Gerrhosaurus* occupy a variety of habitats. Some, like *G. validus,* occur on granitic, boulder-strewn hillsides, rarely leaving their rocky retreats. Like *Sauromalus* (Iguaninae) and many *Cordylus* (Cordylidae), they wedge themselves into crevices and are extremely difficult to remove—though they are not nearly as flattened as many crevice-dwelling cordylids. Others, like *G. nigrolineatus* and *G. flavigularis,* are terrestrial and occupy a variety of habitats, living in holes in the ground. Males of these two species sport brilliant colors during the breeding season. *Gerrhosaurus* feed primarily on arthropods, including grasshoppers, termites, spiders, scorpions, and centipedes. Some species, such as *G. validus,* also eat fruit.

An interesting gerrhosaurid, *Angolosaurus skoogi,* lives in sandy habitats in northwestern Namibia. *Angolosaurus* has converged on the same shovel-nosed snout and fringed toe morphology adopted by sand dwellers in other families, such as the lacertid *Meroles anchietae* of southern Namibia, the phrynosomatine *Uma* of the southwestern United States, and the lacertid *Acanthodactylus boskianus* and the scincid *Scincus philbyi* of northern Africa. When disturbed, they literally dive into the sand, disappearing under the surface.

Tetradactylus has extremely reduced limbs, an elongate morphology, and moves in a serpentine fashion similar to the cordylid *Chamaesaura.* The five species within this genus are endemic to South Africa; absent from more xeric regions, they are locally known as "seps" and can be quite common in coastal mountain forests on hillsides exposed to sun (FitzSimons 1943). They typically deposit clutches of two eggs.

Male *Gerrhosaurus nigrolineatus* are larger and more brightly colored than females. (Laurie Vitt)

ANGUIDAE

The heavily armored anguids originated in Laurasia. When the Atlantic Ocean opened 100 M.Y.B.P., ancestors of the subfamilies Gerrhonotinae (the New World alligator lizards) and Anguinae (the Old World glass lizards) became separated, with subsequent diversification occurring in each area (Macey et al. 1999). *Ophisaurus* later colonized the New World across the Bering Strait, whereas Anniellinae (California legless lizards) arose in North America. Gerrhonotinae, a mostly temperate group, and Diploglossinae (galliwasps), a mostly tropical group, probably diverged in response to Miocene climatic change. Associated geological changes in North America essentially split ancestral groups.

Anguids are characterized by a body covered with large, nearly nonoverlapping scales. Osteoderms underlying the dorsal and ventral body scales add to this effective armor, which in most anguids is separated by well-defined ventral-lateral folds. As anyone bitten by a *Diploglossus* or *Gerrhonotus* knows, most have powerful jaws as well. Two anguine genera and the anniellines *(Anniella)* are completely limbless. Diploglossines and gerrhonotines are elongate with reduced limbs, though some diploglossines are nearly limbless.

Anguids are widespread, occurring in Europe, Asia, Southeast Asia, North America, the Caribbean, Central America, and South America. Richard Estes (1983), in a study of the fossil record, suggested that ancestors of anguimorphans probably arose on the Laurasian plate, with anguids most likely getting their start in North America, where both the diploglossine and gerrhonotine groups appeared early, perhaps as early as the Late Cretaceous. Anniellines also arose in North America but never dispersed. Some ancestral anguids dispersed to Asia during the Eocene, probably via North Atlantic land bridges. Fossil ancestors of *Ophisaurus* appear earlier in Europe than in

Like other limbless anguids, *Ophisaurus harti* superficially resembles a snake. (R.D. Bartlett)

North America, and a fairly good fossil record suggests that these were Eurasian derivatives from an anguioid stock. If so, *Ophisaurus* must have reinvaded North America from Eurasia. These ancestors probably brooded their eggs. When diploglossines reached South America remains unknown. Future work on anguid biogeography and phylogeny will be of considerable interest.

The subfamily Anguinae is represented by 2 genera, the monotypic *Anguis* and *Ophisaurus,* which contains 16 species. European "slowworms" *(Anguis)* are slow moving and extremely long lived (in captivity in Denmark, a forty-five-year-old male successfully mated with a twenty-year-old female!). Slowworms are usually active at night (females sometimes bask during the day), feeding primarily on slugs and snails. They are often quite easily found by turning over rocks. Unlike other anguines, the slowworm is viviparous.

Species in the legless genus *Ophisaurus,* which occur in both the New World and the Old, are known commonly as glass "snakes" or glass lizards—an allusion to their response when threatened of autotomizing their tail, which fragments into several thrashing pieces. The autotomized tail is indeed impressive, representing two-thirds of the

Close examination of the head of an *Ophisaurus attenuatus* reveals eyelids and external ear openings, neither of which occur in snakes. (Paul J. Gier)

lizard's total length. Lost tails are regenerated, but the new tail is generally shorter than the original. A common myth in the southeastern United States is that thrashing tail segments continue moving until after dark, at which time they reunite with the body and the collective whole goes on its merry way. This ridiculous myth attests to the importance of quality biological training in primary and secondary education. The largest and most impressive anguine is the "scheltopusik," *Ophisaurus apodus,* found in southeastern Europe and Southwest Asia; this glass lizard, which eats mice, reaches a total length of over a meter, and its body is nearly as thick as a man's wrist! Four *Ophisaurus* species are currently recognized in the United States: the eastern *(O. ventralis),* slender *(O. attenuatus),* island *(O. compressus),* and mimic *(O. mimicus)* glass lizards; others occur in Mexico. *Ophisaurus* are terrestrial, often found in grassy fields. All *Ophisaurus* are oviparous, depositing relatively small eggs. Like skinks in the genus *Eumeces,* females brood their eggs until they hatch.

Anniellines are represented by a single genus, *Anniella,* of which only two species are known (Hunt 1983). These fossorial, legless lizards live on the coast of California and western Baja. Much of their life is spent underground in coastal sand dunes or in relatively loose soil in inland valleys. Their lower jaw is countersunk, which aids in "swimming" beneath the sand. California legless lizards spend much of their time in the soil under the crown of surface vegetation. Around Riverside, east of Los Angeles, they are common in litter under juniper trees, where they likely feed on termites. These viviparous lizards produce litters of three to four relatively large offspring, similar to Kalahari burrowing *Typhlosaurus* skinks.

Galliwasps or diploglossines superficially resemble skinks, having smooth, glasslike scales, elongate bodies and tails, and short limbs. A ventrolateral fold is generally not present. For the most part, these lizards are terrestrial, and many live among rocks. Genera include Central and South American *Diploglossus,* the Caribbean and Mexican *Celestes,* Caribbean *Sauresia* and *Wetmorena,* and South American *Ophiodes,* which are nearly legless. *Ophiodes* live in open grassy areas in South America where the water table is relatively high; here, they are strikingly convergent with glass lizards *(Ophisaurus).* Some diploglossines lay eggs, but others give birth to living young. In some species of *Diploglossus,* juvenile coloration is distinctly different from that of adults, resembling the color pattern of local toxic millipedes. These are all no doubt long-lived, late-maturing lizards, but little is known about their natural history.

Gerrhonotines, collectively called alligator lizards *(Abronia, Barisia, Elgaria, Mesaspis,* and *Gerrhonotus),* resemble small crocodilians because of their large plate-like scales, especially in *Abronia.* Although most species are subtropical, one, *Elgaria coerulea,* reaches British Columbia. This species spends winter in cliffs along the Pacific, lining up in crevices to bask in what little direct sunlight they can find. Females bear live young, no doubt enhancing embryonic development by basking. When parturition occurs, females eat the remains of the yolk-sac placenta, thereby helping the young to escape from their maternal membranes. Many gerrhonotines deposit eggs.

California legless lizards, *Anniella pulchra,* spend most of their lives beneath sand. (Laurie Vitt)

The brightly colored *Diploglossus fasciatus* of Brazil has a color pattern similar to that of toxic millipedes. (Otávio A.V. Marques)

Elgaria cedrosensis searches during the day for its insect prey.

(L. Lee Grismer)

Elgaria kingi **is among the prettiest of North American alligator lizards.** (Laurie Vitt)

Mesaspis moreletii **of Honduras is very similar to North American alligator lizards.** (Laurie Vitt)

Tails of gerrhonotines are long and muscular and in arboreal genera such as *Abronia* are no doubt used for climbing. Although many *Gerrhonotus* and *Elgaria* are rather drab in coloration, *E. kingi* is beautifully banded. The banding pattern interferes with the general outline of the lizard, aiding in crypsis via disruptive coloration. Most gerrhonotines are secretive, spending much of their time in crevices or under surface material. The largest gerrhonotine is the Texas alligator lizard, *G. liocephalus,* which reaches at least 200 mm SVL. Pianka gets phone calls all the time from people who have come across one of these impressive lizards and want to know what they've seen.

Although most *Gerrhonotus* and *Elgaria* are terrestrial, Mexican and Central American *Abronia* are arboreal—and most are endangered due to habitat destruction. More than 25 species have been described and others wait to be described, comprising more than half of known gerrhonotine lizards. Most *Abronia* occur at high elevations from southwestern Tamaulipas, Mexico, to Guatemala, southern Honduras, and northern El Salvador. The distribu-tions of individual species are both small (some are known only from the type specimen, that is, the individual on which the species description is based) and disjunct. Only a single case of documented sympatry exists: *Abronia gaio-phantasma* and *A. fimbriata* occur together in one portion of the cloud forest of the Sierra de las Minas in Guatemala (Campbell and Frost 1993). Some have spectacular coloration, resembling lichens found on rocks and tree trunks. All species for which data exist are viviparous. Litter size is five or fewer in most species, but *A. vascon-celosi* (erroneously assigned to *A. aurita* by Campbell and Frost 1993) can produce up to twelve young. Although little is known about social behavior of *A. vasconcelosi,* the male bites the female's head during copulation. Because of their restricted habitats, small distributions, and continued development throughout Mexico and Central America, the future of *Abronia* appears dismal. Jonathan Campbell and Darrel Frost (1993) predict that at least 13 species (nearly one-third of the anguine diversity) will go extinct within the next thirty to fifty years. Some species may well have vanished already.

The arboreal anguid *Abronia vasconcelosi* is cryptic against the moss- and lichen-covered trees in which it lives. (D.G. Barker)

Shinisaurus crocodilurus **is an ancient relictual xenosaurid species from China.** (Louis Porras)

XENOSAURIDAE

Xenosaurids, or knob-scaled lizards, include only 2 genera: *Xenosaurus,* with 4 recognized species in southern Mexico and Guatemala (though an additional 8 species are currently being described; J. Campbell, pers. comm.); and the monotypic *Shinisaurus,* found only in southern China. Xenosaurids probably arose in North America, then dispersed along land bridges across Beringia to Asia during the Late Cretaceous. Both genera are apparently very old and may ultimately be placed in separate families (Macey et al. 1999). Although they appear quite different superficially, with *Xenosaurus* being dorsoventrally flattened and *Shinisaurus* compressed laterally, they share many important derived characteristics that tie them together as sister taxa (McDowell and Bogert 1954; Estes et al. 1983).

Xenosaurus are diurnal rock dwellers with spiny tails, having converged on saxicolous iguanians and cordylids. They inhabit narrow crevices, which accounts for their dorsoventrally flattened morphology.

Xenosaurus lives in crevices, usually in rocks, but also on human habitation and possibly under loose bark of trees. Some species, such as *X. rectocollaris* and *X. platy-*

Like other *Xenosaurus, X. platyceps* **is flattened dorsoventrally, which facilitates the use of narrow crevices.** (Laurie Vitt)

ceps, live in rock crevices in open habitats, whereas others, such as *X. newmanorum* and *X. grandis,* live in crevices in tropical forests (Lemos-Espinal et al. 1996, 1997a). Body temperatures of these lizards are quite low, averaging about 22°C (Lemos-Espinal et al. 1998). All species are viviparous, producing from two to seven offspring, typically in June through August (Fritts 1966; Ballinger et al.

2000). Preliminary observations indicate that newly born young remain with the female for some time after birth (Lemos-Espinal 1997b). Similar to other anguimorphans, *Xenosaurus* discriminate prey chemically; however, adults curtail tongue-flicking when outside their home crevices, suggesting that a trade-off exists between feeding and defense (Cooper et al. 1998). Although sexual dimorphism exists in some species, it is subtle, usually involving slightly larger heads in males (Smith et al. 1997).

Shinisaurus crocodilurus is a handsome semiaquatic diurnal lizard found along small ponds and slow-moving streams in a small, very wet limestone region in southern China (Zhang 1985). The area, usually shrouded in mist, freezes during winter, when these lizards hibernate. These sluggish lizards bask on branches overhanging water and escape trouble by dropping into the water. They prey on earthworms, insects (especially caterpillars and aquatic dragonfly larvae), tadpoles, small frogs, and fish (Hofmann 2000), sometimes foraging underwater. Males have larger and deeper heads than females, and heads of most males are red, whereas those of females are usually pale (Hofmann 2000). Males are aggressive toward other males. Mating is said to occur during late summer in the wild (Zhang 1985), but several spring matings have been observed in captivity (Hofmann 2000). Like all xenosaurids, *Shinisaurus* bear live young after a long 8–12-month gestation. Litter size varies from two to fifteen. Neonates are tan colored and average 89 mm in total length (Hofmann 2000). They grow slowly and are long lived (about fifteen years or more); in captivity they reach sexual maturity in two to three years. The tiny geographic range of *Shinisaurus* is shrinking as the area is deforested and streams and ponds dry up, placing this ancient monotypic genus at risk of extinction in its natural habitat (Zhang 1985). Fortunately, these lizards thrive and breed in captivity, so the species might not disappear entirely.

MONSTERS AND DRAGONS
OF THE LIZARD WORLD

A Mexican beaded lizard,
***Heloderma horridum,* in typical thorn**
forest habitat. (Cecil Schwalbe)

As a group, beaded lizards, earless monitors, and monitors cannot be mistaken for any other lizards. All are elongate with extended necks and walk along at a slow pace, swaying their heads and bodies from side to side and flicking their long, forked tongues in and out like snakes. Their gait, compared with that of other lizards, seems almost mammal-like, and they appear to progress with a well-founded sense of confidence. Some are large, and some are venomous; a gigantic fossil monitor probably preyed on Australian aboriginal humans. Stories of fire-breathing dragons no doubt had their origins with these lizards.

Varanoid lizards form a natural monophyletic group consisting of three extant families: earless monitors (Lanthanotidae), beaded lizards and Gila monsters (Helodermatidae), and monitor lizards (Varanidae). Some scientists have placed earless monitors and monitor lizards in a single family (Varanidae) composed of two subfamilies. Varanoidea also contains a number of fossil taxa, including very large marine mosasaurs (Lee 1997). Varanoids, anguids, and xenosaurids make up a natural group known as Anguimorpha (= snake form) (McDowell and Bogert 1954; Estes et al. 1988), which stands out as the evolutionary connection to snakes. Although snakes remain *incertae sedis* within Scleroglossa, they could be nested within varanoids or they could be affiliated with dibamids and amphisbaenians (Wu et al. 1996; see chapter 14). Varanoid affinities are better supported (McDowell and Bogert 1954; Lee 1997).

LANTHANOTIDAE

One of the strangest and least known of all lizards is *Lanthanotus borneensis,* commonly called the earless monitor lizard (the generic name means "concealed ear") and found only in Sarawak on Borneo. These medium-sized secretive lizards (about 42–43 cm in total length as adults) have relatively long cylindrical bodies and long necks

(Harrisson 1966). They have short legs and long, curved, sharp claws, and the way they wrap their muscular bodies and prehensile tails around branches suggests they might climb (Proud 1978). Most of their scales are small, but six longitudinal rows of enlarged scales run from the head down the back, while two central rows extend onto the tail. *Lanthanotus* tails do not regenerate. Similarities with snakes—to which they may be closely related (McDowell and Bogert 1954)—are numerous: They shed their skin in one piece, like snakes and some other anguimorphans (J. Arnett, pers. comm.). The brain case is more solidly encased than it is in varanids, more as it is among snakes. *Lanthanotus* is the only species among anguimorphans with translucent windows in its lower eyelids—possibly a precursor to the "spectacle" covering the eye of snakes. Like snakes, too, *Lanthanotus* have teeth on their palatine and pterygoid bones, a hinge in the middle of the lower jaw, no external ear openings, and they have forked tongues.

Unfortunately, virtually nothing is known about the natural history of *Lanthanotus*. We do know that *Lanthanotus* burrow and are aquatic, and they seem to prefer cool moist habitats. A number were collected after severe flooding in Sarawak in 1963 (Harrisson 1963; Sprackland 1970, 1972), possibly having been flushed out from underground retreats, and one was captured "while hiding

Little is known about ecology of the earless monitor, *Lanthanotus borneensis,* from Borneo. (Alain Compost, with thanks to Frank Yuwano)

in a cave" at 8 A.M. The only eyewitness account of *Lanthanotus* in the wild states: "Sometime in the month of July 1961, while burning the area for the farming season, I suddenly saw two lizards come out from the earth, walking through the hot ashes. The bigger lizard followed about two feet behind the smaller. I quickly caught it: it struggled and coiled its tail. But it soon died on the spot. I caught the smaller lizard and took them both home to our long-house" (H. Harrisson 1961). *Lanthanotus* dig tunnels in banks along water and retreat to the water when threatened (Robert Murphy, pers. comm.). Earthworm setae have been found in stomachs of some presumably wild-caught individuals (Greene 1986).

Some of what we know about *Lanthanotus* derives from observation of these lizards in captivity. Such individuals seem to prefer relatively low ambient temperatures of about 24–28°C; they are sluggish, spending most of their time lying in water, seldom moving; they have eaten squid, small bits of fish, and liver. In captivity, skins are shed infrequently, less than once per year. Several reports of captives suggest that *Lanthanotus* could be nocturnal (Proud 1978; J. Arnett, pers. comm.). Such observations, however, could be mere artifacts of the unusual environmental conditions in captivity and hence largely irrelevant to free-ranging wild lizards.

A fossil lanthanotid from the Cretaceous (75 M.Y.B.P.) found in the Gobi Desert, Mongolia, was described and named *Cherminotus* (Borsuk-Bialynicka 1984); this creature is similar to *Lanthanotus,* suggesting that lanthanotids may have changed relatively little over a vast time period. However, these putative affinities were questioned by Kequin and Norrell (2000). *Lanthanotus* may be a "living fossil" (like the tuatara *Sphenodon*) that could provide valuable insights into our understanding of the ancestor of all anguimorphan lizards (anguids, xenosaurids, helodermatids, lanthanotids, and varanids) and perhaps of snakes as well. A field study of the natural history and ecology of *Lanthanotus* is greatly needed.

HELODERMATIDAE

The two species of *Heloderma* (= nail skin, for scales that are like nail heads) are the world's only truly venomous lizards. (Three fossil genera of extinct helodermatids are also known.) Gila monsters, *Heloderma suspectum,* are found primarily in the Sonoran Desert of the southwestern United States and Sonora, Mexico. These are large lizards,

The beadlike skin of the Gila monster, *Heloderma suspectum,* caught the interest of some early Native Americans. (Laurie Vitt)

from 300 to 500 mm in total length. Beaded lizards, *Heloderma horridum,* are larger, usually about 700 mm but occasionally reaching nearly a meter in total length. They occur at low elevations along the extreme western coast of Mexico from southern Sonora to Chiapas, with one isolated subspecies occurring in southern Guatemala (Campbell and Lamar 1989). In Chiapas, beaded lizards extend inland to near the crest of the continental divide. Fossils from Nebraska, Wyoming, and Colorado indicate that helodermatids once had a much wider distribution (Pregill et al. 1986). Beaded lizards are more streamlined in morphology, and have longer tails than Gila monsters. They are also aptly named: their scales are beadlike, rounded and convex upward; they do not overlap and are underlain by bony osteoderms (also the case in Gila monsters). Belly scales are rectangular.

Most *Heloderma* occur in areas with substantial warm-season precipitation, which probably translates into an adequate food supply. Gila monsters live mainly in dry scrub habitats of the Sonoran Desert, but also occur in the Mojave Desert of northwestern Arizona, southwestern Utah, and southeastern California as well as far western portions of the Chihuahuan Desert in southeastern Arizona and southwestern New Mexico. Beaded lizards occur primarily in dry forest of western Mexico. It is not known whether the two species occur together in southern Sonora, where the two habitats interdigitate.

Gila monsters spend more than 95 percent of their time underground in rocky shelters or mammal burrows (Beck 1990). During spring and early summer—the driest time of year in the Sonoran Desert—they are active in the morning. Later in summer they are also often nocturnal on warm nights, and are occasionally active after summer thunderstorms as well. Beaded lizards are also active in spring and summer (likewise the driest season), with peak activity during May, when they move on average more than 200 m during each one-hour bout of terrestrial activity (Beck and Lowe 1991). Their retreats include cavities in banks of small drainages and at the base of tree trunks, though they sometimes remain inactive in crevices in trees.

Both species are widely foraging lizards and use their large forefeet and sharp claws for digging. While active on the surface, they maintain body temperatures of about 29–30°C. *Heloderma* feed on eggs of ground-nesting birds (they also climb up into bushes and trees for bird eggs), juvenile rodents captured in nesting burrows, as well as eggs of various snakes and lizards. Beaded lizards frequently eat eggs of the large iguanine lizard *Ctenosaura*. Adult lizards are sometimes eaten, but only rarely (these are probably captured in burrows at temperatures below their normal activity). *Heloderma* possess acutely developed olfactory abilities—using their tongue and vomeronasal system, they locate and dig up hen's eggs buried 15 cm deep! Once they detect potential prey with their chemosensory system, they initiate more intense searching behavior, which includes an elevated frequency of tongue flicking (Cooper 1989; Cooper et al. 1994).

Helodermatids can swallow very large prey. Small mammals may be grasped and crushed to death (perhaps with injection of venom), while small rabbits are often swallowed alive. Prey are usually swallowed head first, a series of muscular contractions and bending of the head and neck essentially forcing food items down the throat. Often prey are temporarily blocked by the pectoral girdle, but downward flexing of the neck forces the prey through the esophagus and the pectoral girdle and into the stomach. After a large prey item has been swallowed, the lizard immediately reinitiates tongue flicking and search behavior. Most likely, this behavior results from a long history of finding clumped prey such as eggs and babies in nests of mammals, birds, and reptiles.

Heloderma can withstand extended fasts during long periods of dormancy because of low metabolic rates and availability of stored fat in their tails (which do not autotomize and are not regenerated if broken). Spring and early summer activity may take maximum advantage of availability of reptile and bird eggs as well as juvenile

mammals. A spectacular example of these lizards' capacity to go dormant took place in Arizona during the 1960s, when an instant city called "Sun City" was built in the Sonoran Desert. For a couple years this small town received no appreciable rainfall, and residents irrigated their lawns and trees with groundwater pumped from deep below. When massive August rains finally did fall, Gila monsters began popping out of the ground in people's yards: they had been buried underground, inactive, living off stored fat in their tails during the entire first few years of the city's existence!

Heloderma have long recurved, grooved teeth, which constitute a surprisingly effective venom delivery system. Venom glands are modified salivary glands in the lower jaw. Venom flows along grooves in their teeth by capillary action up into whatever they are biting and chewing (whether prey or enemy). Unlike venomous snakes, which have control over venom injection and often deliver venomless "dry" bites, helodermatids nearly always deliver venom, at least during defensive bites as they hang on with a tenacious bulldoglike grip (Lowe et al. 1986). When cold, *Heloderma* move sluggishly, but when they are hot they can move with astonishing speed, slashing out sideways to bite. Unless molested, these lizards are not dangerous to humans, but they should be treated with great respect. When disturbed, they often hiss with their mouth open and turn sideways to present maximum body area toward an intruder.

Like many other poisonous or dangerous animals, Gila monsters exhibit warning coloration, being brightly colored with pinks, reds, oranges, and yellows. Mexican beaded lizards are not as brightly colored, being black with yellow spots and yellow bands around their tails (though some are uniformly black). Color patterns are more vivid and contrasty in juveniles, becoming paler and more reticulated as a lizard ages.

Helodermatids shed their skin in relatively small patches, such that individual animals seem always to be in the process of shedding. Most likely, a complete shedding sequence occurs only once per year (Bogert and del Campo 1956).

Little is known about the social behavior in helodermatids. However, Gila monsters and beaded lizards observed in the field do engage in male-male combat (Beck and Ramírez-Bautista 1991). The dominant male straddles the subordinate, lying on top of it and holding it with its front and hind limbs. Both lizards arch their bodies, turn

Figure 13.1 Male-male interactions in the beaded lizard, *Heloderma horridum,* involve considerable physical contact in what appear similar to wrestling matches. (Illustration by Kim Duffek, courtesy Daniel Beck)

heads and tails away from each other, and push against each other. In beaded lizards, this results in a full body arch, at which point the dominant lizard forces the other to the ground on its back, collapsing its arch posture—the apparent goal of such an encounter (fig. 13.1). The dominant lizard may bite the upside-down subordinate on its jaw. Male-male combat in Gila monsters is similar, but the body-arch position is never achieved. As a consequence, beaded lizard combat resembles more closely combat in *Varanus.*

Although *Heloderma* have one of the lowest recorded sprint speeds among lizards, aerobic scope values are among the highest measured in lizards (Beck et al. 1995). Even though they cannot run rapidly, they can thus engage in intense behaviors for extended periods. Sexual dimorphism exists in aerobic scope as well, with males having higher aerobic capacities than females. Moreover, prolonged combat behavior in males appears to have resulted in selection for metabolically supported activity in males above and beyond that required for normal activity, an example of sexual selection of a physiological trait.

VARANIDAE

Varanids, also known as monitors, are often large. About 50 species of *Varanus* are currently recognized worldwide; all occur in Africa, Asia, Southeast Asia, New Guinea, and the Australian region, though varanoid fossils are known from North America. New species of *Varanus* are still being discovered in Australia and Southeast Asia.

Monitors live in a wide variety of habitats, ranging from arid deserts to savannas to dense forests to mangrove swamps. Some species are aquatic or semiaquatic, others terrestrial, while still others are saxicolous, semiarboreal, or truly arboreal. Impressive adaptive radiations have occurred in Australia, where about 30 species occur, including a very interesting clade that has evolved dwarfism. The world's smallest varanid is Australia's *V. brevicauda* (about 17–20 cm in total length and only 8–17 g in mass). The largest living species is the Indonesian Komodo "dragon" *(Varanus komodoensis),* which attains lengths of 3 m and weights of 150 kg. Yet even Komodo monitors are dwarfed by a closely related, gigantic varanid, *Megalania prisca* (originally placed in the genus *Varanus*), an extinct Australian Pleistocene species that is estimated to have reached 6 m in length and weighed over 600 kg. Dated at 19,000–26,000 years before present, and hence contemporary with aboriginal humans, *Megalania* probably ate *Homo sapiens.* These large monitor lizards probably evolved large body size in response to availability of very large prey, including rhinoceros-sized diprotodont marsupials for *Megalania* (Rich 1985) and, for *V. komodoensis,* pygmy elephants (Auffenberg 1981; Diamond 1987). At the end of the Pleistocene, however, much of the megafauna went extinct, and without such large prey to support them, large predators also died out. Luckily for varanophiles, *V. komodoensis* managed to survive on smaller prey until the present (deer and pigs that now support these lizards were introduced by humans).

The varanid lizard body plan appears to have been exceedingly successful, as it has been around since at least the

An Australian perentie, *Varanus giganteus,* sampling the external environment with its tongue. (Eric Pianka)

late Cretaceous (80 м.ʏ.в.ᴘ.). *Varanus,* though morphologically conservative, vary widely in size, making this genus useful for comparative studies of the evolution of body size (Pianka 1995). Small body size has evolved at least twice, once in Australia and once in an Asian clade. Large body size has evolved in several lineages of varanids.

Monitor lizards adopt characteristic defensive postures, flattening themselves from side to side and extending their gular pouches, presumably to make themselves appear as large as possible. Often they hiss loudly and flick their tongues. Big species lash their tails like whips with considerable accuracy. Some species stand erect on their hind legs during such displays.

Male monitor lizards engage in ritualized combat, fighting over females. Larger species wrestle in an upright posture, using their tails for support, grabbing each other with their forelegs and attempting to throw their opponent to the ground. Blood is sometimes drawn in such battles. Smaller species grapple with each other while lying horizontally, legs wrapped around each other as they roll over and over on the ground. The victor then courts the female, first flicking his tongue all over her and then, if she concurs, climbing on top of her and mating by curling the base of his tail beneath hers and inserting one of his two hemipenes into her cloaca. (Male varanids have a unique cartilaginous, sometimes bony, support structure in each hemipenis, called a hemibaculum.)

All monitor lizards lay eggs. Clutch sizes vary widely among species, from two to three in the smallest such as *V. brevicauda,* to thirty-five to sixty in large African species like *V. albigularis* and *V. niloticus.* Some monitors *(V. rosenbergi, V. niloticus,* and *V. varius)* excavate termitaria to make their nesting burrows. Termites close off the entrance, sealing the eggs inside in an almost ideal, protected environment of nearly constant temperature and high humidity. Sometimes females return to the termite nest nine months later to open it and free their hatchlings (Carter 1999). In some species of monitor lizards *(V. dumerilli, V. griseus, V. rosenbergi,* and *V. tristis),* hatchlings are much more vividly marked and colorful than adults.

Varanids differ from other lizards in several ways: they have more aerobic capacity and a greater metabolic scope; most range over larger areas; and they appear to be much more intelligent than most lizards. Like many autarchoglossans, monitor lizards have forked tongues, which are used extensively by many species to locate and discriminate among their prey by scent (vision and sound are also used). Most are active predatory species that eat large vertebrate prey (James et al. 1992; King and Green 1999). Many monitor lizards are top predators. Although most varanids are wide foragers, Komodo monitors have secondarily evolved to become ambush predators, lying in wait along trails for large prey such as small deer or wild pigs (Auffenberg 1981; Ciofi 1999). When foraging for smaller prey such as mammals and snakes, *V. komodoensis* forage widely, like most other monitors.

Teeth of many large *Varanus* are curved, with the rear edge serrated for cutting and tearing skin and flesh of prey as these powerful lizards pull back on their bite. Often they grab their prey by a hind leg, severing tendons and hamstringing the prey. Several authors have suggested that *Varanus komodoensis* and *Megalania prisca* are/were ecological equivalents of large saber-toothed cats, using a slashing bite to disembowel large mammals (Akersten 1985; Auffenberg 1981; Losos and Greene 1988). One Komodo monitor killed a water buffalo by ripping a hole in its abdomen, effectively eviscerating the buffalo. Another is reputed to have eaten a small Indonesian boy. A 50-kg female consumed a 31-kg boar in just seventeen minutes. After following Komodos in the field using radiotelemetry for over a year, Auffenberg summed up their ambush technique thus: "when these animals decide to attack, nothing can stop them." He followed one lizard for eighty-one days, during which time it made only two verified kills. Auffenberg himself was attacked by a "maverick" *V. komodoensis,* from which he barely escaped by climbing a tree. Juvenile Komodos are highly arboreal, which may protect them from being eaten by their larger, less agile brethren.

Because of their size, large monitor lizards retain body heat in their nocturnal retreats and can emerge the next morning with body temperatures well above ambient air temperatures. Their mass thus confers a sort of "inertial homeothermy" on them that may have implications for our understanding of the evolution of endothermy (McNab and Auffenberg 1976).

An unusual large climbing monitor lizard, the so-called artrellia or crocodile monitor *(V. salvadorii),* occurs in lowland rain forest habitat in southern New Guinea (Bayless 1998). New Guinea natives consider the artrellia to be an "evil spirit that climbs trees, walks upright, breathes fire, and kills men," and not surprisingly, local hunters are extremely loath to help capture these animals. Characterized by an extremely long tail and a distinctive

During five years of wandering around in the deserts of Australia, I experienced a number of events but once. I only wish I had been able to capture them on film. If one could wander fifty years or five hundred or five thousand, what amazing sights one would see! One of my fantasies is to stand at a single spot for a century to estimate "point diversity." Over such a long time period, one would surely see all species present at that site wander past, not to mention invasions of new species and extinctions of existing ones.

One such once-in-a-lifetime event occurred when a small individual of the large varanid species *Varanus gouldii* attempted to subdue and eat a large individual of the pygmy monitor species *Varanus gilleni*—the first specimen of the latter that I had ever seen in the wild. It was truly an epic battle, two monitor lizards about the same size, tails entwined, rolling over and over in the dust. The *V. gouldii* had the *V. gilleni* by the nape of its neck, and the *V. gilleni* was wrapped around the *V. gouldii*, struggling to free itself.

We came upon this prehistoric scene while driving slowly down a track deep in the Great Victoria Desert 13 km west of Neale Junction. I had recently collated all museum locality data for desert *Varanus* for papers I intended to write, and I was acutely aware that *V. gilleni* was known from only a handful of places in central Western Australia. None had ever been collected from the Great Victoria Desert: this would be about a 400–500-km extension of its known geographic range. Thus, I had to collect this individual as a voucher specimen and a permanent record. However, the fighting monitors were also a sight that I wanted very much to preserve on film. Remaining in our vehicle (monitors usually run from humans on foot but will sometimes hold their ground before a vehicle), my ex-wife, Helen, kept an eye on the two struggling lizards while I frantically tried to dig out the telephoto lens and get it onto the camera. Before I could do so, however, they took off, and we jumped out to follow them. I ran along wringing my hands at losing both the picture and the rare specimen! Luckily, the *V. gouldii* released the *V. gilleni*, and Helen saved the day when she stepped on the stunned *gilleni* and grabbed it. This beautiful maroon and gold lizard, which I quickly photographed, became the first validated record of that species from that part of Western Australia. Years later, I collected two more *V. gilleni* at Yamarna about 200 km west of this locality, documenting the extension of its known geographic range still farther west. *(PIANKA)*

This monitor, *Varanus jobiensis,* occurs in coastal rain forest and mangrove habitats of Irian Jaya in the Indo-Papuan Archipelago.
(Louis Porras)

Some monitors, like this *Varanus pilbarensis* from the Pilbara region of Western Australia, live in rocks. (Brad Maryan)

arched muzzle and bulbous nose, crocodile monitors might be allied with the Australian lace monitor, *V. varius,* also a large climbing monitor. Crocodile monitors use their long prehensile tails as counterbalances when climbing in the canopy, and as effective whips when threatened. Their teeth are very long and fanglike (they probably prey on birds). Ritualized combat, the clinch-bipedal stance, courtship, and copulation have been observed in captive crocodile monitors, which have been successfully bred in Germany several times. Hatchlings are large, measuring half a meter in total length. The largest reliable measurements of adult crocodile monitors had snout-vent lengths of 78–85 cm and tails from 153 to 166 cm, giving total lengths of 231–255 cm, although anecdotal reports of much larger lizards exist.

Some varanids, such as *Varanus niloticus* and *V. mertensi,* are highly aquatic, leaving water only to bask nearby or to dig nest burrows and lay their eggs. Others, such as *V. indicus, V. mitchelli,* and the so-called water monitor, *V. salvator,* are more terrestrial, though they are also strong swimmers, very much at home in the water. Komodo monitors also swim well, which may facilitate colonization of islands.

One of the most devoted students of varanids, Walter

Auffenberg of the University of Florida, spent many years in the field following monitor lizards. He published a monumental trilogy of books providing a vast amount of information about the ecology and behavior of three species: *V. komodoensis, V. olivaceus,* and *V. bengalensis* (Auffenberg 1981, 1988, 1994). We have already alluded to his study of Komodo dragons. Using high-tech radio-telemetry, Auffenberg followed movements and body temperatures of the previously unstudied *V. olivaceus* in the Philippines—and found that this large, rare, arboreal monitor feeds primarily on fruit! His similarly extensive study of *V. bengalensis,* conducted primarily in India and Pakistan, showed that this wide-ranging terrestrial monitor feeds primarily on a wide variety of arthropods (earthworms, crustaceans, and snails) as well as many vertebrates, including their eggs and young.

Varanus griseus is one of the most widespread monitors, living in some of the world's most inhospitable desert regions. They range across the northern Sahara from Morocco through the Arabian Peninsula and into Asia as far east as western India and as far north as the Kyzylkum Desert in Uzbekistan. These handsome golden and maroon monitors are terrestrial; powerful diggers, they eat a wide variety of large prey, primarily other vertebrates.

One of the most arboreal monitor lizards is the beautiful green *V. prasinus* from New Guinea and Cape York, Australia. These small climbing monitors have strongly prehensile tails and spend most of their lives in trees, where they feed on katydids and other arthropods (Greene 1986). Eggs are laid in termitaria.

In the Australian desert, as many as 6–7 species of *Varanus* occur together. All are exceedingly wary, essentially unapproachable and unobservable lizards. Fortunately, however, they leave fairly conspicuous tracks—each species' being distinct (Farlow and Pianka 2000)—and one may deduce quite a lot about their biology from careful study of such spoor. The largest species, the perentie *(V. giganteus),* reaches 2 m or more in total length, whereas some smaller "pygmy goannas," such as the ubiquitous and very important lizard predator *V. eremius,* achieve total lengths of only about 40 cm. Two other species, *V. gouldii* and *V. tristis,* are intermediate in size. Individuals of all four of these species range over extensive areas and consume very large prey items, particularly other vertebrates (especially lizards). Daily forays typically cover distances of a kilometer or more. *Varanus tristis* and two other little-known small semiarboreal species, *V. caudolineatus* and

V. gilleni, have strongly curved sharp claws. Four terrestrial species, *V. brevicauda, V. eremius, V. gouldii,* and *V. giganteus,* have less curved claws.

Varanus brevicauda, the smallest monitor lizard, reaches a snout-vent length of 90–110 mm as an adult, while hatchlings are only about 45 mm SVL and weigh a mere 2–3 g. The smallest male with enlarged testes was 82 mm SVL and the smallest gravid female was 94 mm SVL.

Reaching only about 30 cm in total length, the pygmy monitor from Australia, *Varanus storri,* inhabits rock crevices. (R.D. Bartlett)

The arboreal emerald monitor, *Varanus prasinus,* lives in rain forest habitats of New Guinea, islands of the Torres Strait, and the northern tip of Cape York Peninsula, Australia. (R.D. Bartlett)

Despite their diminutive size, they eat relatively large prey: one female weighing 9.1 g contained a 1.5-ml lizard—an adult *Ctenotus calurus,* constituting a full 16.5 percent of the *brevicauda*'s body weight.

This tiny varanid is seldom encountered active above ground; indeed, the vast majority of specimens have been collected in pit traps. Mark-recapture studies, moreover, show that *V. brevicauda* do not move very far (James 1996). Two were dug up in shallow burrows during August (one must have been active immediately prior to being exhumed, as crisp fresh tail lash marks were at the burrow's entrance and the lizard had a body temperature of 35.4°C, ten degrees above ambient air). Pianka has pit-trapped dozens on a flat sandplain covered with large, long-unburned clumps of spinifex, possibly their preferred habitat (Pianka 1996).

The typical monitor lizard threat posture and behavior has been conserved in the evolution of these tiny monitors, which hiss and lunge with throat inflated as if they are a serious threat. Their tails are very muscular and pre-hensile, and they "hang on for dear life" when inside a spinifex grass tussock.

The Australian *Varanus caudolineatus* is a small, semi-arboreal monitor preferring habitats with mulga trees, which offer small hollows that provide tight-fitting, safe diurnal and nocturnal retreats. While some foraging takes place in trees—stomach contents of one sample contained intact arboreal geckos *(Gehyra)* as well as their tails (Pianka 1969a)—*V. caudolineatus* do descend to the ground to forage as well: three of thirteen active specimens observed in another study were on the ground when first sighted, and one stomach contained a ground-dwelling gecko, *Rhynchoedura ornata,* as well as their tails (*V. caudolineatus* and *V. gilleni* actually "harvest" exceedingly fragile tails of geckos too large to subdue intact), while gut contents of a sample from Atley Station in Western Australia consisted largely of scorpions and ground-dwelling spiders (Pianka 1969; Thompson and King 1995). These monitors forage on the ground searching for prey by going down into their burrows (Thompson 1993). Move-

TRACKING AND CATCHING MONITOR LIZARDS IN AUSTRALIA

Sand constitutes a natural event recorder, leaving a record of what creatures have moved past. Strong winds regularly dull and erase all tracks. Tracks are difficult to see during midday when the sun is high or on overcast days. Morning and afternoon of bright, relatively still days are therefore prime times for tracking, when the sun is low in the sky and shadows are long. Tracks are best seen by looking into the light.

My first experiences with tracking were more than a little frustrating: the animals I was following had made large loops, and the track frequently wound back to its point of origin. Being relatively unskilled, I couldn't determine for certain exactly where a track began or ended. I had followed many sets of one very distinctive track without ever finding an animal, but I was not about to give up: I had to learn the identity of this mysterious beast that left a bold, broad tail

drag mark. It might be a large skink, or perhaps a monitor.

I followed one such fresh track to a fallen, burned-out eucalyptus branch. The track went directly to the hollow end of this small log, but no exiting track was evident. Circling carefully around and around, I determined that, indeed, the lizard had not departed. I picked up the branch and peered in but couldn't see very far because of its curvature. So I plugged the ends with other smaller branches and took my "surprise package" back to camp. Breaking the log apart, I found a large jet-black racehorse monitor lizard, *Varanus tristis,* an arboreal forest dweller, previously unknown in desert habitat. In the end, however, these cryptic lizards turned out to be ubiquitous at all my study sites throughout the Great Victoria Desert, which was a major discovery in itself.

One can easily learn to judge the "run" of a track, that is, where the ani-

mal is headed. It is almost like becoming the lizard yourself. This allows one to move ahead, cutting the track at intervals, to find the lizard rapidly. A track can be aged by its crispness and whether or not other tracks, say those of nocturnal species, cross over it. Nothing is more exciting than finding a crisp new track less than an hour old, for you know that the maker of that track is close by. When hot on the trail, walk as quietly as possible, barely breathing, scouting ahead to spot the lizard.

Tracking large lizards across sandy areas is one of my favorite pastimes. A great deal about wary unobservable species such as *Varanus* can be learned this way. It is an incredible thrill when a track suddenly ends in a magnificent animal, captured in midstride and frozen in time. More often than not, however, before you see it, the lizard breaks into a run and dives down a hole or escapes into a hollow of a tree. *(PIANKA)*

Diggings of a sand monitor, *Varanus gouldii*. The lizard is just behind the upright snag. (Eric Pianka)

ments of *V. caudolineatus* marked with a radioactive tracer are not nearly as extensive as movements observed in other varanids, suggesting that these pygmy monitors may be fairly sedentary (Thompson 1993). Both sexes appear to mature at about 91 mm SVL, the size of the smallest male with enlarged testes and the smallest gravid female. One specimen with an SVL of 111 mm, estimated to weigh 15 g, contained an intact 3-ml *Gehyra* (20 percent of its mass).

Varanus eremius are fairly common in Australian sandy deserts, judging from the frequency of their unique, conspicuous tracks (Pianka 1968). Unlike larger monitors such as *V. giganteus, V. gouldii flavirufus,* and *V. panoptes,* which walk with their legs beneath them and their bodies and tails elevated above ground, *V. eremius* walk in the typical primitive tetrapod stance with their bodies close to the ground, dragging their tail straight behind. They leave a distinctive tail drag mark, which loops around when the lizards turn back.

Although this beautiful little red *Varanus,* active all year long (unlike larger goannas), is extremely wary and very

Eye contact by monitors like this Australian *Varanus gouldii* is so mammal-like that it can present the impression of emotions, reflecting the advanced state of neural integration achieved by these lizards. (Eric Pianka)

seldom seen, a great deal about its activities can be inferred from its tracks. Individuals usually cover great distances when foraging. Fresh tracks often cover distances of up to a kilometer. Tracks indicate little tendency to stay

A terrestrial Australian *Varanus panoptes* in typical monitor walking stance. (Eric Pianka)

within a delimited area, which suggests that home ranges of these lizards are extremely large. These pygmy monitors are attracted to fresh holes and will often visit any digging within a few days after it is made. In contrast to *V. gouldii,* which locate most prey by scent, *V. eremius* do not dig for their prey but rather rely on vision to catch it above ground. Once a *V. eremius* ambushed a small blue-tailed skink *(Ctenotus calurus)* when it came within a few centimeters of the edge of a loose *Triodia* tussock—a few centimeters too close. Sometimes an *eremius* track intercepts the track of another, smaller lizard, with evidence of an ensuing tussle. One such pair of tracks came together, rolled down the side of a sandridge leaving a trail of big and little tail lash marks, and finally became one track, dragging away a fat belly.

Over 70 percent of the *V. eremius* diet by volume consists of other lizards; large grasshoppers plus an occasional large cockroach or scorpion constitute most of the remainder (Pianka 1968, 1982, 1994a). Nearly any lizard species small enough to be subdued is eaten. In a typical foraging run an individual *V. eremius* often visits, and goes down into, several burrows belonging to other lizard species, especially complex burrow systems of *Egernia striata.* These activities could be related to the search for prey or to thermoregulatory activities, or they might simply involve escape responses. Certainly a *V. eremius* remembers exact positions of burrows it has visited, since it almost inevitably runs directly to the nearest one when confronted with the emergency of a lizard collector.

The distinctive tracks of *Varanus tristis,* which consumes other lizards as well as baby birds and probably bird eggs, typically run more or less directly from tree to tree (it climbs them looking for food). Its activity is highly seasonal, and lizards rely on building fat reserves during times of plenty to get them through lean periods (Pianka 1971d). A survey of stomach contents revealed that 70 percent of the diet of *V. tristis* by volume consists of other lizards; large grasshoppers, cockroaches, and lepidopteran larvae are also eaten (Pianka 1994a). The 75 stomachs with food contained 137 large prey items, including 35 lizards from some 11 species: 7 species of skinks *(Ctenotus brooksi, C. colletti, C. grandis, C. helenae, C. pantherinus, C. quattuordecimilineatus,* and *Lerista bipes),* two species of agamids *(Pogona minor* and a small *Moloch horridus),* one gecko *(Gehyra variegata),* and one pygmy monitor *(Varanus caudolineatus).* Most likely, *V. tristis* would eat any lizard species small enough to be subdued.

Varanus tristis also raids bird nests and eats eggs and baby birds (Pianka 1971d). On one occasion, a Galah cockatoo was screeching as if in distress. The bird was on the ground, its crest held high and wings partially outstretched. It flew up onto a fallen log under a marble gum tree, and then into the tree—which proved to be its nesting tree. A large *V. tristis* was clambering over the same log toward the tree. The cockatoo continued to screech, and then began to harass the lizard. When the lizard had climbed about 3 m up the tree, it disappeared from view around the other side, whereupon the Galah attacked and actually drove the monitor back down the tree. The bird's mate was also present. These large, climbing, predatory lizards must constitute a potent threat to hole-nesting parrots.

Based on enlarged testes of males and yolking ovarian eggs of females, both sexes of *V. tristis* appear to achieve sexual maturity at about 200 mm SVL (Pianka 1994a). Individuals are most active during the austral spring, October through November, when mating occurs. Clutch sizes are large, from 5 to 17 eggs, with a mean of 10.1. The relative clutch mass of eleven females with eggs in their oviducts averaged 16.2 percent of female body weight.

TRACKING DOWN DRAGONS

By far the most impressive desert lizard in Australia is the enormous—and exceedingly unapproachable—perentie, *Varanus giganteus,* which attains a total length of more than 2 m. Highly prized as food and hence sought after by Aborigines, these large mammal-like predators have evolved mammalian-level intelligence. I once found a perentie track that intercepted my own track then turned immediately back on itself, suggesting that these lizards possess incredible olfactory sensitivity in addition to acute intelligence. Their food must once have been small wallabies and mid-sized marsupials, many of which are now extinct. Now perenties feed largely on introduced European rabbits (though when an assistant and I flushed the stomach of one, we found that it had eaten another monitor lizard, a large *V. gouldii*).

I wanted very much to see, and to photograph, a perentie during my first stay down under, but couldn't seem to find one. One day a station owner told me that, just a few weeks before, his dogs had had one at bay, which he had killed down in the south paddock, very close to one of my study areas. When I asked why he had killed it, he said that "perenties were simply vermin." I insisted that he take me to the spot; its battered head seemed worth taking to try to salvage a skull for a specimen. Its long, sharp, serrated cutting teeth were most impressive.

While in Australia, I wanted to become familiar with as many different species of lizards as possible. In 1968, on the last bush trip of my first great expedition down under, Helen and I therefore decided to do some incidental collecting on a rocky area (or "tor") with granitic outcrops—a relatively rare feature in the sandy desert, but a good habitat for lizards. Sitting in our Land Rover Matilda away from the bushflies (I had fitted screens to some windows), we ate our standard lunch of Vitawheat crackers and cheddar cheese with Vegemite, washed down with a cool drink. After a quiet lunch I got out of the car to head toward the rocky outcrops behind. As I came around the back I saw, just a few meters away, the horselike head, long neck, and shoulders of a huge perentie! Instantly this enormous lizard dove, disappearing down a hole beneath a small tor. This was the first live perentie I had seen. Naturally, I had to try to capture this magnificent beast. I spent the better part of a week doing just that, but came up empty-handed.

Tracks of this big lizard, as well as several more deep burrows, were all around the outcrop. The first thing I tried to do was smoke the lizard out. I shoved a long length of dynamite fuse as far as it would go deep down into the hole and lit the end: whoosh! Sputtering and smoking, great wisps of acrid smoke poured out. But no lizard. Next, I decided to try to capture the perentie

using nylon snares. Dozens of nooses soon festooned all the major entrances and exits underneath the rock. We then moved half a kilometer away so as not to disturb the perentie. After a day had passed I crept up to the tor to check, and found my nooses undisturbed, exactly as I had left them. The big lizard had not budged. I left quietly and returned the next day: same thing. On the third day, I found one of my nylon nooses frayed and broken. The perentie had been snared but had rubbed the nylon line against the rough granite until it broke.

The next morning very early, I crept up to the tor and positioned myself a few meters above the main entrance. For several hours nothing moved except arthropods and an occasional small lizard. There wasn't a sound, except for the ever-present bushflies. Finally it grew hot. By about noon my legs were going to sleep and I was thirsty, stiff, and cramped, so I slowly got up and moved toward the hole. As I did so, I heard a loud noise: the perentie—looking to be a full two meters long—was charging right at me, diving for its hole. In an instant, before I could respond, the big lizard had gone right between my legs. It must have emerged from another entrance.

The following morning I tried again, creeping slowly behind bushes up toward the perentie's lair. From a few *continued on next page* ▶

Eggs are laid in underground nests dug by females in October and November, and hatchlings emerge in February (incubation time is short, probably around 114–117 days). Hatchling snout-vent lengths are about 72–73 mm, with weights about 4.3 g. Hatchlings have a checkered pattern, quite different from that of adults (Thompson and Pianka 1999).

Varanus tristis was studied in the Great Victoria Desert using radiotelemetry (Thompson and Pianka 1999; Thompson, de Boer, and Pianka 1999; Thompson, Pianka, and de Boer 1999). Temperature-sensitive transmitters were surgically implanted in twelve lizards, which were then released at the site of capture. They were followed daily for varying periods of up to six weeks. One female dug an oviposition burrow and laid a normal clutch of eggs while carrying a transmitter internally. Males moved

▶ *continued*

hundred meters away, I scanned the outcrop with binoculars and saw the magnificent beast, head held high, surveying its kingdom. But as I tried to sneak closer, it vanished down a hole. I never saw that perentie again.

A decade later, in early 1979, I came upon a large perentie's track at a study area far from rocks. Surprised, for I had thought these animals never strayed very far from rock outcrops, I knew I simply *had* to catch this lizard. In the end, the job took over two weeks. Following its tracks that first day, I quickly found where the lizard had first spotted me and broken into a run, heading directly to a large burrow in the bank of a sandridge. I tried snares to no avail and then sat all day long for several days, waiting in ambush, but the lizard never left the burrow.

After waiting patiently for an opportunity to ambush the perentie, I decided I had no choice but to dig the big lizard out. Using a mirror, I learned that the tunnel went straight into the sandridge, then turned sharply to the right. After measuring the distance to the turn with a stick, I dug a pit about a meter deep, and found that the burrow turned sharply again. Making a second stick measurement, I then dug an adjacent pit and peered with the mirror into the tunnel beyond the second turn. It plowed ahead another couple of meters before turning again. So I dug still another pit, deeper this time, down to the burrow, where the

mirror inspection was repeated two or three more times. Finally, hours later, I was surprised to find that the tunnel turned abruptly upward, heading directly toward the far side of a large clump of shrubs. The side of the sandridge was by now covered with large piles of red sand next to several-meter-deep holes (these "digs" were still evident over a dozen years later). Climbing out of my pit, I went around to the place the tunnel seemed to be heading and found tracks: the perentie had exited via a back door!

I followed that perentie's tracks again another kilometer or so along the same sandridge to another burrow system, this time a rabbit warren. Again I set snares, but this time wire ones. I caught a rabbit, but no perentie. The next time the perentie moved, it took off cross-country, going about 3 km in yet another direction straight to a large burrow under a small bush. After I had convinced myself that it definitely had not left this third burrow, I set another batch of snares. For a full week I checked the nooses daily, but the big lizard did not move. At last, I decided once again that the time had come to try to exhume the lizard: another daylong effort, shining mirrors, poking sticks, and carefully excavating. Finally, after following the tunnel for almost 10 m with a series of pits, I spotted the perentie's yellow-tipped black tail, only a meter away. Carefully, I resumed digging. When I reached its

hind legs, I hog-tied them together (I wasn't about to take any chances at this point!). I then put on a long-sleeved shirt and leather gloves in case the perentie got aggressive and fought back. At last I had the great beast.

Back at camp several kilometers away, I chilled the perentie down overnight in a wet cloth bag, and the next day, after weighing and measuring him (he had everted his hemipenes, allowing me to determine his sex), I photographed him and finally released him. But he would not make a break for it, and just lay as if dead, even after he had warmed up, under the shrub where I put him. Eventually I left him there playing possum. A few hours later I returned to find him gone: and his tracks headed right back toward where I had exhumed him. I got my camera and followed the track to take pictures of his trail. Just for the heck of it, I tracked him down again, this time catching him above ground away from the burrows. He had traveled about 2 km from the point of release in a few hours. Following an abortive attempt to outrun me, he went into a state known to physiologists as "oxygen debt" and held his ground at bay, this time inflating his throat and hissing, flicking his great long tongue at me. A great photo op—after a few more pictures, I finally left the great beast to his own devices. With his tail, this lizard measured 1.64 m in length. *(PIANKA)*

A laid-back *Varanus gouldii* sprawled out in typical resting posture. (Eric Pianka)

more often and farther than females, though they also spent time in the same hollow trees with females. Tracks overlaid other tracks, probably a sign of males following females or their scent trails. One male traveled 723 m in an almost straight line directly into the wind and was found the next day in a hollow, dead but upright marble gum tree with a female; most likely he had used an airborne scent trail to find her.

When foraging, Australian *Varanus gouldii* hunt by smell, swinging their long necks and heads from side to side, constantly flicking their long, forked, very snakelike tongues, searching for scent trails over as big an arc as possible. Upon detecting a scent signal, these monitors follow the trail to the source, usually a burrow, and dig up the intended prey. Digging is methodical, using the forearms and sharp claws of the forefeet, with the pointed snout, mouth, and sharp teeth right in between, ever ready to snatch up prey as it dashes to escape. Geckos are important prey (dug up in their dead-end diurnal retreats), but many diurnal species of lizards are also eaten (probably any they can catch), including *Ctenophorus, Ctenotus, Lerista, Lialis, Menetia, Moloch, Pogona,* and even other *Varanus,* such as *V. brevicauda, V. caudolinea-*

tus, V. gilleni, and *V. gouldii*; they also eat reptile eggs, baby mammals, and baby birds (Pianka 1970b, 1994a). Among specimens of *V. gouldii* examined, the largest relative prey mass was a *Pogona minor,* estimated to weigh about 25 g—about 13.9 percent of the varanid's 180-g mass.

Helodermatids, varanids, and *Lanthanotus* now occupy only a portion of their historical geographic ranges. Although the historic diversity of these three advanced lizard groups will never be fully appreciated, taxonomic and ecological diversity of *Varanus* species attests to their ability to persist and diversify even as radical changes have occurred across the planet. Whereas in Australia habitats have remained fairly stable for most *Varanus* and they are protected, in Africa and Asia skin trade poses a threat. Helodermatids and *Lanthanotus* have been less fortunate. *Lanthanotus* will unquestionably soon join the ranks of its extinct ancestors as deforestation of Borneo robs it of its last outpost. Meanwhile, habitat for Gila monsters is rapidly diminishing not only as urban sprawl and agriculture eliminate their habitat, but also as redirected water threatens to permanently affect the hydrological cycles that support the Sonoran Desert vegetation.

Chamaeleo chamaeleo overlooking massive
habitat destruction of the Riff Mountains
in Morocco. (L. Lee Grismer)

Part Three

SYNTHESIS

CHAPTER 14

HISTORICAL PERSPECTIVE

The arboreal teiid *Kentropyx striata* is common in Amazon savanna habitats. (Laurie Vitt)

Taken in its entirety, the evolutionary history of lizards is among the most fascinating examples of natural selection in action among vertebrates. Key evolutionary innovations caused several explosive adaptive radiations leading to today's lizard and snake fauna. It began with lizardlike ancestors, the lepidosaurians, giving rise to rhynchocephalians, which lacked a male sex organ but had many features that led to the success of squamates. Basic lizard morphology was already established, with well-developed optic systems for detecting moving prey; prey capture by tongue prehension; and rudiments of a potentially powerful

This *Uroplatus phantasticus* from Madagascar mimics dead leaves. (Bill Love)

chemosensory system including taste, olfaction, and vomerolfaction.

At the same time, world environments contained a rapidly diversifying invertebrate fauna that had responded first to the earlier, remarkable diversification of gymnosperms and then vascular plants and second to invertebrate diversification itself. Spiders, for example, responding early on to the diversification of arthropods, underwent their own adaptive radiation, such that today virtually all spiders are predators, mostly on arthropods and secondarily on small vertebrates.

The first lizards no doubt entered a world teeming with available arthropod prey, and they were well equipped to take advantage of those resources. Within lizards, at least five spectacular events occurred: (1) diversification within Iguania, (2) the profound evolutionary shift that produced Scleroglossa, (3) diversification within Gekkota, (4) diversification within Autarchoglossa, and (5) evolution of snakes from within Autarchoglossa (likely within Anguimorpha). At various junctures in this evolutionary history, limbs were highly modified or lost repeatedly and tongues changed both shape and function. Both eyes and external ear openings were reduced or lost again and

again, with visual signals now augmented or replaced by detectable thermal and chemical signals. Some lizards developed the ability to detect and use ultraviolet light in social communication. One group, snakes, lost their limbs, eyelids, and ear openings and reduced their eyes as their bodies elongated for subterranean existence; then they resurfaced and rebuilt their eyes to become one of the most successful groups of terrestrial vertebrates. The key innovation enabling snakes to return to the surface may have been the subterranean banquet they found in narrow channels of social insect nests, which set snake ancestors free from physical constraints of living in friable soils because tunnels already existed and food was highly concentrated. Many snakes came back to haunt descendants of their lizard ancestors as highly specialized and efficient predators on lizards or their eggs. Some snakes developed a sixth sense, infrared heat sensors mapped directly onto their optic systems, allowing detection of endothermic prey in total darkness. Others reverted to the iguanian sit-and-wait mode of foraging, exploiting their heightened chemosensory ability to locate ideal ambush sites. Although snakes may be "mysterious" (Greene 1997), they aren't nearly as diverse as lizards.

To appreciate this remarkable history, we first examine aspects of the fossil record. We review numerous morphological, behavioral, physiological, and ecological aspects of lizards within the context of their evolutionary history. We examine the evolution of major lizard clades and appraise diversification within each. Finally, we speculate on global scenarios and suggest a working hypothesis for evolution of diversity within lizards.

Because of their small size, delicate bones, and terrestrial habits, lizards fossilize poorly and relatively few have been described. Most lizard fossils are mere bits and pieces, but some relatively complete fossils have been found. Five late Jurassic (about 150 M.Y.B.P.) fossil lizards represent ancient extinct lineages; because these are scattered across the phylogenetic tree, we can conclude that much early evolution into major clades (Iguania, Gekkota, Scincomorpha, Anguimorpha) had occurred by the end of the Jurassic (Estes 1983). The common ancestor ("stem group") of all lizards must have existed during the Upper Triassic to the Lower Jurassic (about 200 M.Y.B.P.), but no fossils have yet been found (Carroll 1988). Fossils of two lizardlike offshoots from this stem branch, Paliguanids and Kuehneosaurids, from the Lower Triassic of Australia and South Africa (Molnar 1985; Frey et al. 1997), suggest that lizards originated in the part of Pangaea destined to become Gondwana. When Pangaea broke up 200 m.y.b.p., some lizard ancestral stocks had also established themselves on the plate later destined to become Laurasia. No true lizards existed yet.

A fossil from 125 m.y.b.p. found in China is thought to be an ancient offshoot of Gekkota known as an ardeosaurid. A late Cretaceous fossil site in Mongolia contains many spectacularly well preserved dinosaur and lizard fossils (perhaps the animals were buried by sandstorms?). An 80-million-year-old fossilized skull of a varanoid from this site (named *Estesia mongoliensis* in honor of Richard Estes) looks very much like a modern-day *Varanus*, demonstrating that the very successful varanid body plan has been around for at least that long. *Estesia* may have been venomous, as it had grooved teeth like those used for

Fossilized skull of *Estesia mongoliensis*, 80 million years old, from Mongolia. (From Norrell et al. 1992; courtesy of the American Museum of Natural History)

venom conduction in *Heloderma;* however, the grooves are broader in *Estesia,* perhaps suggesting a different function. *Estesia* probably raided dinosaur nests and ate dinosaur eggs and neonates, among other things.

Fossil scleroglossans reveal that an explosive diversification occurred during the Cretaceous. Whereas North American teiids became diverse and widespread (Gao and Fox 1991, 1996; Nydam 2000), Asian teiids were restricted to the Polyglyphanodontinae from the Upper Cretaceous of Mongolia (Sulimski 1975). Evolutionarily, teiids appear to have originated with relatively simple dentition consisting of unicuspid, conical teeth similar to those in gekkotan lizards (Winkler et al. 1990). By mid-Cretaceous, several distinct teiid genera had appeared. Some, such as *Socognathus* and *Sphenosiagon,* retained unicuspid, conical teeth; others, including *Chamops* and *Leptochamops,* had typical teiid teeth with bicuspid or tricuspid crowns oriented anteroposteriorly; yet others, such as *Peneteius* and *Polyglyphanodon,* had teeth transversely oriented with bicuspid crowns (Nydam 2000). Although all these went extinct in North America, some persisted in South America, but they did not exhibit the tooth diversity seen in fossil North American teiids. By the end of the Cretaceous, migrants from South America had replaced extinct teiids in North America, with reinvasion being restricted, however, to two genera, *Ameiva* and *Cnemidophorus,* of which only *Cnemidophorus* reached northern temperate latitudes. High tooth diversity in present-day South American teiids (e.g., *Crocodilurus, Dracaena*) is therefore a relatively recent event.

Other scleroglossan lizards were evident in the Cretaceous as well. Cordylids, for example, are known from the late Cretaceous in Madagascar (Gao 1994), and one helodermatid-like platynotan is known from the early

Fossil of an ancient gekkotan from China, *Yabeinosaurus tenuis* (Ardeosauridae), 125 m.y.b.p. (Courtesy of the National Museum of Natural Science, Taiwan)

Cretaceous in Utah (Cifelli and Nydam 1995; Nydam 2000). Because both are highly derived, we know that scleroglossan evolution had been well under way much earlier. Mosasaurs, large marine varanoids, are the best-known Cretaceous lizards, partly because they left such dramatic fossils. They have been extinct since the end of the Cretaceous (Estes 1983).

Even snakes had begun diversifying by the mid-Cretaceous. The recent discovery of *Coniophis* from the Cretaceous of Utah indicates that snakes likely occurred on Gondwana before North America drifted to the northwest (Gardner and Cifelli 1999).

Another spectacular lizard fossil was found from the Miocene (about 20 M.Y.B.P.) of East Africa: a chameleon's head encrusted inside a calcite crystal (Estes 1983)! Yet another is a complete *Succinilacerta succinea* (Lacertidae) embedded in Baltic amber that dates back 40 million years to the middle Eocene (Borsuk-Bialynicka et al. 1999).

EVOLUTION AT HIGHER TAXONOMIC LEVELS

To appreciate the dramatic sequence of events that led to present-day lizard diversity, we need to examine lizard history on two levels: (1) that of broad patterns, reflecting roots deep in the higher taxonomy of lizards, and (2) that of more detailed patterns—or perhaps more accurately, deviations from patterns—reflecting more recent evolutionary events. We therefore begin by examining evolution of higher taxa but will return to what might appear to be exceptions.

Let us start again with *Sphenodon,* the only living genus of Rhynchocephalia, the sister taxon to Squamata. *Sphenodon,* though lizardlike in overall morphology, differs from modern lizards in several important ways. *Sphenodon* lives in cool environments and operates at low body temperatures. Many aspects of its life history are therefore extended because of a short growing season. Whether *Sphenodon* ancestors differed dramatically from this ecological profile remains unknown.

Sphenodontids (tuatara and their fossil allies) had large adductor jaw muscles and a powerful bite, but a limited gape and slow speed of jaw closure. The rigid tuatara skull with its two temporal arches (diapsid) has a fixed quadrate bone. Iguanian skulls are much like those of *Sphenodon* except that the lower temporal arch is absent and the posterior margin of the lower temporal fenestra has a ro-

tating quadrate bone, which articulates with the temporal arch above it. This new hanging jaw setup (known as streptostyly) increased both gape and speed of jaw opening and closure (Carroll 1988; Smith 1980). With evolution of new musculature, mechanical advantage and biting force were also increased (Smith 1980, 1982). Streptostyly was probably the key innovation that allowed lizards and snakes to diversify by developing new feeding strategies and a wide variety of dietary specializations. Scleroglossan skulls, though less robust than iguanian skulls, are more flexible, and many are highly modified for fossoriality. The temporal region is narrower due to reduction or loss of the upper temporal fenestra. Many scleroglossan lizards have two additional points of potential flexibility in their skulls, a condition known as cranial kinesis or mesokinesis (fig. 14.1). In lizards with this ability, the upper jaw bends and better conforms to prey, presumably enhancing feeding success and allowing capture of more agile prey. Streptostyly not only opened up new possibilities for prey capture and handling but, when combined with other changes that occurred in scleroglossans, allowed some lizards to respond evolutionarily to important habitat shifts (e.g., development of fossoriality) not once but multiple times.

Iguanians (3 families and about 1,230 species) retained numerous ancestral states—again, shared with tuatara—including a fleshy tongue used to capture small prey (a "feeding-type tongue"). For all practical purposes, the tongue simply was ejected from the mouth, pressed on a

An Australian pygopodid, *Lialis burtonis,* eating a gecko. Notice how cranial kinesis allows its jaws to wrap around its prey.

(Steve Wilson)

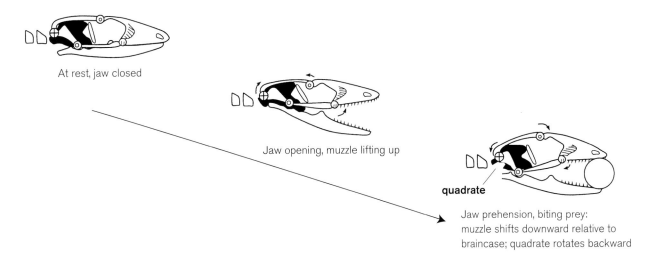

At rest, jaw closed

Jaw opening, muzzle lifting up

quadrate

Jaw prehension, biting prey:
muzzle shifts downward relative to
braincase; quadrate rotates backward

Figure 14.1 Lizard mesokinesis. Three points of flexibility in lizard skulls allow the upper jaw to bend and better conform to prey, presumably enhancing feeding success. (Redrawn from Frazzetta 1983 by permission of the Museum of Comparative Zoology, Harvard University)

prey item, and retracted, pulling the prey into the oral cavity, a process known as tongue prehension or lingual feeding. Most iguanian tongues are short and rounded off at the tip, but some are as much as twice as long as the lizard's body. Even those with long tongues retain lingual feeding. Chameleons took lingual feeding to its logical extreme, evolving a sophisticated ballistic tongue-projection apparatus.

Although iguanian olfactory abilities remained relatively rudimentary (with the exception of herbivorous species, a point to which we will return), they did include both nasal olfaction and vomerolfaction, as in *Sphenodon*. For these sit-and-wait foragers, however, visual cues were most important for detecting prey. Some species, such as chameleons, perfected optic prey detection by evolving telescopic vision to magnify distant prey, whereas others, like anoles, evolved mechanisms for detecting even the slightest movement of prey. Over time, this method of finding food caused many iguanians to become even more sedentary than their ancestors.

Because sit-and-wait foragers must remain in one spot for extended time periods, natural selection favored background matching and disruptive patterns and colors to reduce detection by potential predators. As a group, therefore, iguanians are among the most cryptic of lizards, equaled only by some geckos (also sit-and-wait predators). Such ambush predators watch for moving prey but do not detect sedentary prey. Because iguanians sit and wait for

prey, they forage and bask for thermoregulation simultaneously. Autarchoglossans, which move while searching for prey, must cease foraging to bask. Autarchoglossans in deserts may bask very little because sun availability and environmental temperatures are high. Tropical autarchoglossans, however, spend considerable time basking to gain heat.

Finally, iguanian social systems center on territoriality, defense of well-defined pieces of real estate. Although territoriality is a logical evolutionary pathway for cryptically colored lizards that remain in one spot for extended periods waiting for prey, it is not, however, entirely restricted to sit-and-wait foraging species.

This suite of characteristics—sit-and-wait foraging, camouflage, territoriality—has many implications, some of which have never been explored. For example, predaceous iguanians should, given these characteristics, be microhabitat specialists. And indeed, natural selection has led to a correlation between preference for a given microhabitat patch and fitness within that patch type. Just imagine an iguanian like *Stenocercus fimbriatus* that resembles dead leaves perching on a large green leaf!

Crypsis in iguanians also releases them from some predation sources, particularly predators that detect prey with vision, thus favoring evolution not only of morphological diversity but also of reproductive options. Because lizards that remain motionless and match their backgrounds can be more robust, they can also hold a larger

The Brazilian tropidurine *Stenocercus fimbriatus* is nearly invisible on dead leaves. (Dante Fenolio)

clutch without costs of reproduction associated with motion. Thus some horned lizards produce a clutch of eggs that weighs almost as much as the female. At the other end of the morphological spectrum are anoles that make maximal use of arboreal habitats, including grass stems. Their streamlined morphology may be partly responsible for a fixed clutch size of a single egg, which not only reduces the female's weight but also interferes minimally with her ability to match a narrow perch. To offset production of a single egg, anoles produce many eggs in rapid succession throughout the breeding season.

From a higher taxonomic perspective, then, let us summarize: Iguanians, as sit-and-wait foragers, are masters of crypsis that detect their prey visually and are supported by low to moderate metabolic rates and a wide range of active body temperatures. They live in a mosaic of defended territories, each individual with its own piece of real estate and in many cases, because they slice their territory along different niche axes, not even recognizing that

other species of lizards live in the same place. Many iguanians remain close to a site safe from predators. When crypsis fails, explosive bursts of short-range activity provide access to nearby refuges. Because making it to a refuge determines escape and exercise recovery times can be extended with minimal additional cost, little selective pressure exists for increased endurance. Nevertheless, some iguanians in open habitats, such as *Callisaurus* and long-legged *Ctenophorus,* can run, zigzagging moderately long distances.

Morphological ornamentation in sexual dimorphism reaches its peak in iguanians. The same logic applies here. Extreme ornamentation, such as the horns on male Jackson's chameleons, doesn't interfere with locomotion because they move very little, and it may actually enhance crypsis by helping to break up a lizard's body shape, which a predator might otherwise use as a cue for detection. Such ornamentation in an active autarchoglossan would incur high costs.

At some critical point in evolutionary history, as yet unknown selective pressures produced the ancestor to scleroglossans, and lizard evolution took off explosively in a brand new direction. This is the deepest split in the phylogeny of lizards. Ancestors to scleroglossans used their jaws rather than tongues to capture prey, a process known as jaw prehension (Schwenk and Throckmorton 1989). Although this may seem like a rather minor change, its ramifications cannot be understated. Jaw prehension freed scleroglossans to evolve new, harder tongues that could be devoted to chemoreception for prey discrimination, among other things. This shift from visual prey detection facilitated exclusion of noxious or toxic prey and allowed choices based on prey quality. Morphological ornamentation as a mode of sexual dimorphism now disappears, though head and body size dimorphism as well as sexual differences in coloration persist in some groups.

Coincident with divergence in skulls within ancestral lizards, therefore, was divergence in tongue morphology, sensory receptor systems, feeding mechanics, foraging mode, and a host of other attributes. As Scleroglossa diverged from Iguania, it underwent its own split into Gekkota and Autarchoglossa. Gekkotan tongues were a compromise between feeding-type tongues of iguanians and more highly developed chemosensory tongues of more derived lizards; they were no longer used to stick to prey and bring them into the mouth, but rather aided in moving prey *through* the mouth.

Gekkotans remained largely sit-and-wait foragers, but with numerous exceptions. Like iguanians, many are sedentary, have social systems centered on territoriality, and use their tongues and vomeronasal systems very little for prey discrimination—though they do to some extent use their nasal olfactory system (Schwenk 2000). They have broad, short tongues with a rounded apex, again similar to iguanian tongues, but not used for prey adhesion. For gekkotans that have eyes covered by transparent spectacles, freedom of tongue movement allowed a rather unusual innovation: the ability to clean the spectacle, much as a windshield wiper cleans a window. This strange behavior shows up again in xantusiids and has been used to suggest closer affinities with geckos than with scincomorphans. Sexual dimorphism in coloration persists in many diurnal geckos, but size dimorphism, although it exists (as in *Nephrurus,* where females are considerably larger than males), is usually limited.

Gekkotan evolution took some other interesting turns.

One clade, the pygopodids, became snakelike, with one genus, *Lialis,* apparently losing typical lizard feeding mechanics altogether. Like snakes, it ingests other lizards whole, using skull kinesis to help subdue and swallow its prey. Other geckos evolved adhesive toe pads and, in some cases, adhesive tail pads, giving them access to smooth vertical surfaces. Some of these developed expanded vocal cords producing sounds for social communication much like frogs.

Autarchoglossans followed a different evolutionary trajectory. Involvement of the tongue in vomerolfaction became better developed in autarchoglossans (although scincomorphan families of the Autarchoglossa continued to use the tongue to some extent for manipulating prey), playing important roles in social behavior and prey discrimination. The vomeronasal organ in the roof of their mouth grew more acutely sensitive, expanding their world into one filled with plumes of chemical signals. Development of such an acute chemosensory mechanism went hand in hand with acquisition of a new, more active, mode of foraging, which in turn had a cascading effect on morphology, physiology, behavior, and ecology.

Tongue elongation enhanced these lizards' ability to sample their external environment. Anguimorph tongues, particularly, became longer and more sophisticated, with the anterior portion functioning primarily to pick up chemical cues and the posterior portion still involved in feeding; however, unlike scincomorphans, the hindtongue and foretongue are distinct, with the foretongue a slender, hydrostatic extension. These are the superficially snake-

An Australian agamid, *Ctenophorus caudicinctus,* lifts its toes as it stands on a hot surface. (David Pearson)

like tongues of large teiid and tiny gymnophthalmid lizards. Tongues of teiids often have deep notches, giving the appearance of forked tongues, but structurally they are different from the forked tongues of varanoids and snakes, where the feeding function of the tongue has been lost entirely. Forked tongues are used as edge detectors to locate scent trails (Schwenk 1994b, 1995, 2000).

These novel adaptations allowed autarchoglossans, particularly those within the Anguimorpha, access to a wide variety of previously unavailable prey resources and microhabitats, and in turn enabled an extensive adaptive radiation to take place (there are nearly 3,000 species of autarchoglossan lizards plus almost 3,000 snakes). Chemical discrimination also allowed autarchoglossans to avoid noxious and toxic prey. Just as chameleons took tongue feeding to its logical endpoint, varanids and snakes took prey prehension by jaws and chemical prey discrimination to their extremes. Many varanoids developed the ability to swallow enormous prey, in a manner similar to snakes.

Other radical changes occurred among autarchoglossans in parallel with the evolution of feeding mechanics and use of the vomeronasal system. These lizards in general shifted to a highly active lifestyle, moving about in search of prey. The payoff for increased mobility must have been substantial, considering that movement not only requires energy but also attracts attention. Visually oriented predators that had difficulty detecting nonmoving, cryptically colored sit-and-wait lizard species would have no problem spotting moving lizards. Increased alertness and decreased response times were tightly coupled with evolution of active lifestyles, and neural integration developed to a level providing some lizards with an almost mammal-like awareness. Anyone who has observed teiid or varanid lizards in the field is immediately struck by how these lizards gauge every move an intruder makes, maintaining safe distances while continuing to forage; at some critical distance, the lizard will dart off to a new safe distance and resume its activities. Predation costs are offset at least partly by these lizards' watchfulness and responsiveness. This more active lifestyle is supported by higher metabolic rates, higher activity body temperatures, increased endurance, and perhaps most important, reduced exercise recovery times.

Actively foraging lizards tend to have relatively large home ranges through which they move in search of food. For the most part, defense of these areas is inversely related to activity levels. Most teiids, many skinks, and most varanoids, which are among the most active foragers, do not appear to defend territories; instead the male actively guards his mate as they move through their relatively large home ranges. Coupled with the shift to active foraging is a major shift in mating systems. No longer are portions of the home range defended, largely because females cover large areas while foraging, areas too large for a male to effectively defend. The mating system shifts from the typical resource defense polygyny found in iguanians to sequential female defense polygyny.

Major payoffs for the switch from ambush foraging to active foraging include access both to a wider spectrum of prey and to a greater number of potential mates. Foraging and social behavior often occur simultaneously. Unlike iguanians, widely foraging lizards that require extended basking to gain heat (such as teiids in tropical forest) cease foraging and curtail social behavior while basking, since these activities compete directly with thermoregulation. Sexual dimorphism is common in these lizards; it does not involve morphological ornamentation but frequently affects color, head size, and body size.

Finally, within the varanoid clade of actively foraging lizards, an unknown limbless ancestor with reduced eyes and an increased number of vertebrae associated with its long, flexible body gave rise to snakes—a point to which we will return.

Major innovations driving the evolution of various squamate clades are summarized in table 14.1.

DIVERSIFICATION WITHIN EACH MAJOR CLADE

We now return to each of the major clades—Iguania, Gekkota, the *incertae sedis* dibamids and amphisbaenians, Scincomorpha, and Anguimorpha (Estes et al. 1988)—to examine some of the fascinating diversification that has occurred within each.

IGUANIA

In many respects, evolution within Iguania has been relatively conservative, possibly a consequence of constraints associated with sit-and-wait foraging. Even though body sizes, shapes, colors, ornamentation, and patterns are remarkably diverse, this diversity tends to be driven by one of two things: crypsis and sexual selection. Because natural

TABLE 14.1

Innovations driving the evolution of squamate clades

INNOVATION	CLADE
Streptostyly	All squamata
Jaw prehension	Scleroglossa
Mesokinesis	Scleroglossa
Wide foraging	Scleroglossa
Vomeronasal prey discrimination	Iguaninae, some Agaminae, most Scleroglossa
High aerobic capacity (endurance)	Autarchoglossa (with some reversals)
Nocturnality	Most Gekkota, with secondary reversions to diurnality
Leglessness (including near leglessness)	Some Scincomorpha, some Gymnophthalmidae, some Cordylidae, some Gerrhosauridae, some Anguidae, Pygopodidae, and all snakes
Fossoriality	Some Pygopodidae, many Scincidae, some Anguidae, some Gymnophthalmidae, Dibamidae, all amphisbaenians, and some snakes
Herbivory	All Iguaninae, some Agaminae, some Scincidae, some Tropidurinae, *Cnemidophorus* (rare)
Adhesive toe lamellae	Some Diplodactylidae, some Gekkonidae, most Polychrotinae, some Scincidae
Deeply forked tongues ("edge detectors")	Teiidae, Varanoidea, snakes
Fringed toes	Some Gekkota, some Iguanidae, some Scincidae, some Lacertidae, some Gerrhosauridae, some Teiidae
Arboreality	Some Iguania, some Gekkota, and some Autarchoglossa
Aquatic	One iguanian (*Amblyrhynchus*),* some Scincidae, some Teiidae, some Gymnophthalmidae, and some Varanidae
Saxicolous	Many Iguania, some Gekkota, some Autarchoglossa
Viviparity	Some Iguania, some Gekkota, Trogonophiidae, many Scincomorpha, some Anguimorpha (some Anguidae, Xenosauridae, many snakes)
Spiny body armor	Iguanidae, Agamidae, Cordylidae
Ballistic tongues	All Chamaeleonidae
Zygodactyly	All Chamaeleonidae
Telescopic eyes	All Chamaeleonidae
Infrared receptors	Boas, pythons, pit vipers
Venom production	Helodermatidae and many snakes
Venom injection	Many snakes
Constriction	Many snakes
Liberation of mandibles	Snakes
Reversion to ambush foraging	Cordylidae, a few Lacertidae, *Xenosaurus*, boas, pythons, and vipers

Note: Although some evolutionary reversals have occurred, these are usually restricted to a small portion of a given clade.

Amblyrhynchus is terrestrial but forages for algae in the ocean. No other iguanians do anything like this, though several jump into or run across water for escape.

habitats are structurally complex, many opportunities for background matching exist, and iguanians have certainly taken maximum advantage of this diversity. Fringe-toed lizards *(Uma)* and zebra-tailed lizards *(Callisaurus)* match the color of sands on which they live almost exactly; round-tailed horned lizards *(Phrynosoma modestum)* and Australian *Tympanocryptis* are identical to rocks in their habitats; chameleons in the genus *Brookesia* are virtually indistinguishable from dried leaves; and Jackson's chameleons, *Chamaeleo jacksoni,* blend in perfectly with green leaves in their arboreal habitats, even rocking back and forth to give the impression of leaves blowing in the wind.

Nevertheless, males of many otherwise cryptic species are often brilliantly colored during the breeding season, not blending in at all with their microhabitats. Indeed, within iguanians, sexual selection has produced some of the most extreme morphological sexual dimorphisms known in lizards. Huge horns in *Chamaeleo jacksoni,* strange proboscises in *Ceratophora,* brilliant dewlaps in most polychrotines and many agamids, and a diversity of brilliant neck, head, ventral, and ventrolateral schemes advertise an individual male's genetic quality. Although females often have some of these features, most are drab compared with males and could easily be mistaken for different species altogether. While sexually dimorphic characters so prevalent in Iguania were generally not carried through to scleroglossans, some reappear in cordylids, a scleroglossan group that has converged on iguanians in foraging mode, patterns of space use, and behavior (Branch 1988; Cooper et al. 1997).

During breeding season, some males change color rapidly, and many increase their activity while defending territories, becoming quite conspicuous. Although this contrast between crypsis for escaping detection by predators and conspicuousness associated with breeding may appear paradoxical, it represents an evolutionary balance between gains in immediate reproductive success and potential loss of all future reproductive success. An eye-catching male that attracts numerous mates and escapes predation *in spite* of conspicuous coloration and behavior must be doing something right, and he will pass on genes for whatever those successful behaviors are, leaving behind colorful descendants like himself. By contrast, a similar male without the appropriate appearance or be-

The smallest North American horned lizard, *Phrynosoma modestum* resembles a rock when not moving. (Laurie Vitt)

havioral repertoire will likely *not* pass on as many genes, either because he fails to attract mates or because he becomes somebody's dinner. Of course, as in many animals, alternate mating strategies exist that allow some of these "lower quality" males to reproduce successfully.

Great emphasis has been placed on dewlap displays of lizards, particularly *Anolis,* in a social context. These function in intraspecific communication, incurring potential risks associated with exposure during social displays. The displays, however, are usually so brief that they can be considered "flash" displays. From the perspective of a conspecific, to whom the signals are directed, a flash display may be too brief to allow exact location of the displayer to be ascertained, but long enough to provide general directional and distance information. A potential male competitor can thus learn enough to decide whether continued movement in that direction might result in an aggressive interaction, and potential mates can learn enough to proceed in the direction of a good territory holder. From a predator's perspective, the flash is enough to signal that potential prey is nearby but not enough to allow pursuit and capture.

Because diurnal *Anolis* are usually cryptically colored, the flash of a dewlap sends a signal that a predator cannot tie to a particular object because it appears and disappears quickly, leaving no trace of the color that caught the predator's attention in the first place. A male lizard without a territory might pester a male territory holder into presenting his flash display just often enough to give a predator the critical sensory input needed to locate the lizard. The payoff to the non-territory-holding male would be high (acquisition of a territory and resulting mating opportunities) and would represent a fascinating alternative mating strategy. We know of no studies that have considered this possibility, nor studies that have measured the total time during a day that brightly colored male dewlaps are exposed. We suspect that time period is very short.

In spite of constraints within Iguania, several rather exceptional evolutionary trajectories stand out. Herbivory has arisen at least three times, twice in Iguanidae and once in Agamidae. Associated with evolution of herbivory in iguanians is enhancement of the vomeronasal chemosensory system to discriminate among plant foods. Although the largest iguanians are herbivores, associating size with herbivory is risky at best (see chapter 3). Nevertheless, large body size of many iguanines (*Cyclura, Iguana, Am-*

A Galápagos marine iguana, *Amblyrhynchus cristatus,* surveys its kingdom. (Eric Pianka)

blyrhynchus, Conolophus) releases adults from predation by many typical lizard predators: no cuckoos or shrikes are large enough to take an adult iguana! As a result, costs of many movements to and from foraging areas and of high activity levels during courtship and male-male aggressive interactions are likely to be relatively low. Of course, juveniles are much more vulnerable to predators.

Another exceptional evolutionary trajectory is evident in horned lizards, *Phrynosoma.* Dietary specialization on ants has occurred repeatedly within Iguania (e.g., *Moloch horridus,* Amazonian *Tropidurus*), but only in *Phrynosoma* has it resulted in a highly successful diversification event producing the dozen species seen today. If the bad-tasting blood of some species turns out to be derived from ants they eat (see chapter 4), then horned lizards have developed a chemical arsenal from natural products in their own environment without ever having sat through organic chemistry, and they did so without wasting a single moment they could have spent reproducing.

Clockwise from top: **Australia's thorny devil, _Moloch horridus_, is the most extreme example of morphological evolution among agamid lizards.** (Brad Maryan)

Some southern Asian agamids, like this _Ceratophora stoddartii_, are adorned with surreal-looking head ornamentation. (Christopher Austin)

The Australian agamid _Chlamydosaurus kingii_ erects its frill as a display to startle and fend off predators. (Dr. Hal Cogger)

Because iguanians move very little while foraging, morphological adaptation has not been constrained evolutionarily as it has been in actively foraging autarchoglossans. The thorny devil, *Moloch horridus,* is a particularly radical extreme. Such monstrosities deserve additional emphasis because they provide a yardstick against which morphological evolution in other lizards can be compared. Considering the success of horned lizards in deserts of the American West, one wonders why a swarm of *Moloch*-like species did not evolve in Australia. Perhaps they have and just remain undiscovered. Cryptic species could occur within *Moloch* based on variable phalangeal formulae in their reduced digits (Greer 1989).

Chlamydosaurus represents another extreme, with its huge neck frill that is erected when the lizard is disturbed. Not only does this frill make the lizard's head appear enormous, but instantaneous erection of the frill also serves as a flash display that would catch any potential predator off guard. Think of the scene in the movie *Jurassic Park* where the clumsy, overweight computer technician, having just stolen a set of dinosaur embryos, is confronted by a *Dilophosaurus* that erects its frill, setting off a terrifying escape attempt. The moviemakers' special effect may be traceable to *Chlamydosaurus,* since no evidence exists that *Dilophosaurus* had an erectible frill.

Another extreme in agamid morphological evolution is represented by strange *Ceratophora* species that have evolved outlandish appendages on their snouts (larger in males). Still more extreme ornamentation occurs on heads of male chameleons. An especially remarkable appendage is that of the rare Amazonian anole *Anolis phyllorhynus,* whose scaly proboscis extends forward from its snout nearly 70 percent of total head length (Avila-Pires 1995)!

While maintaining all the standard iguanian characteristics, chameleons have bent, twisted, and stretched iguanian morphology virtually to its limits, with their strange, often ornamented heads; turretlike eyes; leaflike, clumsy bodies; zygodactylic feet; coiled, prehensile tails; and psychedelic color patterns, which are further confounded by remarkable color change. This animal, moreover, can sit for an eternity, ever so slowly stretching out its body to get a better shot at an insect. Then in a millisecond, this un-

CONVERGENT EVOLUTION

I studied North American desert horned lizards in detail between 1962 and 1965. These lizards were so different from other desert lizards that I suggested a new integrated perspective to explain various aspects of their biology, including their anatomy, behavior, and ecology. These ant specialists have tanklike bodies covered with spines, which protect them from potential predators. They cannot run very fast and rely on crypsis to avoid detection by predators. Their body temperatures are more variable than in other species, and they are active over a long time period. They also have much larger stomachs and clutch sizes than other lizards. All these unusual features are interrelated and work together to make horned lizards effective ant specialists.

One cannot help but be struck with the morphological similarity between horned lizards and thorny devils. I went to Australia in 1966 as a postdoc hoping to document evolutionary convergence between these lizards and also perhaps contribute new ideas about the vital process of natural selection. After traveling more than 12,000 km during several months in the arid interior of the continent without seeing a single thorny devil, I began to despair that I would ever find one. Finally I got lucky and spotted a thorny devil crossing the road. Like a horned lizard, this thorny devil was fairly immobile and proved easy to catch. I took it to some soft sand and backed off a ways and waited, allowing it to walk and leave a trail, which I studied with care. Eventually I learned how to find these elusive lizards by following their delicate footprints. This is far from an easy feat, and even now I often cannot locate individuals of this species. Tracks that lead into a large bush or a messy area of litter and bushes cannot be followed, and one must then try to find the animal by sight.

Once, at a very remote site deep in the Great Victoria Desert where thorny devils had never before been recorded, I had spent most of the day trying to find an animal whose track went around and around in a large figure eight. We were running out of food and water and were scheduled to return to civilization the next day to resupply. It was late in the day and the sun would soon set. Helen and I were standing a couple of meters apart at the center of the figure eight, shaking our heads, wondering where that thorny devil was, when suddenly both of us saw the lizard at the same instant: in its dark green color phase, it was hunkered down right between us. Even when you have the correct search image in mind, its camouflage works amazingly well! (PIANKA)

This chameleon, *Bradypodion damaranum*, is among the strangest of all known lizards. (Wolfgang Schmidt; Chimaira Edition)

likely assemblage of strange features hurls its projectile tongue through space like a missile, sticking to a fly in midair, and hauls the fly back into its mouth with the grace of a fly fisherman reeling in a cutthroat trout.

As strange as iguanians seem, their diversity alone attests to their evolutionary success. What iguanians have not done is lose limbs and develop elongate, flexible morphology. None live a truly subterranean existence. And relatively few have evolved viviparity.

GEKKOTA

Windshield-wiper tongues, adhesive toe and tail pads, skin that feels like rubber, lightning-speed color change, fixed small clutch size, and chirps in the night all provide clues to the many directions gekkotan evolution has taken. Indeed, gekkotans have experimented with evolution in a nearly surreal fashion. If current understandings of gekkotan relationships are correct, the ancestor to this clade was a nocturnal lizard with velvety skin. Today, velvety-skinned eublepharids move about on the ground in slow motion, tails often waving above their backs, presumably retaining many features of ancestral scleroglossans. Some

terrestrial diplodactylids, too, have feet like those of ancestral scleroglossans and velvety skin, while others have evolved toe pad lamellae and are arboreal. Even though most diplodactylids are nocturnal, some display a kaleidoscope of color patterns.

Gekkonidae diverged into two groups: Gekkoninae and Sphaerodactylinae. Most Gekkoninae are nocturnal lizards with adhesive toe pads that allow them access to a habitat dimension seldom exploited by other lizards: smooth, vertical surfaces. A few, however, like *Heteronotia,* lack toe pads and are terrestrial. Sphaerodactylines are small and diurnal and include some of the tiniest lizards. *Coleodactylus, Pseudogonatodes,* and *Lepidoblepharis* are so small that one almost needs reading glasses to see them. Their world is filled with frightening predators: spiders weighing a hundred times as much as they do, scorpions four times their length, amblypygids (tailless whip scorpions) with giant lancelike spikes on their mouthparts, centipedes that could easily trample them to death, and troops of ants that can dismember a lizard in minutes. As they feed on tiny springtails and mites, their small body size allows them access to interleaf spaces in leaf litter, an extensive and structurally diverse microhabitat.

Strangest among gekkotans are pygopodids, snakelike lizards that we are just beginning to understand ecologically. Within pygopodids, one, *Lialis burtonis,* is for all practical purposes a "snake," swallowing its lizard prey whole. Others, though snakelike in morphology, feed on more typical lizard prey, arthropods.

The several directions that gekkotan evolution has taken are so divergent that speculating on logical endpoints is impossible. If anything, gekkotans represent experimental forays into the unknown, with unlikely results. The simple fact that these various trajectories have been so successful confirms how truly complex and little understood natural environments really are.

Sexual dimorphism in gekkotans is much less extreme than in iguanians. Most species are slightly dimorphic in size, but the larger sex differs among species (Doughty and Shine 1995; Tokunaga 1984; Vitt, Souza et al. 2000). In diurnal species, such as some sphaerodactylines and *Phelsuma,* males are usually brilliantly colored, though some can change from drab to brilliant in seconds. Nocturnal geckos typically show no sexual dimorphism other than size. Most striking among gekkotans as a group is the complete absence of morphological ornamentation so common in iguanians.

DIBAMIDS AND AMPHISBAENIANS

Although their affinities remain uncertain, dibamids and amphisbaenians probably arose within Scleroglossa as sister taxa to gekkotans. These animals are so strange and so different from all other lizards that it is difficult even to speculate on their origins. Their only obvious connection to other lizards is the presence of hind limbs in male dibamids and molelike front limbs in a single genus of amphisbaenians, *Bipes.* Given that amphisbaenians are pantropical in distribution and dibamids occur only in Mexico and Southeast Asia, both must be very ancient, dating back to Gondwana.

All aspects of the morphology of these two groups have been highly modified for subterranean existence. Rather than swimming through soft soil or sand like many other legless lizards, amphisbaenians push their way through the soil and under leaf litter using rectilinear locomotion for forward thrust and head movements to widen their burrows as they move forward. Their vomeronasal system is well developed, whereas eyes are reduced, serving primarily as light detectors.

SCINCOMORPHA

Among the most active and alert lizards are scincomorphans: teiids, gymnophthalmids, lacertids, gerrhosaurids, cordylids, xantusiids, and scincids. Present-day teiids and gymnophthalmids dominate tropical South American lizard assemblages, while skinks and lacertids dominate both tropical and desert lizard assemblages in the Old World. Teiids are strikingly similar in morphology, all with long, powerful hind limbs, long tails, and streamlined bodies. These alert, fast-moving predators, adept at finding hidden prey, will often dig into termite runs to feast on hundreds of termites at a single sitting. The smallest teiid, *Cnemidophorus inornatus,* only about 55 mm in snout-vent length, races among grass clumps in sandy habitats of the Chihuahua Desert, whereas one of the largest, *Dracaena guianensis,* swims around in flooded forest and swamps in the Amazon River system, feeding on mollusks that it crushes with its large, molarlike rear teeth.

Teiids achieve some of the highest body temperatures known in lizards—up to about 40°C for some species. The ability to maintain activity for extended periods at high body temperatures has allowed teiids to do very well in open habitats such as deserts, thorn forests, and beaches. In tropical forests, teiids such as *Ameiva* are concentrated along woodland edges—along trails, in disturbed forest, or in large treefalls—where they gain access to direct sunlight for thermoregulation. Larger teiids in tropical rain forest are associated with edges along waterways, where they, too, bask to gain heat. Species in only a single genus, *Kentropyx,* have moved up into vegetation; most of their arboreal activity is relegated to low shrubs, particularly those along edge habitats such as streams and lakes, where they gain access to sun on exposed leaves and branches. Although a vast majority of teiids are masters at finding and pursuing animal prey, herbivory has evolved independently in two genera, *Dicrodon* and *Cnemidophorus,* and several other teiids include some plant parts (usually fruits) in their diets.

The most striking divergence between teiids and gymnophthalmids is in body size: nearly all gymnophthalmids are smaller than the smallest teiid. Gymnophthalmids dominate the leaf-litter lizard fauna of Amazonian rain forest as well as slopes and valleys of the Andes. Their small body size allows access to interleaf spaces and tiny refuges among roots and along stream banks. Gymnophthalmids are teiid-like in that they appear to have high activity levels, yet unlike teiids they maintain relatively low body temperatures (28–31°C). Several genera—*Neusticurus, Alopoglossus,* and *Leposoma*—frequently live near water, often diving in and swimming away like salamanders. Scales of semiaquatic species are highly ridged or keeled, whereas scales of many leaf-litter species (e.g., *Gymnophthalmus* and *Iphisa*) are smooth and glasslike, resembling the cycloid scales of skinks. Few gymnophthalmids are arboreal, but limblessness or near-limblessness is common. One nearly limbless genus, *Bachia,* is wormlike, dragging its elongate body around through leaf litter and soil with tiny front limbs. When disturbed, it can move quite rapidly by means of undulatory movements of the body and tail or even saltate. *Calyptommatus,* which have only tiny vestiges of limbs, literally swim through sand in Brazilian sand dunes. Most gymnophthalmids have transparent scales on their lower eyelids, allowing them to see with their eyes closed. No known gymnophthalmids are herbivorous.

Most lacertids, Old World counterparts to teiids, are active foraging heliothermic lizards common in microhabitats that receive direct sunlight. Most are relatively small, but a few, such as *Gallotia,* are large. One genus, *Meroles,* has diversified into a swarm of sand lizards with toe fringes and the ability to breathe beneath the sand. Desert species such as *Nucras* resemble the teiid *Cnemidophorus* in foraging behavior and morphology. As in teiids, herbivory and parthenogenesis occur in some lacertids. In contrast to teiids, however, viviparity has arisen once within lacertids, allowing a single species, *Lacerta vivipara,* to extend into the higher latitudes in Europe.

The most successful scincomorphans in terms of species diversity are skinks, with more than 1,400 species. No other lizard family has as complete a worldwide distribution, and nearly every evolutionary trajectory taken by autarchoglossan lizards has occurred at least once within Scincidae. Sandfish run across sand using their fringed toes like snowshoes, diving into sand and disappearing beneath the surface. Skinks have lost their limbs many times and taken up subterranean existences in a variety of soil types. Although most skinks have relatively smooth, glasslike scales, some have keeled scales or scales with so much architecture that they appear spiny, almost like reptilian porcupines. Skinks live in the most extreme deserts, the wettest tropical forests, at high elevations and latitudes in cool environments, and even in water. Most are small to moderate in body size, fast moving, and carnivorous,

though some, such as *Corucia* and *Tiliqua,* are large, slow moving, and herbivorous.

Among lizards, skinks exhibit the greatest diversity in reproductive modes and patterns of social behavior. Many species are oviparous, with some depositing only one clutch a year and others depositing many. Likewise, while some oviparous species simply lay their eggs and leave, others brood their eggs until they hatch. Even viviparity takes various forms, with some species having no placental connection between mother and offspring, while others have the most highly developed placental arrangements known in any reptile. Social behavior varies from relatively simple polygynous mating systems with no parental care to highly complex monogamous systems and extended parental care. Most skinks do not defend territories, but some do.

Cordylids as a group have not diversified much—most species are flattened in morphology and live in rock crevices. These characteristics have reached their extreme in *Platysaurus,* which differ from other cordylids in other features as well: *Platysaurus* are oviparous, whereas all other cordylids are viviparous, and *Platysaurus* are highly dimorphic in coloration, unlike many cordylids. Some species of *Cordylus* have scales with extreme architecture. Members of one genus of cordylids, *Chaemasaura,* are elongate with vestigial legs.

Many gerrhosaurids are crevice dwellers as well, a characteristic probably traceable to the common ancestor of cordylids and gerrhosaurids. Morphologically, gerrhosaurids have diversified considerably more than cordylids, with some species in the genus *Tetradactylus,* for example, converging on anguid glass lizards, others (in *Angolosaurus*) converging on fringe-toed lizards and sandfish, and yet others (e.g., in *Gerrhosaurus* and *Zonosaurus*) resembling large skinks.

The fact that cordylids and gerrhosaurids are sister taxa gives their divergent evolutionary trajectories a particular fascination. For whereas cordylids have converged on iguanians, gerrhosaurids have retained skinklike features: cordylids are typically sit-and-wait foragers, gerrhosaurids are active foragers; cordylids do not discriminate prey on the basis of chemical cues, gerrhosaurids do; and cordylids are territorial, gerrhosaurids are not (Cooper 1992; Cooper and Van Wyk 1994; Cooper et al. 1997; Martin et al. 1997; Van Wyk 2000; Mouton et al. 2000). The fact that a major evolutionary reversal has taken place in cordylids involving foraging mode, chemosensory abilities, and behavior shows how tightly these characteristics are intertwined. It also illustrates the evolutionary success of the iguanian body plan and ecology, especially when combined with scincomorphan features.

Xantusiids have followed an altogether different evolutionary direction. Most are small, and all are highly secretive, viviparous, and long lived. Morphologically and ecologically, they vary greatly, some being dorsoventrally flattened and living in crevices (e.g., *Xantusia henshawi*), while others are nearly cylindrical and live within debris under Joshua trees (e.g., *X. vigilis*).

As in other autarchoglossans, sexual dimorphism in scincomorphans is restricted to differences in body size and does not include morphological ornamentation. However, many scincomorphan males do have relatively larger heads and often have enhanced coloration either on the head or on the jaws, neck, lateral body surfaces, or portions of the front limbs. Male teiids and skinks, for example, often have larger heads than females, and they may be brightly colored (Anderson and Vitt 1990; Vitt and Cooper 1985b). In many gymnophthalmids, males have targetlike markings on the sides of their bodies (e.g., *Neusticurus ecpleopus* and *Cercosaura ocellata*), often enhanced by brilliant background coloration (e.g., *Prionodactylus oshaughnessyi*). The most striking sexual dimorphism in scincomorphan lizards occurs in cordylids, the autarchoglossans that have converged on iguanians. Males of *Platysaurus*, for example, are brilliantly colored dorsally and ventrally and perform spectacular displays during social interactions (Broadley 1978; Whiting 1999).

ANGUIMORPHA

Diverging radically, anguids have taken high activity levels and active foraging behavior of their ancestors and put them in slow motion. The underpinnings of this change could be as simple as an evolutionary shift in the thermophysiology of ancestral anguids, giving them an edge in relatively cool environments. Most anguids are active at body temperatures below those of teiids, skinks, or lacertids, and most species live in relatively mesic habitats or microhabitats. They are secretive, and sexual dimorphism is limited to relatively minor differences in body and head size.

Anguids have taken four evolutionary trajectories. In one, Anguinae, limbs have been lost, body and tail have elongated more than in any other lizards, and scales are like those of skinks. The long tails of anguines autotomize

easily and break apart, hence the common name "glass lizards." Although their morphology appears snakelike, it differs from that of true snakes in two important ways: the tail comprises 60 percent or more of total length (usually less than 20 percent in snakes), and jaw structure is similar to that of other anguids. Glass lizards are terrestrial and diurnal. The second trajectory can be found in *Anniella* (Anniellinae), which, though also limbless and elongate, has developed a fossorial lifestyle, living in—and swimming through—sand or friable soils. Its tail is relatively short compared to other anguids. Diploglossines or "galliwasps," representing the third evolutionary offshoot, resemble smooth-scaled skinks at first glance: limbs are reduced but well developed, although one genus, *Ophiodes*, has nearly lost its limbs, converging on the anguine *Ophisaurus* in both morphology and ecology. Finally, gerrhonotines are mostly robust lizards with tails of variable length and short but well-developed limbs. Most have an alligatorlike aspect—the result of heavy armor on the head and body, usually consisting of scales with some vertical architecture—hence the name "alligator lizard" for some species. Tails of gerrhonotines are strong, and in the arboreal genus *Abronia* are considered truly prehensile. Anguids represent a radiation successful in mesic habitats and microhabitats where diversity of other autarchoglossans is low.

If Xenosauridae is a distinct evolutionary group (i.e., monophyletic) as generally depicted (Pough et al. 1998; Zug et al. 2001), its two distinct evolutionary trajectories—*Xenosaurus* and *Shinisaurus*—do not make a lot of sense. The approximately twelve species (six described) of *Xenosaurus*, flat-bodied lizards with broad heads and powerful jaws that live in tight places (rock crannies and crevices in trees), are restricted to Mexico and Guatemala. The single species of *Shinisaurus*, which does not have a dorsoventrally flattened morphology but rather is compressed from side to side, is restricted to mesic montane forests of southern China. As such, about all that can be concluded about either genus is that neither has diversified much and both can be considered relics of relatively unsuccessful evolutionary experiments. Perhaps *Xenosaurus* have converged on saxicolous iguanians and cordylids. Both groups, in any case, appear to be very old and may be distinct families (Macey et al. 1999).

If a logical endpoint to limbed autarchoglossan evolution exists, the varanoids (Helodermatidae, Varanidae, and *Lanthanotus*) have achieved it. The self-confident gait of an active *Heloderma* or *Varanus* leads one to believe that even they recognize their status among lizards. Their vomerolfaction system is surpassed only in snakes (their descendants); they achieve phenomenal body size *(Varanus komodoensis),* and advanced neural integration in combination with superior tooth, jaw, and neck morphology, which has allowed them to join the ranks of the most impressive top predators on Earth.

The natural history of *Lanthanotus,* the sister group to varanids, remains virtually unknown. Found only in northern Borneo, these cryptic lizards are seldom encountered. A Cretaceous (75 M.Y.B.P.) fossil varanoid from Mongolia similar to present-day *Lanthanotus* suggests that the latter may be a "living fossil," and as such could provide valuable insights into varanoid ancestors.

Beadlike yellow, orange, or red skin with a black background and relatively large body size distinguish helodermatids, the only truly venomous lizards, from all other lizards. These and varanids reflect the ultimate in foraging ecology. Using their highly developed chemosensory forked tongues and vomerolfaction system to locate and discriminate prey, they frequently feed on large vertebrate prey—including other lizards, as well as eggs and juveniles found in bird and mammal nests—thus gaining the maximum amount of energy in each foraging bout. Helodermatids are currently represented by just two species in arid and thornscrub habitats in western North America, but numerous fossils indicate a much wider former distribution.

Varanids, more than any other lizards, have developed a neurophysiology that makes them seem nearly mammal-like. Just look into the eyes of a Komodo dragon if you are not convinced: these lizards make eye contact in a way that would put most young lovers to shame, but make no mistake—the image a Komodo derives in such an exchange has nothing to do with love. One wonders whether

An Australian monitor, *Varanus panoptes,* in a defensive display. (Eric Pianka)

their advanced neural systems did not free them from many attributes other lizards use to escape predators, such as cryptic coloration, limited activity periods, and a heavy reliance on established escape routes for predator escape.

These top predators are the most intelligent of all lizards. At the National Zoo in Washington, D.C., individual Komodos exhibit curiosity about their keepers, whom they appear to recognize; one lizard approached its keeper and gently climbed up on him, then, with its mouth, extracted the man's notebook from his pocket, descended, laid it down on the ground, and tongue-flicked it (James Murphy, pers. comm.)! Recent experiments on captive *Varanus albigularis* conducted by John Phillips at the San Diego Zoo suggest that some varanids can even count. Lizards were fed a certain number of snails placed in separate compartments, which were opened one at a time; upon finishing the last snail, the lizards were allowed into another chamber containing the same number of snails. After such conditioning, one snail was removed from some snail groups; the lizards searched extensively for the missing snail, even when they could see the next group. These experiments showed that these varanids can count up to six, but more than that and they seemed to stop counting and simply moved on to the next chamber (King and Green 1999). Such an ability to count probably evolved as a consequence of raiding nests of reptiles, birds, and mammals, since average clutch or litter size would often be around six.

In overall morphology, most monitors are similar, with elongate bodies, long necks and tails, and long powerful limbs. Varanids, however, exhibit the greatest range in body size of any lizard genus, or family for that matter: from a mere 17–20 cm in total length and a weight of only 10 g, to 3 m in length, with a mass of 150 kg. An extinct Australian varanid from the Pleistocene *(Megalania)* was positively monstrous, reaching 6 m and 600 kg. The success of the varanid body plan—which, indeed, works well in a variety of habitats, from terrestrial to arboreal to aquatic—is attested by the fact that it has endured for at least 80 million years.

Snakes represent a putatively monophyletic adaptive radiation of exceedingly specialized, but nevertheless highly successful, anguimorphan lizards. Like their ancestors, which were probably fossorial (burrowing lizards with reduced appendages or limb loss), snakes rely heavily on chemosensory cues to locate prey. However, not all snakes are active foragers; some, such as boas, pythons, and viperids, have reverted to the iguanian sit-and-wait mode of ambush foraging. Many snakes are dietary specialists, most eating only a narrow range of various vertebrate prey. However, different snake species have specialized on different prey, including amphibian and reptilian eggs, avian eggs, snails, frogs, toads, lizards, other snakes, birds, and mammals (Greene 1983). Like most lizards, however, some snakes also consume certain arthropods, including ants, termites, spiders, centipedes, and scorpions, as well as other invertebrates such as earthworms, slugs, and snails. Snake skull morphologies and dentitions have evolved along a wide array of pathways, each presumably adapting its bearer to efficient exploitation of its particular prey. Many snakes will not even eat anything outside of one of these (their own) particular categories. Why do so few lizards eat mammals and birds, whereas so many snakes do? Why do so few snakes eat invertebrates, whereas a majority of lizards do? One reason might be the generally larger size of snakes than lizards. Another is the snake jaw structure, allowing larger items into and through the mouth, and the lack of a pectoral girdle, allowing large items into the body.

Snakes are a relatively specialized clade due to lack of limbs; hence, their "diversity" is restricted by morphology. Despite their limblessness, snakes have solved many of the same problems as lizards, sometimes in similar ways, sometimes using different means. Some snakes, for example, pull snails out of their shells; snail-eating lizards, in contrast, crush snail shells with molariform teeth. And whereas those lizards that are termite specialists have developed various modes for finding their prey (certain geckos, for instance, catch termites at night when they are active above ground; others, like lacertids and teiids, break into termite mud tunnels during the day; and still others, like some fossorial skinks, are more snakelike, finding termites in tunnels and termitaria below ground), all termite-specialized snakes locate termites below ground, and many even live inside termitaria.

Snake eyes appear to have been rebuilt after degenerating during an extensive subterranean existence in the ancestral, fossorial form. All other tetrapods focus by changing lens curvature using muscles within the eye, but snakes have no such muscles and focus instead by moving the lens with another set of muscles in the iris (Bellairs and Underwood 1951; Rieppel 1988).

Though severely limited by their limbless body form, snakes have nevertheless undergone a massive adaptive radiation (nearly 3,000 species). Many snakes kill their prey by constriction, which requires short vertebrae; heavy, supple bodies; and slow movements. Very fast snakes like cobras and racers have elongate vertebrae, with musculature extending considerable distances between vertebrae; such snakes are slender, not as supple, and seldom can constrict their prey. About 20 percent of snakes envenomate prey, a trait that has evolved repeatedly. A friend of ours recovering from the bite of a large Costa Rican fer-de-lance wrote a poem likening his nurse who was injecting him with antivenin to the pit viper that bit him—"the inventor of the hypodermic needle"!

Snake skulls have diversified widely. Although they are still diapsids, because they are descendants of diapsids, snakes have lost both temporal arches and apertures, allowing greater independent movements of head bones. They have also carried cranial kinesis to a higher level than their lizard ancestors, possessing numerous flexible joints in their skulls as well as the liberation of the mandibular symphysis (tendons connecting the two lower jaws together). Coupled with streptostyly, these adaptations allow snakes to swallow exceedingly large prey (Gans 1961; Cundall and Greene 2000). Independent movement of the two sides of a snake's upper jaw and the complex musculature of a snake's head essentially allow snakes to "walk" their way down large prey, opening first one side of their jaws, extending the jaw bones forward, biting down, and then repeating on the other side (Gans 1961). Skulls of burrowing snakes are secondarily compacted.

Hypotheses for evolution of snakes have been highly controversial. One posits that snakes are the sister taxon to mososaurs and thus of marine origin and only secondarily terrestrial (Caldwell and Lee 1997; Lee and Caldwell 1998; Lee 1998). This hypothesis rests largely on a mososaur-like fossil, *Pachyrachis problematicus,* believed to represent an ancient snake. A second hypothesis posits that snakes arose within varanoids, but from a gape-limited subterranean ancestor (Rieppel 1988; Cundall and Greene 2000; Greene and Cundall 2000). The recent discovery of a fossil, *Haasiophis terrasanctus,* combined with a reanalysis of *P. problematicus,* however, places both fossil snakes within macrostomate snakes, a derived clade nested within Alethinophidia (Tchernov et al. 2000), making a marine origin of snakes highly unlikely.

A useful approach to understanding snake evolution is to compare the two major sister clades of snakes, Scolecophidia (blind snakes in the families Leptotyphlopidae, Typhlopidae, and Anomalepididae) and Alethinophidia (all others). Blind snakes have solid blunt skulls and are nearly toothless. Even so, considerable variation exists in scolecophidian skulls: leptotyphlopids manipulate and transport prey with the mandible, whereas typhlopids and presumably anomalepidids rake prey into the mouth by rapidly protracting and retracting maxillary teeth (Cundall and Greene 2000). In the end, which clade of snakes—scolecophidian or basal alethinophidian—is most like their lizard ancestors is hard to judge because living examples are highly derived and appropriate lizard and snake fossils are lacking. Indeed, snakes might not even be monophyletic (McDowell and Bogert 1954)! Snake maxillae are extremely variable and movable: these are the bones in the upper jaw to which viperine fangs are attached, which hinge through almost a full 90 degrees from the folded-back closed-mouth position to the fully erect stabbing position. Mouths of Alethinophidia, meanwhile, are filled with hundreds of sharp recurved teeth arrayed along their mandibles, maxillaries, premaxillaries, pterygoids, and palatine bones.

Let us consider another possibility. If snake ancestors were subterranean (Greene 1997), they were clearly the most successful among many scleroglossans that experimented with fossoriality. Yet why was this particular evolutionary experiment so successful, given the many times limblessness has arisen? The answer perhaps lies among varanoid lizards, for they share a combination of characteristics that may have opened up a unique opportunity for a fossorial varanoid precursor, one that other subterranean lizards would never have been able to tap into. First and foremost, possession of a forked tongue would have allowed keen chemosensory discrimination of prey and detection of airborne chemical signals, including the critical ability to determine the direction of those signals. A fossorial varanoid encountering termites could thus identify them and feed on them, *and* it could trace their chemical trail back to the colony, something other fossorial termite-eating autarchoglossans would be unable to do. Because all fossorial lizards are relatively small, moreover, a fossorial varanoid would likely have been small and thin-bodied, with a correspondingly small head (perhaps similar to a small *Lanthanotus*), which would allow it access to the passageways of social insects such as termites. Mandibular prey manipulation, present in varanoids, would facilitate ingestion of termites in restricted passageways, while travel through these narrow spaces would favor concertina locomotion, which in turn would select for longer bodies (as opposed to longer tails) in these snake ancestors. This set of traits describes fairly accurately the three primitive snake families within Scolecophidia: Typhlopidae, Leptotyphlopidae, and Anomalepididae.

Elongation of the body most likely preadapted these animals for a return to the surface; once there, these snake ancestors underwent selection for increased ability to ingest large prey—amphibians, lizards, birds, and mammals—and evolved larger body sizes. They also evolved a loose mandibular symphysis, which allows the two lower jaw bones to spread apart, facilitating ingestion of large prey (Gans 1961).

One question remains: Why did snakes return to the surface? Or perhaps better phrased: What innovation gave snakes an edge in a terrestrial world? The answer to this lies in overall body plan. Tetrapods must expend considerable energy working against gravity as they move their body mass up and down with each step; snake locomotion, in contrast, is much more efficient, relying as it does on an elongate and very flexible body with a high degree of trunk control. Not only can snakes move more rapidly than tetrapods, but by making an S-shaped loop in their neck, they can strike more quickly to capture prey. After returning to the surface, therefore, snakes could eat large things relative to their body and head diameter, thanks to their jaw structure, and they could move their highly flexible body around in ways that no other elongate lizards could. Any crevice, hole, or passageway they could get their head into was accessible. Increased number of vertebrae and associated musculature facilitated swimming, climbing, and types of locomotion either poorly developed or nonexistent in other lizards, including rectilinear, concertina, and sidewinding undulation. A fourth type of locomotion used by snakes, lateral undulation, is found among present-day fossorial lizards and many with reduced limbs, as well as by aquatic lizards such as *Crocodilurus.*

A key difference between snakes and most legless lizards is relative tail length. Most fossorial lizards live either in sand (the skink *Typhlosaurus,* for example, and the gymnophthalmid *Calyptommatus*) or at relatively shallow depths in humus on the floor of tropical forests (such as the gymnophthalmid *Bachia* and the skink *Larutia*). These microhabitats may well restrict the evolution of such snakelike characters as extremely elongate bodies and the neural integration to move them because of drag during locomotion in loose soil. For although most sand-swimming lizards are elongate compared to their lizard relatives, they are short compared to blindsnakes. Even snakes that have returned to the sand (e.g. *Chilomeniscus, Simoselaps*) have converged on lizard morphology (shorter and stouter) rather than retaining snake morphology.

But did snakes evolve only once? Pygopodids provide an extremely interesting case for exploring this question, for following the same logic, if pygopodids evolved snake-like morphology from a terrestrial ancestor, they should have long tails like *Ophisaurus, Ophiodes,* and some of the other "glass lizards." And indeed, terrestrial pygopodids do have extremely long tails (three to four times SVL in some *Delma*), such that tail movement rather than trunk movement as in snakes largely drives their locomotion. Fossorial pygopodids, in contrast, have elongate bodies and relatively short tails. (*Anniella,* the anguid that has gone back into the sand, also has a relatively short tail and long body.) Because fossorial pygopodids (*Aprasia*) are well embedded within the pygopodid phylogeny, we conclude that fossoriality is derived in these lizards, not ancestral (Bryan Jennings, pers. comm.).

The ability to eat large prey evolved (at least functionally) independently and via a different pathway in pygopodids. Most pygopodids possess typical gekkotan jaw structure and feed on relatively small invertebrates. *Lialis,* however, have extremely long jaws and kinetic skulls and swallow lizards whole—just like snakes. Add to that pygopodids' snakelike morphology (especially in fossorial species), and it seems quite reasonable to say that "snakes" didn't evolve only once, they evolved twice: we simply gave each a different name!

ESCAPING AUTARCHOGLOSSANS

Throughout this book, we have pointed to remarkable differences between iguanians and scleroglossans, especially autarchoglossans. Jaw prehension of prey, increased use of the tongue and vomeronasal system to discriminate prey, and a much more active lifestyle would appear to provide autarchoglossans a competitive advantage over more sedate, visually oriented iguanians and even gekkotans. In addition to being formidable competitors, moreover, autarchoglossans often eat other lizards, especially iguanians, and some even specialize on lizards as prey

TABLE 14.2

Relative percentages of Iguania, Gekkota, and Autarchoglossa at study sites in various world habitats

HABITAT	NO. LOCALITIES	IGUANIA	GEKKOTA	AUTARCHOGLOSSA
NEW WORLD				
Amazon rain forest	15	37 ± 2	17 ± 2	46 ± 2
Amazon savanna	2	34	10	56
Caatinga	6	28 ± 1	26 ± 2	45 ± 1
Brazilian cerrado	6	34 ± 4	13 ± 7	54 ± 4
Central America	3	54 ± 2	18 ± 1	28 ± 2
Caribbean	1	63	10	26
North American deserts	12	74 ± 2	9 ± 2	17 ± 2
OLD WORLD				
Australian desert	14	18 ± 2	31 ± 2	51 ± 4
Kalahari	10	8 ± 1	35 ± 1	58 ± 1

Note: New World iguanians are all Iguanidae, whereas Old World iguanians are mostly Agamidae; New World autarchoglossans are mostly Teiidae, whereas Old World autarchoglossans are mostly Scincidae and Lacertidae. Mean percentages ± one standard error are shown.

(note that a few iguanians specialize on lizards). Nevertheless, most lizard assemblages in the world contain mixtures of iguanians, gekkotans, and autarchoglossans. Most strikingly, in lizard assemblages containing a substantial number of autarchoglossans, most of the iguanian and gekkotan fauna is arboreal, saxicolous, nocturnal, or active in the shade. In lizard assemblages lacking autarchoglossans, iguanians in particular occupy many of the microhabitats occupied by autarchoglossans in mixed assemblages. The present ecological and geographical distribution of these lizards may be explained by two factors: displacement of iguanians and gekkotans by autarchoglossans throughout their evolutionary history; and an inability on the part of autarchoglossans to persist in some environments and microhabitats that iguanians and gekkotans dominate, owing to the very set of traits that gave them a competitive advantage in the first place. Global distribution and microhabitat affinities of extant lizards support this idea (tables 14.2 and 14.3).

With their tremendous ecological and morphological diversity, iguanians certainly demonstrate the ability to adapt to nearly every imaginable microhabitat. The remarkable radiation of *Liolaemus* in the southern Andes is a clear example of their almost universal success in a single environment, with most species being terrestrial, using open microhabitats. The same is true for the huge radiation of *Anolis* lizards in the Bahamas, except there many species are arboreal. In the Sonoran Desert we find

a much less diverse lizard fauna, composed of a mere eleven species: eight iguanians (*Callisaurus draconoides, Dipsosaurus dorsalis, Uta stansburiana, Gambelia wislizeni, Phrynosoma platyrhinos, Sceloporus magister, Urosaurus ornatus, U. graciosus*), one teiid (*Cnemidophorus tigris*), one eublepharid (*Coleonyx variegatus*), and one xantusiid (*Xantusia vigilis*)—but again, it is an iguanian-dominated lizard fauna, with several terrestrial species using open spaces (Pianka 1986). In these places, significantly, a diverse autarchoglossan fauna is lacking.

Lizard assemblages in tropical South America are very different. In an Amazon savanna area in northern Brazil, open habitat contains only two iguanians, *Tropidurus hispidus* and *Anolis auratus;* one native gecko, *Hemidactylus palaichthus;* four teiids, *Tupinambis teguixin, Ameiva ameiva, Kentropyx striata*, and *Cnemidophorus lemniscatus;* and one gymnophthalmid, *Gymnophthalmus leucomystax* (Vitt and Carvalho 1995). In this fauna, one iguanian is arboreal on grass and small shrubs, and the other sits on top of termite mounds or lives in trees. The single gekkonid is nocturnal. In the Amazon rain forest, lizard faunas at the local level contain twenty to thirty species. Nearly every fauna contains at least three teiids and three to nine gymnophthalmids, most of which are terrestrial. In contrast, only one of the six to eight iguanians, *Anolis nitens,* is terrestrial, and the only terrestrial geckos—*Coleodactylus, Pseudogonatodes,* and *Lepidoblepharus*—are much smaller (by mass) than any gymnophthalmids or

TABLE 14.3

Relative percentages of Iguania, Gekkota, and Autarchoglossa in various geographic areas

AREA	IGUANIA	GEKKOTA	AUTARCHOGLOSSA	TOTAL NO. SPECIES
NEW WORLD				
United States	48.4	6.3	45.2	95
Mexico	53.9	10.1	36.0	336
Yucatán	47.0	19.6	33.0	51
Belize	50.0	20.5	29.5	44
Guatemala	50.5	11.5	37.9	87
Puerto Rico	51.5	30.3	18.2	33
Amazonia	32.2	14.4	53.3	90
Argentina	73.3	8.1	12.5	86
Mean U*	50.9 ± 1.4	15.1 ± 1.0	33.2 ± 1.7	
Mean W*	51.8	12.0	35.5	
OLD WORLD				
Iran	18.4	33.6	48.0	125
India	26.8	34.1	39.0	165
North Africa	14.7	27.4	57.9	95
South Africa	10.9	27.9	61.1	241
Madagascar 1[†]	32.5	37.0	30.5	197
Madagascar 2[†]	32.9	36.9	30.2	225
Southeast Asia	31.6	27.9	40.5	192
New Caledonia	0.0	38.2	61.8	68
Australia	11.4	27.5	64.7	553
Mean U*	19.9 ± 1.3	32.2 ± 0.5	48.2 ± 1.5	
Mean W*	19.6	30.8	49.6	

Note: Total number of New World species = 822; total number of Old World species = 1,880.

*Means, ± one standard error: U = each region weighted equally; W = regions weighted by number of species.

[†]Two sources for Madagascar, both underestimates of total numbers of species.

teiids. Remaining geckos are arboreal. Moreover, iguanids and gekkonids usually live in shady environments less accessible to heliothermic teiids.

Neither teiids nor gymnophthalmids are diverse at high latitudes in the New World (gymnophthalmids drop out totally), and teiids usually are absent in well-shaded forest, suggesting that some correlate of high activity (e.g., temperature) limits their ability to persist (a climatic or physiological constraint). In habitats where gymnophthalmids and teiids do not exist, iguanians dominate lizard faunas, and many species live in open habitats used by autarchoglossans elsewhere.

In the Old World and Australia, patterns are similar but the players differ. In habitats with high skink and varanid diversity such as Australia, geckos are nocturnal, pygopodids have diverged to become functional snakes,

and many agamids are arboreal or saxicolous. Several terrestrial iguanian species manage to coexist with many autarchoglossans, perhaps by virtue of high body temperatures. In South Africa, where skink, lacertid, cordylid, and gerrhosaurid diversity is high, most gekkonids are nocturnal and very few agamids are present. Some cordylids, particularly *Platysaurus,* have converged on iguanians and inhabit rock crevices in the same manner as agamids in the Middle East and iguanids in rocky habitats of the New World. In habitats with low autarchoglossan diversity, such as Madagascar and New Zealand, many geckos are diurnal.

Given this scenario, it could be difficult—perhaps impossible—to reconstruct the phyloecology of tropical and midlatitude iguanians, because the terrestrial ancestors of arboreal iguanians were outcompeted by autarchoglossans long ago, leaving few survivors. If autarchoglossans are responsible for driving many gekkotans into the trees and others into the night (which must have happened long ago), reconstructing the phyloecology of gekkotans proves difficult as well. This scenario also suggests that evolution of diversity in lizards and their descendants (e.g., snakes) was constrained by the very characteristics that made iguanians so successful: the sit-and-wait foraging mode, low activity levels, and crypsis to escape detection. By breaking free of these constraints, scleroglossan ancestors not only gained many advantages—higher activity levels, enhanced chemosensory ability, and a much improved prey ingestion system—but they also were able to diversify to high degrees. This diversity was facilitated by a key fact: scleroglossans shared most of their evolutionary history with iguanians. Their "bag of tricks" thus included all those of iguanians plus novel ones added as a result of divergence, making them—especially the autarchoglossans—the most formidable competitors iguanians had ever faced.

CHAPTER 15

LIZARDS AND HUMANS

Karen and Gretchen Pianka are captivated watching two Australian thorny devils feeding on ants. (Eric Pianka)

Lizards command our attention: appealing to look at, they appear as motifs for all sorts of products and businesses, from bicycles to database programs to insurance companies to climbing walls to graphic design firms. Perhaps more important, they are model organisms for ecological and evolutionary studies and, as such, can teach us volumes about the living world that surrounds us. Without lizards, our lives would be sadly impoverished.

Human fascination with lizards has a long history. Certainly nearly every child is mesmerized when offered the opportunity to handle his or her first lizard. Quality time spent with an individual lizard offers a glimpse into the fascinating world of evolutionary biology: the high species diversity of lizards, as well as their abundance in many world habitats, is a sure testament to their evolutionary success. As animals whose most relevant adaptation is a highly developed integrative nervous system, we humans would do well to look to lizards to learn what they did right. (After all, if the number of species is any indication of success, primates have done poorly compared with lizards.) Given the many insightful contributions to our understanding of ecology, physiology, behavior, functional morphology, and evolutionary biology that studies on lizards have made, these animals have proven their "value" many times over.

LIZARDS IN NATURAL ECOSYSTEMS

Ecosystems have many emergent properties, which are often difficult to connect directly to particular animal or plant groups. Nevertheless, these properties can be measured, monitored, and used to determine an ecosystem's overall health. Take nutrient cycles, for example (other examples would be carbon cycles, nitrogen cycles, and water cycles). In parts of the world where lizard diversity and abundance are high (most intermediate to low latitudes), carnivorous lizards consume enormous numbers of in-

The spark of wonder in Amy Couper's eye as she marvels at a gecko walking up a pane of glass is a testament to children's fascination with the natural world. (Steve Wilson)

sects on a daily basis. These insects contain nutrients and stored energy directly or indirectly derived from plants. In turn, higher-order consumers such as birds, snakes, mammals, and even other lizards feed on lizards, thereby transferring nutrients and energy up food chains. Stability of food chains requires adequate energy and nutrient transfer across levels, and healthy ecosystems require relatively stable food chains. Complex ecosystems with large numbers of connections are usually stable over time, whereas simple ecosystems with fewer connections are more susceptible to catastrophic collapse following relatively minor perturbations.

Lizards, like all life forms—including humans—contribute to the complexity that sustains stability. Although we often forget our connections to the natural world as we drive through our paved-over landscapes, make no mistake: the future of humans, as of all life, depends on the preservation and maintenance of complex ecosystems. Understanding ecosystems so that we can continue to manage them thus remains one of our greatest challenges.

LIZARDS IN CULTURE AND MYTHOLOGY

Lizards have been woven into human culture throughout our evolutionary history. Human primate ancestors in tropical Africa certainly knew that large varanids were dangerous and some smaller lizards were edible. Even ignoring these practical concerns, just think how such animals as Komodo monitors, male Jackson's chameleons, or horned lizards must have played on aboriginal peoples' imaginations in their respective corners of the world. To the naive and rapidly developing human mind, these bizarre creatures must have conjured up powerful images of gods and the forces under their control: lightning and thunder, drought and rain, volcanoes and earthquakes. We sample a few such stories in what follows, but we encourage interested readers to track down others, especially ones not yet recorded, for they represent a fascinating aspect of the history of our own species.

The role of lizards in native cultures of South America is poorly known. Because native South Americans have been continually displaced throughout their history, first by other indigenous groups and later by a combination of colonists and other displaced Indians, it is difficult to determine how recently lizards appeared in their art

Petroglyph of a chuckwalla, *Sauromalus obesus*, from the White Tank Mountains of central Arizona. (Jim Rorabaugh)

or culture. Making the situation even trickier is the fact that few anthropologists or historians have asked for information about lizards. As a result, nearly every book on South American Indians contains data on mammals, birds, and fishes, but rarely anything on lizards. Yet considering the abundance and visibility of large teiids and iguanids throughout the continent, and given that most indigenous cultures hunted to at least some degree, native peoples must have recognized and perhaps developed taxonomies for these animals, especially those that they ate or were threatened by.

The Canela Indians of Maranhão, Brazil, present a good example of how South American Indians use taxonomic systems. Mammals and birds, which are hunted, are often recognized to the species level, whereas most lower vertebrates are not (Vanzolini 1956). However, when shown seven lizard species occurring in the area, most Canelas recognized and had names for five: *Phyllopezus pollicaris* (Gekkonidae), *Iguana iguana* and *Tropidurus torquatus* (Iguanidae), and *Ameiva ameiva* and *Tupinambis nigropunctatus* (Teiidae). The other two, *Mabuya nigropunctata* (listed as *M. mabouia* by Vanzolini) and *Micrablepharus maximiliani* (Gymnophthalmidae), were considered adults and juveniles of the same species by some Canelas. However, an old and apparently knowledgeable Canela recognized them as distinct, calling *M. nigropunctata* "ti-glo-ti" and *M. maximiliani* "ti-glo-ré." Clearly, the Canela were in tune with the natural world surrounding them, including the lizards.

The information about Canela lizard awareness was gleaned by Paulo Vanzolini, a world expert on South American lizard taxonomy. Anthropologists have been less successful at extracting such information from South American Indians. David Maybury-Lewis, for example, studied the Shavante societies, at one time distributed between the Araguaia and Tocantins Rivers in the central plateau of Brazil, and observed that they hunted, captured, and ate everything from deer, peccaries, and anteaters to macaws and even beetle grubs. Yet, he said, "I have never seen Shavante eat reptiles. They are not in any case very plentiful on the savannah, and the Shavante certainly do not seek them out at the water's edge" (Maybury-Lewis 1967). We may never know whether the Shavante included lizards in their diets, but we do know that Brazilian "savannah" (correctly, cerrado) contains both high lizard diversity and abundance, and that the water's edge is not the best place to find them (Colli et al., in press).

Similarly, in a discussion of the Yanoama (Yanomamö) of the Parima highlands of Venezuela, which borders Roraima in northern Brazil, William Smole (1976) claims that "salamanders and certain toads are eaten," when in fact no salamanders occur anywhere near the region! The only "salamanderlike" animals would have to be small lizards.

Not all anthropologists spread misconceptions about native people's relationship to the natural world, however. Indeed, some have made highly insightful contributions to our understanding of the roles lizards play in native cultures. For example, thanks to the work of anthropologist Ellen B. Basso (1973) we know that the Kalapalo Indians of the upper Xingu region of central Brazil classify lizards within a broad group of "land animals" *(nene),* giving them status equal to cats, deer, and armadillos. Except for monkeys and coatis, they eat no "land animals," so lizards are relatively unimportant, at least as food, in their culture. In sharp contrast, the Archuar of the upper Amazon in Ecuador have named about 600 animal species that live around them, including 48 species of reptiles, many of which are lizards. This comprehensive taxonomy is not dependent on species that figure in their diet: the only reptiles that appear in summaries of their hunts, for example, are caimans, and even these are rare (Descola 1994). The Suya of the upper Xingu have a different way of classifying animals: by smell. Animals with strong smells are considered dangerous, whereas those with weak smells are considered harmless. In their "smell scale," lizards (probably *Ameiva ameiva,* which they called by the Portuguese term *calango rabo verde,* and tropidurines, which they called *lagartixa*) are rated "bland." Nevertheless, the eating of lizards is forbidden among the Suya (Seeger 1981). Although the origin of their taboos has not been identified by anthropologists, we now know that tropidurines eat lots of ants, which contain noxious chemicals, and as a result, these lizards might make a human who eats them sick.

Throughout much of Amazonia, both Indians and rural Brazilians believe some lizards to be venomous. The large gecko *Thecadactylus rapicauda* is particularly singled out on this count. Bill Lamar informs us that the Barasana of the Rio Papuri call geckos *anya rihoa,* which means "snake-head." They also think that geckos climb on a person's body and inject venom with their feet. The thornytail, *Uracentron flaviceps* (called *rima yua*), is believed to sting as well. And in fact, Laurie Vitt once did receive multiple stings while climbing trees in pursuit of *U. flaviceps*—not by lizards, but by ants living in the trees!

The place of lizards among the Mayans of Mexico and Central America is well summarized in an outstanding treatise by Julian Lee, *The Amphibians and Reptiles of the Yucatán Peninsula* (1966), from which most of the following is derived. As local environments deteriorated fol-

lowing the demise of the Mayan empire, due to both overpopulation and overexploitation, Mayans appear to have incorporated more and more reptiles into their diets, particularly turtles, iguanas, and spiny-tailed iguanas. During colonial times Mayans continued to harvest iguanas for food, and some present-day Mayans eat not only these large lizard species but also *Ameiva undulata* and *Corytophanes.* As in some other Indian cultures, *Iguana* fat was considered to have medicinal value (Thompson 1963).

A few lizards are considered dangerous by Mayans. Severe headaches purportedly result from even glancing at brightly colored adult male *Ameiva undulata. Laemanctus serratus,* called *yaxtoloc,* is believed to inflict a bite that heals very slowly. Like many other New World Indians, Mayans have special superstitions about geckos. Some geckos, for example, are called *escorpión,* from a belief that they deliver venom by throwing their tails. This belief could have its origins in the similarity between eublepharid geckos, which walk with their tails raised, and scorpions, a possible example of mimicry (Parker and Pianka 1974); mistaken identification would indeed result in a painful sting. The strangest belief attributable to

While conducting fieldwork in the northern Brazilian state of Roraima during the summer of 1991, my wife, Jan, and I were offered an opportunity to make some reptile and amphibian collections for the Museum of Zoology at the University for São Paulo through a Brazilian colleague, Celso Morato de Carvalho. Gaining access required permission from the local Indian agency, Fundação Nacional do Índio (FUNAI), as well as from the Catholic bishop for the region, since a Catholic mission was in the reserve. Once we had jumped through these hoops, we were on our way.

We were met at a river by the mission padre, Guilherme Damiol, who helped load our equipment into his truck. We then sped down an arrow-straight dirt road (a portion of the abandoned trans-Amazon highway), leaving the colonized

part of Roraima in the rearview mirror. I rode in the back of the truck with a Yanomamö chief and a Macuxi Indian who worked for the padre. After stopping at several small Yanomamö villages along the 100-km drive, we arrived at Missão Catrimani. The following morning, virtually every Yanomamö in the area passed by to check out the pale gringos from the north. We then spent nearly two weeks collecting lizards, snakes, and frogs, some of which the Indians brought to us.

Having accomplished our collecting goals, we decided to create a laboratory exercise to determine how well the Yanomamö understood their own fauna. We learned through Padre Guilherme that the chief was the wisest man in the village and he would take our little exam.

We laid out an array of freshly preserved specimens, including several du-

plicates to check for consistency. We then showed each one to the chief and asked him a series of questions: what is it called, what does it eat, where does it live, is it active in day or night, and so on. We also asked some general questions about the Indians' beliefs with respect to amphibians and lizards. Padre Guilherme, who spoke both Portuguese and the local indigenous language, served as our intermediary. We taped the entire session.

Our findings were interesting. All *Anolis* lizards were called *oramisiparoa* (the local name, but using Portuguese phonetic spelling), where, according to the chief, *orami* means throat, *si* means surface, and *paro* means pain. The first part of the name clearly refers to the dewlap, and the last part may refer to dewlap expansion, which could appear painful. The chief did not have a name

Coleonyx, called *ix-hunpekin,* is that it throws its tail when covered by a person's shadow, resulting in a severe headache. The headache can apparently be cured by a plant with the same name. *Coleonyx* is believed by many to be so poisonous that simply brushing up against one can cause death in a short time.

In myth, too, lizards play a major role among the Mayans. They, together with some other Central American Indians, believed, for example, that the world was formed from the back of a lizard or crocodilian of immense proportions. One name for this deity, Itzamná, translates to "Iguana House." Itzamná was believed to exist in quadruplicate as four lizards, each facing a cardinal direction, that formed the four sides of the "iguana house." Iguanas appear in other Mayan legends as well. Finally, lizards, most likely iguanas and spiny-tailed iguanas, were sacrificed in some Mayan rituals and thus held special significance.

The Anasazi, Hohokam, Mimbres, and Mogollon Native American cultures used horned lizard images on pottery, petroglyphs, effigy bowls, figures, and shells, and the Hopi, Navajo, Papago, Pima, Tarahumara, and Zuñi cultures portrayed horned lizards in their ceremonies and stories as symbols of strength. Piman people believed that horned lizards would cure them of a persistent illness if they appealed to the lizard's strength and showed them respect. Cures were effected by singing songs describing the animals and placing a lizard fetish on the patient's body (Sherbrooke 1981). Native Mexican people also respect horned lizards, attributing to them the words "Don't tread on me! I am the color of the earth and I hold the world; therefore walk carefully, that you do not tread on me" (Lumholtz 1902). A Mexican common name for horned lizards is *torito de la Virgen,* or "the Virgin's little bull," both because of their horns and because of their blood-squirting behaviors, which are likened to weeping tears of blood (Manaster 1997).

Horned lizards were first introduced to people in Europe in 1651 by the Spaniard Francisco Hernández, who observed one squirt blood from its eyes; he marveled at this behavior in his report on the first scientific expedition to Mexico by Spain. Over a century later, in 1767, a Mexican cleric of Spanish descent, Francisco Saverio Clavigero, described his wonder at horned lizards in his illustrated volumes of Mexican history (Manaster 1997).

---

for *Coleodactylus* or *Gonatodes,* although he had seen them. *Thecadactylus* was called *koikoiank; Mabuya* (two species) were called *haremukeruk; Ameiva ameiva* was called *waiashi; Kentropyx calcarata* was called *shamamuk;* and *Tupinambis teguixin* was called *oasukar-a.* Surprisingly, he had names for the four tiny gymnophthalmids that we showed him as well: *Arthrosaura reticulata* was *ihuruamish-i; Cercosaura ocellata* was *reorea; Gymnophthalmus underwoodi* was *haremukeruk;* and *Leposoma percarinatum* was *temtem.* According to the chief, *Leposoma* makes a call that sounds like the Yanomamö given name Temtem, hence its name, though in fact *Leposoma* produces no sounds. Use of the same name, *haremukeruk,* for *Mabuya* and *Gymnophthalmus* is not surprising, considering that *Gymnophthalmus* is very skinklike. The Yanomamö in this group were afraid of gymnophthalmids because they believe that they hold spirits *(reore)* associated with magic. If one is bitten by a gymnophthalmid, necrosis supposedly results. *Cercosaura,* which the chief had identified as *reorea,* seemed to have special significance; the added *a* on the end of the word for spirits identifies it as "the" *reore* or spirit.

Even though he had no general name to cover all lizards, the chief believed that lizards (the species we showed him, at any rate) were "friends" with other lizards, but lizards and snakes were enemies. This was based on his observation that some snakes ate lizards. The exact meaning of "friends" was not clear, but it seems to be used in a much more general way than we might use it.

When it came to easily observable natural history information (where the species lives, when it is active, what it eats), the chief performed quite poorly, answering more than 90 percent of our questions wrong. This was surprising, considering that he had names for many species. Perhaps, even though as chief he was considered wise, he was simply a very poor naturalist. Then again, since the Catrimani Yanomamö had historically lived in some low mountain ranges (the largest of which they called Opúktheri) prior to moving into the lowlands, perhaps this herpetofauna was simply foreign to him. This might explain why some of the information was so general and other information was wrong. *(VITT)*

In 1897 in Eastland County, Texas, at a dedication ceremony for a new county courthouse, a horned lizard was placed, along with a Bible and other suitable objects, in a time capsule to add appropriate flair to the occasion. Thirty-one years later the courthouse was demolished, and three thousand people gathered to see how the lizard had fared. A judge removed the contents of the time capsule and saw the lizard; an oilman retrieved the dust-covered lizard and passed it to a reverend, who in turn handed it over to another judge. To the crowd's delight, the horned lizard awoke! No one expressed any doubts about its longevity. Over the years, however (not to mention a beer or two), some have claimed that the original dead lizard was palmed and a new, live one substituted (Manaster 1997; Welch 1993). We'll never know if it was the judge, the reverend, or the oilman!

Gila monsters played a role in many Native American myths and appear as pictographs on rocks and pottery. Some Indians wore sections of the beaded skin of a Gila monster's tail as decorative finger rings (perhaps giving rise to the widespread Indian predilection for beaded decorations). For Apaches, to see a Gila monster was good luck: it meant rain was imminent (Brown and Carmony 1991). Members of a tribe in Utah thought that these lizards could produce good or bad weather at will and so should not be molested (Bogert and Del Campo 1956).

On the other side of the world, Australian Aborigines have an exceedingly rich mythology of "dreamtime" stories about the origins of the sun and moon, the creation of various topographic features, the causes of weather, the ecological role of humans, as well as innumerable, very interesting myths about all sorts of animal species, including lizards. Many of these myths have been depicted by the Australian artist Ainslie Roberts in fanciful, sometimes spectacular, illustrations (Roberts and Mountford 1965, 1969, 1971; Roberts and Roberts 1989).

Australian Aborigines are known for their artistic talents, using red and yellow ochers, white clay, and black charcoal mixed with fat for paints. Elaborate rock paintings are found across the continent. Bark paintings of northern tribes depict legends, often showing "x-ray" versions of both animals and humans, accurately depicting bones, muscles, and internal organs. Eric Pianka acquired three such paintings in Darwin, painted by Lipundja at the Milingimbi Mission in Arnhem Land. The largest one, measuring 60 cm x 125 cm, depicts a scene at a waterhole with a water monitor lizard (or goanna:

Varanus mertensi, a species described scientifically only in 1951!), a snake, an anhinga (a predatory aquatic bird, also known as a "diver" or a darter), plus many fish and yams. The following description is pasted on the backside:

Djambarrpuynu Mortality Rites
The above cycle tells in song and dance the habits of the various flora and fauna found in and around a certain waterhole on the nearby mainland. Here a large fresh-water goanna is attacking a grass snake, whilst nearby a diver prepares to plunge upon unsuspecting fish. Around the waterhole grow many yams.

Aborigines believe that the dead go on living as spirits who must be appeased lest they pester or even kill the living. A persistent watering hole, for example, is imagined to be blessed with the friendly spirit of a lizard; if that lizard's spirit is offended and driven away, the waterhole will dry up. In the Numarika Swamp on the island of Groot Eylandt in the Gulf of Carpentaria lives Ipilya, the spirit of an enormous gecko about a hundred meters long! When the wet season is about to start, Ipilya eats large quantities of water-grass and drinks enormous quantities of water from the swamp, which he squirts up into the sky. The water quickly forms thunderclouds, which the grass binds together. Soon monsoon rains begin to fall and lightning strikes the ground. Then the giant gecko roars with the voice of thunder, delighted with his work and the benefits he has bestowed on the earth and all its creatures. After the rainy season is over Ipilya returns to his swamp, where, except to punish intruders, he rests quietly until the next wet season is due (Roberts and Mountford 1969). This myth is appealing because, indeed, geckos are one of the few groups of lizards with a voice—though it is more of a squeak than a roar.

Another dreamtime story involving lizards explains the origin of day and night (Roberts and Mountford 1971). When the world was young, light was provided by the great fire of a cannibalistic sun-woman, Bila, who cooked human victims. Lizard-man Kudnu, a famous boomerang thrower, and gecko-man Muda went to visit their neighbors, the euro-people. Upon discovering that all the euro-people had been killed by the sun-woman's dogs and dragged to her camp to be devoured, Kudnu and Muda decided to kill Bila in revenge. When the sun-woman saw the lizard-men approaching, she howled with rage and grabbed her boomerang to throw at them, but she was too slow: before she could hurl her weapon, Kudnu's boomerang wounded her badly. She then transformed herself into a ball of fire and disappeared over the horizon, leaving the world in complete darkness. The two lizard-men were terrified by the calamity they had caused. Kudnu decided to try to bring the light back with his remaining boomerangs. He threw one north, but the darkness continued; he then threw two more, one south and one west, but still it remained pitch black. When he threw his last boomerang east, however, a great ball of fire arose and traveled slowly across the sky, eventually disappearing below the western horizon—and thus were created day and night. After this, no Aborigine would kill a monitor lizard or a gecko, for these creatures had saved humans from destruction by creating the day, a time for gathering food, and the night, for rest and sleep.

Still another engaging Australian legend tells how two species of monitor lizards—*Varanus tristis* and *V. giganteus,* the perentie—got their coloring, by painting each other. The perentie went first, and did a beautiful, artistic job painting scales in a sort of rosette pattern on the smaller *tristis* (an accurate description of that lizard's coloration). But when the *tristis*'s turn came, about halfway through it grew tired of painting the big perentie, and finished up by throwing the bucket of paint on the other lizard's back half (again, perenties really are colored in such a pattern). The legend finishes with the statement that to this day, perenties keep *tristis* in the trees on account of their ancient betrayal (also an accurate statement about the habits of these lizards!).

Medieval cartographers marked their maps of the Sunda Shelf with an intriguing warning, "Here be dragons," a reference to giant Komodo monitor lizards. In nearby Indonesia, Komodo dragons have impressed all inhabitants of the Lesser Sunda Islands, past and present. The original inhabitants of the Komodo, the Ata Modo, believed themselves to be kin to the giant lizards. One day, it was said, Najo, a man living on the island, encountered a beautiful spirit woman named Putri, whom he married. Soon they had twins, but only one was a human child; the other was a Komodo dragon. Devastated and humiliated, Putri left the village and Najo, and raised her children in a cave. Eventually the Komodo dragon went its own way, founding the dragon population, and the child left to start the Ata Modo clan. The clan therefore believed that they must take care of their "brother," the Komodo dragon; not to do so would endanger the clan because it would anger the powerful spirit Ina Babu, who protected both. Consequently, portions of fish catches were left on beaches for the lizards. Because the Komodo dragon was in essence kin of the clan, their futures were considered interdependent. Extinction of one necessarily meant extinction of the other.

Although very few modern inhabitants of the Lesser Sundas can be traced directly to the Ata Modo, nevertheless, the legend lives on. During the 1980s, the Indonesian government was planning to relocate inhabitants of the national park on Komodo Island. One day, some residents of Sumbawa, an island near Komodo, saw two adult Komodo dragons emerge from the sea and come onto the beach. The arrival of the dragons verified these people's link to the Ata Modo, and convinced them that descendants of the Ata Modo must remain on the island to ensure that the dragons—and so their human cousins—did not go extinct.

LIZARDS TO EAT AND WEAR

Lizards, especially iguanids and varanids, are sold in third world marketplaces around the world. Because little refrigeration is available in such poor countries, lizards are kept alive until it is time to consume them. In Thailand, people eat the agamid *Leiolepis*. In Central America at Lent, great piles of hog-tied *Ctenosaura* (their mouths sewn shut so they can't bite) are offered for sale. Iguanas are sold alive in markets throughout southern Mexico and Central America as well.

Aborigines living in Australian deserts, who have tracked down and hunted lizards for millennia, developed valuable tricks that aid in capturing lizards. A fleeing monitor, for example, can be stopped dead in its tracks by an imitation hawk call, making the lizard an easy target for a throwing stick or spear. A big varanid is viewed as "bush tucker": food for the taking. Aborigines have traditional recipes for preparing and cooking various lizards. Take a monitor lizard, for example: Using a hooked stick, first extract its stomach and intestines through its mouth and discard. Never cut the lizard open: its intact skin acts like an aluminum foil wrapping. Next, flame the lizard over a fire, then bury it under hot sand and ashes for ten to fifteen minutes, until it begins to sizzle. Upon removal from the cooking pit, its skin peels away easily, exposing the succulent white meat (fat bodies—fat storage organs in the body cavity and tail—are preferred). When Aborigines are really hungry, they eat the entire lizard except for its head; even its brains are often sucked out. In times of severe food scarcity, once the soft parts are picked off, even the bony remains, after being crushed with rocks, are consumed.

Large numbers of lizards are harvested for their skins in third world countries, especially in South America (tegus) and southeast Asia (varanids). Lizard (and snake) skin boots fetch exceedingly high prices. Annual harvests of almost three million monitor skins have been estimated (DeLisle 1996). Anyone who has visited Mexico knows that virtually every tourist shop sells tacky stuffed iguanas—an embarrassment in terms of home decor, and a tremendous waste in terms of natural resources.

In South America, several species of *Tupinambis* are exploited for food and commercially for their hides. A vast majority of the trade in tegus takes place in Argentina and Paraguay, where almost two million *T. rufescens* and *T. merianae* (formerly *T. teguixin;* see Avila-Pires 1995) were harvested annually during the 1980s (Hemley 1984; Fitzgerald et al. 1991, 1994), giving these lizards the dubious honor of being the most heavily exploited reptiles for a time. Presumably reduced demand caused the trade to fall off to less than a million per year during the 1990s. Needless to say, Argentina and Paraguay must have had a lot of tegus, and a lot of tegu habitat, to sustain such a harvest level. Trade also occurs in Bolivia and parts of Brazil.

Harvested for food on the beaches where it lives and popular in the pet trade, the butterfly agama, *Leiolepis guttata,* may have a dim future. (R.D. Bartlett)

Most of these South American lizard skins are exported to places such as the United States, Canada, Mexico, and Europe, where they are primarily used for cowboy boots (Fitzgerald 1994b; Rose 1992). Considering that neither of us has ever seen a *real* cowboy wearing tegu-skin boots, we surmise that the primary market is the wanna-be urban cowboy. Unfortunately, markets for lizard hides are driven by fashion rather than need.

The meat and fat of lizards are used by people who harvest them, primarily rural poor and indigenous Argentinians and Paraguayans, who also benefit substantially from skin sales (Donadio and Gallardo 1984; Norman 1987; Fitzgerald et al. 1994). Unfortunately, tegus are secretive and wary, which has made it difficult to obtain life history data to determine the impact of large-scale harvesting of these lizards in the long term. Conservation programs put into place by the governments of Argentina and Paraguay since 1990 require that tegu harvests be monitored, providing data on size distribution, sex ratio, and total number of skins exported from those countries each year. These data will provide information about demographics of harvested tegu populations and will allow population biologists to make inferences about the sustainability of tegu hunting. Harvest limits of 1 million for Argentina and 350,000 for Paraguay were also established.

Computer simulations of population levels performed by Lee Fitzgerald (1994a) reveal that populations of tegus (females of which produce many eggs, with larger females producing larger clutches; Fitzgerald et al. 1993) are relatively sustainable because the long adult life span allows repeated opportunities for breeding even during years when juvenile mortality is high (low recruitment). Exceptional recruitment in just a few years, Fitzgerald found, is enough to result in rapid population recovery. This represents the classical "bet-hedging" life history model (see chapter 5). Of course, this is a mathematical model, and all models are constrained by assumptions—one of which, in this case, is continued high harvest rates. Harvestable lizards always run the risk that a new capture technique will be found or that hunters might become particularly desperate, thus driving the population to zero. Conversely (and on a more positive note), if more census information can be gathered, more precise population models will allow managers to control the resource more efficiently. Because of the economic impact of tegus at the local level, management strategies that use life his-

"Lizarding" can be even more fun than fishing. Indeed, the two activities are somewhat similar, and I suspect that they satisfy our primitive hunter-gatherer instincts. In both cases, some skill is required in pitting one's wits against those of a fast, agile, small animal. I haven't yet attempted to catch lizards with a hook and line, though that might be possible (they sometimes lunge open-mouthed at snares). Instead, I catch lizards by grabbing them by hand or noosing them with a fishing rod and a tiny nylon snare around their neck. Individuals of some species are so slow that they can just be picked up—no skill required. Catching an alert fast lizard by hand, however, requires extreme speed and coordination, as well as some cunning in the approach. Nonherpetological laypeople are often quite impressed, even startled, when watching a herpetologist attempt to capture a snake or lizard. I have seen two-hundred-pound men throw their entire body into a bush in a superhuman effort to catch a lizard or a snake. (Indeed, I do it myself!)

All herpetologists have stories to tell, and some of them sound like "fish tales." One of my own favorites goes like this. In the mid-1960s, I spent eighteen months in the Australian deserts studying lizards. I learned a lot about dozens of species, including how to track down large monitor lizards. I returned State-side, took my job at the University of Texas, and began teaching. After a decade riding the academic merry-go-round, I was eager to return to Oz to resume fieldwork. In 1978, a Guggenheim fellowship allowed me to go back down under. Leaving my family behind (they joined me later), I headed off alone into the outback. My first stop was a former study site where Helen and I had discovered half a dozen undescribed species of skinks a decade before. As I was nostalgically setting up camp late in the day, I saw fresh tracks of a large sand goanna (*Varanus gouldii*). Naturally, I couldn't resist following them. The tracks led me on a circuitous course for nearly a kilometer, ending at a large burrow with a bucketload of sand at its entrance. Crisp imprints of the monitor's belly scales told me it had basked there moments before, and footprints and a tail lash mark indicated that it had gone down into its burrow.

Varanus gouldii dig U-shaped tunnels, with one end of the U being an escape hatch that stops just below the surface (when an enemy threatens, monitors pop out of this escape hole and dart away). Knowing this, I took my shovel and, about 1.5 m away from the open entrance to the burrow, began scraping the sand's surface in a rough circle. When the sand fell in, I knew I had located the escape hatch. Then, judging that the lizard would be in the bottom of the U about halfway between the two openings, I positioned myself in the middle, raised my right foot, and stomped down hard with all my might. I felt the sand collapse into the burrow below. I knelt down and thrust my right arm into the loose sand, wiggling and rotating my hand until my arm was into the sand all the way up to my elbow. Locating the goanna by touch, I grabbed hold of it, then pulled it out of the sand and stood up. Holding the meter-plus lizard by its shoulders, I felt triumphant, like a true master of my domain. (Years later, I learned that Aborigines catch goannas this way, so all I had really done was reinvent the wheel.) Unfortunately, no one was there to witness my Great Moment.

Laurie Vitt thinks that herpetology should be a spectator sport like football. Indeed, we have contemplated making a video for one of the next meetings of the American Society of Ichthyologists and Herpetologists, in which you would see many such amazing events, each followed by a flash to crowds roaring and cheering, then on to another great moment in herpetology, more cheers, etc. Of course, we would expect our salaries to be commensurate with those of coaches and professional athletes!

(PIANKA)

tory information to determine how best to harvest them stand to have a far-reaching positive impact on people whose livelihoods depend on these lizards.

LIZARDS AS SOURCES FOR PHARMACEUTICALS

Scientists have recently discovered useful pharmacological agents in varanoid lizards. Venoms of *Heloderma* are complex mixtures of over a dozen small peptides, neurotransmitters, proteins, and other molecules, which have powerful effects on mammalian physiology (Raufman 1996). These are molecular analogs to important mammalian hormones such as serotonin and secretin. One lowers blood pressure; another regulates the release of insulin; another attacks certain cancers. Such molecules could prove useful in drugs to control hypertension, diabetes, and cancer. Indeed, one drug derived from Gila monster venom is currently being evaluated for treatment of type 2 diabetes (Edwards et al. 2001, Seppa 2001). Another molecule, a peptide called gilatide, improves memory in rats and is a candidate for development of a drug to treat Alzheimer's disease. Some of these molecules are also found in snake venoms, which tend to be far more toxic.

Komodo dragons may possess a natural immunity to bacterial infection. Their saliva harbors more than fifty different strains of bacteria, some of which are highly septic. If a Komodo does not kill its prey outright, its bite introduces germs potent enough to kill its quarry with a massive infection in a few days. Yet when these big lizards fight each other, their bites do not become infected. These observations prompted an investigation of the blood plasma of Komodo dragons. Recent preliminary work by Dr. Gill Diamond at a medical school in New Jersey has identified a powerful antibacterial agent in the plasma, which could be developed as a new antibiotic in our ongoing worldwide battle against the evolution of antibiotic-resistant microbes (Diamond, pers. comm.).

LIZARDS AS PETS

Lizards are sold in many pet stores and are popular as "domestic companions," particularly in the Temperate Zone—possibly because low-latitude enthusiasts can enjoy lizards on their walls and in their yards, even within large cities. Some lizards make great pets, but many species are ill suited for captivity and soon die. Provided that a lizard has adequate fat reserves, it can live for months in a deficient captive environment, giving the false impression that it is doing well. Before buying a lizard as a pet, therefore, it is important to do thorough research: be sure you know what you're getting and, assuming it is suitable for captivity, how to take care of it.

Most professional herpetologists had snakes, lizards, turtles, frogs, or salamanders as pets when we were young, and the experience was pivotal in our becoming professionals. Many of us still maintain a few reptiles or amphibians and no doubt always will. As academics, we could construct an esoteric discourse on why people should not keep these animals, which seem incapable of returning affection or communicating in the ways dogs and cats do. We would be hypocrites if we did. The truth is that for those who want a "pet" lizard and have given the idea serious thought, the experience can be both rewarding and educational.

We certainly don't advocate that everyone go out and buy a lizard or catch one in the wild. Rather, we provide some points to consider before making a decision, and offer a few suggestions. First, if your strategy is to go into the hills and bring one back, we recommend that you keep it in a terrarium for no more than a week and then return it to where you found it. Keeping field-caught lizards for extended periods is difficult and requires a major commitment, and you must know about the species' requirements. Most lizards are insectivorous; where will you get food for it? Most lizards have temperature and light requirements; do you know what they are, and can you reproduce them in a cage? Of course, if you house a wild-caught lizard in a terrarium that offers a range of temperatures, you will rapidly learn a lot about lizard thermoregulation without even reading a book. If you give it a variety of insects, you will learn that it prefers some insects over others. If you hassle it, it may bite, teaching you a lesson in lizard defense. Remember, from its perspective, you are an alien giant. If you decide a lizard isn't for you, let it loose where you found it, and nothing is lost. If, however, you decide to keep this field-caught lizard, it is time to hit the scientific literature and search for the best information available on what the lizard does in its natural habitat. You may learn that this lizard has particular requirements that make it inadvisable to keep it, or you may learn what to do to keep it healthy.

Several species of *Agama* do not appear to distinguish between rock faces in their natural habitats and rock walls constructed by humans. (C. Ken Dodd)

The Mediterranean gecko, *Hemidactylus turcicus,* is so closely associated with humans that it is difficult to imagine what its original habitat was. (Steve Wilson)

If searching the hills is not an option where you live, your decision becomes more difficult. Once you purchase a lizard, you must be committed to it: the option of letting it loose doesn't exist. Choose smart! Juvenile iguanas, for example, are spectacular, tame easily, eat plants, and will feed out of your hand. At first glance, they appear to be perfect pets. But what happens when your cute little iguana reaches 2 m in length? Do you really want to add a large, temperature-controlled greenhouse to your house or apartment? What's more, large iguanas can be downright dangerous. A large, excited iguana can tear open a person's face or arm with its claws, and iguanas are known to have bitten pieces of flesh from their owners. Iguanas, as attractive and interesting as they seem, are bad choices, especially if you are starting out in the lizard-keeping business.

Large teiids and varanids also make spectacular pets when they are small. However, they can be difficult to keep because they are not herbivorous. Moreover, they have high activity levels, thermal requirements, and spatial needs that make them poor choices for anyone not willing to invest seriously in lizard housing. Indeed, as they grow, their spatial requirements increase—and if they should bite, consequences can be serious.

A good choice, especially for first-time lizard keepers, is a leopard gecko. These lizards are not too small and not too big; they do very well on a diet of crickets and meal-worms, which can be purchased at pet stores; and if you lose interest, they are so charismatic that you will have little problem finding them a new home.

LIZARDS AND THE PET TRADE

Wheeling and dealing in lizards can be profitable. Traffic in lizards captured in the wild should be illegal, but in many places, especially in third world countries, it is not. Some dealers import hundreds or even thousands of lizards that have been removed from their natural habitats. Nevada still ignorantly sells licenses to collect lizards: kids earn $0.25–1.00 for each lizard, which commercial merchants turn around and resell for $20–30 to hobbyists. A nice profit margin, but certain death for most lizards collected.

In 1967, the progressive Texas legislature passed protective legislation preventing collection, exportation, or sale of horned lizards from the state. Prior to this legislation, hundreds of thousands of horned lizards were exported—dead and alive—from Texas every summer to tourists, curiosity seekers, and would-be pet owners, leading to the certain demise of the lizards. In many but not all parts of world, lizards are now protected by laws from general collection. However, enforcing those laws is difficult. Thousands of lizards are smuggled out of protected areas to be sold in the pet trade every year.

Some more enlightened pet dealers have set up breeding colonies, focusing their energy on species relatively easy to maintain and breed in captivity. This is far preferable to exploitation of wild-caught animals, for several reasons. First, it has no impact on natural populations beyond initial collection of breeding stock, with occasional recollection or trading needed to maintain genetic diversity. Second, lizards provided to retailers are healthy and free of disease and parasites. Third, most captive-reared lizards are already proven to be amenable to captivity, and instructions on proper care can be provided. Finally, a certain degree of accountability is involved, such that purchasers can have some confidence in the quality of their purchase.

Success of breeding programs for the pet trade has been phenomenal, resulting not only in high profits for breeders and dealers but also in increased awareness of and interest in lizards on a global scale. The downside of this industry is that it provides an avenue for illegal trade and exploitation under the guise of legal trade. There is nothing to keep wild-caught animals from being added to stocks and sold as captive bred. However, considering the cost of energy (gasoline and travel in general), time involved collecting, and risk of collecting illegally, such trade is likely to replaced in the long term by captive-bred supplies.

Another serious downside is introduction of unnatural genetic strains. Captive breeders have discovered, just as domestic animal breeders did long ago, that deviants can be sold for more than normal animals. Consequently, breeders deliberately produce albinos or cross-breed species and subspecies to produce color patterns that can be sold at higher prices. This low-tech genetic engineering has the potential of impacting natural populations negatively should these animals reenter natural systems.

ENDANGERED AND THREATENED LIZARD SPECIES

For most of the world, too little information is available to determine the status of lizard species (Gibbons et al. 2000). At present, six lizard species are listed as endangered or threatened in the United States and its territories, and another twenty-eight are listed worldwide (table 15.1). Not a single lizard species has ever been removed from the list, indicating that efforts at recovery are either not in place or have failed. The list includes many of the largest, and therefore most conspicuous, lizards in certain families: *Varanus komodoensis,* species of *Cyclura, Xantusia riversiana,* and *Gallotia simonyi simonyi.* It does not include the many lizard species on tropical islands that are losing their habitat due to deforestation (Madagascar, Borneo, the Philippines, Indonesia, and many more) or those mainland species that have lost major portions of their geographic ranges to agriculture, deforestation, urbanization, or other human development, including most of the midwestern United States, the cerrado of Brazil, coastal areas of most countries, and riparian habitats worldwide that have been inundated by floodwaters from dam projects.

Although potential loss of lizard species associated with drastic habitat modification would be relatively easy to document, this is rarely done. Paradoxically, current regulations in most countries (including the United States) require that scientists seeking to document diversity obtain permits to collect lizards. Yet very few countries require developers to obtain permits for the many (and usually unknown) lizards that will die as the result of building projects. Developers are held accountable only when species already listed as threatened or endangered are impacted. Cat owners are not held responsible for lizards killed by their cats.

A more subtle threat to lizard populations worldwide is the use of agricultural chemicals. Long-term effects of most such chemicals on lizards remain unknown, largely because research required for approval of agro-chemicals focuses on instantaneous effects and usually ignores life history consequences of sublethal doses (Rose et al. 1999). But think about it: Water uptake is critical to reptilian egg development, influencing hatching success as well as size and quality of offspring (Gutzke and Packard 1986, 1987; Miller et al. 1987; Packard et al. 1982; Packard and Packard 1986; Overall 1994). Because environmental chemicals readily move with water across porous shells, the likelihood that they may interfere with hormone production and regulation is great (Guillette 1995). And if exposure to low levels of such toxicants during development does not kill the embryo, it may well affect a host of important life history traits in juveniles and adults.

Other chemical threats exist as well. Many byproducts of plastics released into the environment are estrogen mimics (Colburn et al. 1996). Endocrine disrupters, leading to a shift in age or size at sexual maturity, would have cascading effects at the population level, as would feminization of males or interference with females' ability to produce eggs as the result of toxicant exposure during development. These and other human-created threats not just to lizards and their reptilian cousins but to the planet as a whole demand attention—before it's too late.

THE FUTURE OF LIZARDS AND HUMANS

Lizards were here long before humans, and they have evolved through many millions of years of environmental changes. Unfortunately, many lizards cannot cope with the rapid, drastic, man-made habitat modification occurring today (Gibbons et al. 2000). An environmentalist might use stronger words—habitat destruction, deforestation, desertification—whereas a politician would gloss over all of these with a single word: development, or even, to invoke that classic oxymoron, "sustainable development."

Whether you look on habitat modification as destructive or desirable, in the end it is tied to one overwhelming force: human overpopulation. Quite simply, we are

TABLE 15.1

Endangered and threatened lizard species

SCIENTIFIC NAME	COMMON NAME	STATUS*
UNITED STATES AND ITS TERRITORIES		
Ameiva polops	St. Croix ground lizard	E
Cyclura stejnegeri	Mona ground iguana	T
Gambelia silus	Blunt-nosed leopard lizard	E
Sphaerodactylus micropithecus	Monito gecko	E
Uma inornata	Coachella Valley fringe-toed lizard	T
Xantusia riversiana	Island night lizard	T
OTHER COUNTRIES		
Brachylophus fasciatus	Fiji banded iguana	E
Brachylophus vitiensis	Fiji crested iguana	E
Cnemidophorus vanzoi	Maria Island ground lizard	E
Conolophus pallidus	Barrington land iguana	E
Cyclura carinata bartschi	Mayaguana iguana	T
Cyclura carinata carinata	Turks and Caicos iguana	T
Cyclura collei	Jamaican iguana	E
Cyclura cychlura cychlura	Andros Island ground iguana	T
Cyclura cychlura figginsi	Exuma Island iguana	T
Cyclura cychlura inornata	Allen's Cay iguana	T
Cyclura nubila caymanensis	Cayman Brac ground iguana	T
Cyclura nubila lewisi	Grand Cayman ground iguana	E
Cyclura nubila nubila	Cuban ground iguana	T
Cyclura pinguis	Anegada ground iguana	E
Cyclura rileyi cristata	White Cay ground iguana	T
Cyclura rileyi nuchalis	Acklins ground iguana	T
Cyclura rileyi rileyi	Watling Island ground iguana	E
Cyrtodactylus serpensinsula	Serpent Island gecko	T
Gallotia simonyi simonyi	Hierro giant lizard	E
Leiolopisma telfairi	Round Island skink	T
Phelsuma edwardnewtoni	Day gecko	E
Phelsuma guentheri	Round Island day gecko	E
Podarcis pityusensis	Ibiza wall lizard	T
Sauromalus varius	San Esteban Island chuckwalla	E
Varanus bengalensis	Indian monitor	E
Varanus flavescens	Yellow monitor	E
Varanus griseus	Desert monitor	E
Varanus komodoensis	Komodo Island monitor	E

*E = endangered; T = threatened

Source: U.S. Fish and Wildlife Service website, http://endangered.fws.gov.

converting the world's biomass into human tissue at an ever-increasing rate, with no apparent end in sight. A vast proportion of these humans, moreover, are now living their entire lives in large, sprawling cities, which means that an ever greater percentage of the world's population is losing sight of the natural systems that support them.

Overpopulation, a growing lack of care for the natural world, the threats of habitat change and ecotoxicants—all these have a tremendous impact on our planet's biota, including hundreds of lizard species. One of many examples is the clade of giant terrestrial iguanas inhabiting the West Indies.

The giant land iguanas *(Cyclura)* have thrived on islands of the West Indies for at least the past 15 million years. Prior to human arrival, their only predators were raptors and snakes. Adult populations must have been very stable over time because of their large body size, which made them unavailable to natural predators. Land iguanas reach sexual maturity at six to seven years of age and can live forty years (Iverson 1979; Iverson and Mamula 1989). They deposit clutches of two to ten eggs once per year, with larger females having larger clutches, and thus contributing most to future generations. Survivorship is age and size related; older, larger animals have a higher probability of surviving to the next year than younger animals. High survivorship of large animals and high fecundity of large females in the face of relatively higher juvenile mortality are life history characteristics that historically determined long-term persistence of land iguana populations.

Enter humans. Indigenous peoples who first settled the islands harvested adults for food. The impact on iguana populations was no doubt significant, for not only did adult individuals disappear, but their entire reproductive future vanished with them (a tenet of conservation theory as it relates to long-lived, late-maturing animal species; Congdon and Dunham 1994; Congdon et al. 1994).

Before indigenous peoples reached a high enough density to decimate land iguana populations, western Europeans discovered and colonized the islands, bringing with them their pets, farm animals, and the general attitude that the world was theirs for the taking. Goats compete with iguanas for food; pigs and cattle disturb their nests; rats and cats eat their eggs and hatchlings; and dogs kill lizards of all sizes (Derr 2000; Iverson 1978; Denney 1974; Christain 1975; Coman and Brunner 1972; Mittermeier 1972). Mongooses introduced to control rat populations also eat eggs and young lizards. More recently, the developing tourist industry has competed directly with iguanas for land, completely wiping out their prime habitat.

On Pine Cay in the Caicos Islands, an adult population of nearly 5,500 *Cyclura carinata* was driven almost to local extinction by construction of a hotel and tourist facil-

Some adaptable lacertids, like this *Podarcis milensis* sitting on Greek ruins, are often equally as common on man-made structures as in their rapidly disappearing natural habitat. (Steve Wilson)

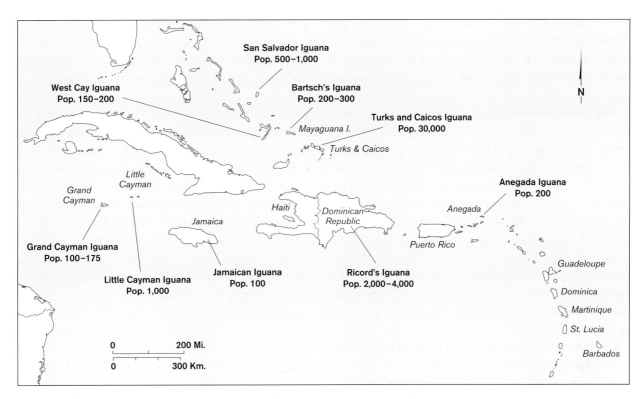

Figure 15.1 Population size for species of land iguanas *(Cyclura)* on various islands of the Caribbean.

ity and introduction of domestic cats and dogs (Iverson 1978). The first impact the cats had was to decimate rat populations. When rats became scarce, the cats moved on to lizards, crabs, and birds. In addition to preying on *C. carinata,* cats decimated populations of curly-tailed lizards *(Leiocephalus)* by digging them from their burrows before they became active. Similar damage occurred on other islands in the Turks and Caicos Banks.

Conservation efforts for ground iguanas were stepped up considerably when a small population of Jamaican ground iguanas was discovered—a species believed to be extinct. A combination of land protection and captive breeding programs in zoos is keeping this and various other land iguana populations from slipping over the precipice into extinction; nevertheless, as human tourism promoters continue to encroach on the islands, the future of these magnificent lizards diminishes. Populations of several species are already so small that long-term survival seems highly unlikely (fig. 15.1).

In the United States, despite protective legislation, horned lizard populations continue to decline and disappear as, every day, hundreds of acres of habitat are con-

A majestic land iguana *(Cylura)* **watches as the habitat its ancestors occupied for thousands of generations is transformed into a tourist facility.** (C. Ken Dodd)

verted into golf courses, strip malls, and agricultural fields in California, Arizona, and Mexico. The Texas horned lizard, *Phrynosoma cornutum,* has disappeared from nearly half its historic range (Donaldson et al. 1994). Populations of many other lizard species are also suffering from dramatic habitat loss due to human encroachment. *Phrynosoma mcallii* and *Uma,* which had narrow ranges even before human encroachment, have lost half their historic sand dune habitat in the last few decades.

Much of this decline is due not to habitat loss per se, but to habitat alteration. After all, housing developments do not have to squeeze out horned lizards if native vegetation is kept intact and cats and dogs are not allowed to run free. Some small Texas towns rightfully boast of their amazing horned lizard populations. Generally, though, development means planting exotic grasses, which choke out native plants and prevent lizards access to dirt to dig their burrows or lay eggs. In addition, along with development typically comes a new set of animals. Ravens and blue jays are not native to the deserts, any more than cats and dogs are. Trash pits become havens and resources for these avian and mammalian predators.

Development also means paved roads and fast cars. Like many other reptiles, lizards often bask on roads, a dangerous predilection for a small animal whose primary defense against predators (cars, in this case) may be to not move. On a small paved military road in Arizona, a full one-quarter of all flat-tailed horned lizards *(Phrynosoma mcalli)* seen were dead, run over by vehicles. Moreover, roads are usually lined with fences and utility poles, great perch sites for avian predators. Road kills may also attract predators, even allowing some, such as roadrunners, to increase in abundance, which leads to a concomitant increase in rates of predation on lizards.

Habitat alteration in Texas and the southeastern United States has promoted the spread of a terrible pest, the red fire ant, *Solenopsis invicta,* accidentally introduced from Brazil. Fire ants can kill almost anything given the chance, and they are fierce competitors against native ants, which horned lizards require for food. (Horned lizards do not eat fire ants, probably because of their different natural history, different venom in the sting apparatus, and different nutritional makeup.) The widespread invasion of fire ants has given all ants a very bad reputation. Since the early 1950s, broad-spectrum pesticides have been sprayed over large areas by air to combat ants and other nuisance insects, including mosquitoes and crop pests. Homeowners regularly use ant insecticides, but in so doing they only make it easier for fire ants to invade because they have no competitors to stop them. Not only do these insecticides deplete the horned lizards' food supply, but ants that are

not killed outright may be eaten by horned lizards, poisoning them. The Horned Lizard Conservation Society was formed in 1990 to try to stop the decline of these charismatic lizards in North America.

We have now come full circle. Like most biologists, we are appalled as we watch the destruction of natural habitats. Humans have now transformed nearly one half of the earth's land surface; not so long ago, we were surrounded by wilderness and wild animals, but now we surround them. As natural habitats vanish, their floras and faunas, many species of which are undescribed, disappear as well. In first world countries, most fertile valleys have filled with urban sprawl and much of the surface has been replaced with concrete, as though we can somehow isolate ourselves from the natural world that constitutes and supports all life on Earth.

As, with increasing technological developments, we have lost contact with our connections to the natural world, we have also lost contact with our hunter-gatherer roots. People no longer know or care where food comes from; instead it is a mere commodity, bought and sold as if it existed in limitless supply. We are expected to think that our resource base will expand indefinitely, when in fact, with continued population growth, it can only contract. Per capita shares of food, water, and land are falling everywhere. All the world's oceans have been heavily overfished, and more than half of all accessible surface freshwater is now used by humans. Freshwater aquatic

While inactive during the day, geckos like this *Oedura* often hide in crevices behind lights, emerging at night to capture insects drawn to the light. (Steve Wilson)

Having successfully survived drastic natural environmental change throughout its evolutionary history in the Middle East, a _Uromastyx_ gazes obliviously into the distance as its natural habitat is converted to agriculture. (L. Lee Grismer)

systems everywhere are polluted, and their resident fishes and amphibians seriously threatened.

Six billion of us currently consume more than half the solar energy trapped by plants, primarily via fisheries, agriculture, pastoral activities, and forestry. More atmospheric nitrogen is "fixed" by humanity than by all other natural terrestrial processes combined.

The frenzy of energy use in which we are engaged contributes dramatically to buildup of greenhouse gases, decay of the ozone layer, and environmental contamination that will be economically infeasible to clean up. Perhaps ironically, however, unlimited cheap clean energy, such as that envisioned with the much-vaunted cold fusion plan, would be one of the worst things that could befall the planet, for it would only enable even more massive energy consumption. Mountains would be leveled, huge water canals dug, ocean water distilled, and water pumped into deserts, which would be turned into green fields of crops—all to further the growth of the human population.

Heat dissipation would of course set limits on this energy consumption, for when more heat is produced than can be dissipated, the resulting thermal pollution would quickly warm the atmosphere to a point where all life is threatened, perhaps the ultimate ecocatastrophe. Die-hard technologists will doubtlessly argue that we will invent ways to shoot our excess heat out into space. Considering current estimates of global warming, they had better get busy.

A large proportion of children growing up today never experience natural environments, or do so in such a cursory way that the connections between us and the natural world cannot be appreciated. Nor are they being educated in basic biological concepts. In some places, attempts are even under way to eliminate evolutionary biology from school science curricula. At the level of higher education, many zoology departments are so intent on supplying the burgeoning health industry (which, rather than addressing *causes* of human health problems, capitalizes on the problems themselves, selling pharmaceuticals that provide symptomatic relief) with doctors, nurses, and technicians that most "real" biology (non-applied, organism-based) has been excised from the curriculum.

Such curriculum planners forget that evolution and systematic biology provide us with the conceptual basis for all areas of biology, including medicine. It is the integrating principle: biology is the science of life. And the science of life gives us the food we eat and it provides us the background for combating new diseases. We desperately need all the basic biology we can get—it is not some luxury that we can live without!

Human survival depends on understanding our natural world in enough depth to manage it in a sustained fashion. Today we have the technology to do science at a rate far and away faster and better than ever before in our history. Now is the best time ever, both at home and in the most remote parts of the globe, to study the vanishing book of life. The battle to keep the planet alive is the greatest challenge that humans will ever face, and we will not win it with arrogant ignorance.

We hope our perspective on the evolution of lizard diversity has provided you with new insights into the fascinating world of lizards. Lizards represent but a single chapter, albeit an important one, in the book of life. Yet all chapters are inherently connected by the systems that organize all life on Earth, including anatomy and physiology, behavior and ecology, and, of course, evolution. If a single piece of information or idea presented in this book adds to your appreciation of lizards as part of our natural heritage, then we will have succeeded in our task. In closing let us just say, if you are afraid of snakes but not of lizards, bear in mind: snakes are lizards too.

APPENDIX

Taxonomic summary of lizard genera of the world (mostly from Zug et al. 2001)

IGUANIA

FAMILY	SUBFAMILY	NO. SPECIES	GENERA
Agamidae	Agaminae	400+	*Acanthocercus, Acanthosaura, Agama, Amphibolurus, Aphaniotis, Bronchocela, Bufoniceps, Caimanops, Calotes, Ceratophora, Chelosania, Chlamydosaurus, Complicitis, Cophotis, Cryptagama, Ctenophorus, Dendragama, Diporiphora, Draco, Gemmatophora, Gonocephalus, Harpesaurus, Hydrosaurus, Hylagama, Hypsilurus, Japalura, Laudakia, Lophocalotes, Lophognathus, Lyriocephalus, Megalochilus, Mictopholis, Moloch, Oriocalotes, Otocryptis, Phoxophrys, Phrynocephalus, Physignathus, Plocederma, Pogona, Psammophilus, Pseudocalotes, Pseudocophotis, Ptyctolaemus, Salea, Sitana, Thaumatorhynchus, Trapelus, Tympanocryptis,* and *Xenagama*
	Leiolepidinae	14	*Leiolepis* and *Uromastyx*
Chamaeleonidae	—	130+	*Bradypodion, Brookesia, Chamaeleo,* and *Rhampholeon*
Iguanidae	Corytophaninae	9	*Basiliscus, Corytophanes,* and *Laemanctus*
	Crotaphytinae	12	*Crotaphytus* and *Gambelia*
	Hoplocercinae	10	*Enyalioides, Hoplocercus,* and *Morunasaurus*
	Iguaninae	29	*Amblyrhynchus, Brachylophus, Conolophus, Ctenosaura, Cyclura, Dipsosaurus, Iguana,* and *Sauromalus*
	Leiocephalinae	28	*Leiocephalus*
	Leiosaurinae	28	*Anisolepis, Aperopristis, Diplolaemus, Enyalius, Leiosaurus, Pristidactylus,* and *Urostrophus*
	Liolaeminae	155	*Ctenoblepharys, Liolaemus,* and *Phymaturus*
	Oplurinae	7	*Chalarodon* and *Oplurus*
	Phrynosomatinae	110+	*Callisaurus, Cophosaurus, Holbrookia, Petrosaurus, Phrynosoma, Sceloporus, Uma, Urosaurus,* and *Uta*
	Polychrotinae	302+	*Anolis* and *Polychrus*
	Tropidurinae	106	*Eurolophosaurus, Microlophus, Plesiomicrolophus, Plica, Stenocercus, Strobilurus, Tropidurus, Uracentron,* and *Uranoscodon*
TOTAL		1,340+	

SCLEROGLOSSA: GEKKOTA

FAMILY	SUBFAMILY	NO. SPECIES	GENERA
Diplodactylidae	—	110+	*Bavayia, Carphodactylus, Crenadactylus, Diplodactylus, Eurydactylodes, Hoplodactylus, Naultinus, Nephrurus, Oedura, Phyllurus, Pseudotheca-dactylus, Rhacodactylus, Rhynchoedura, Saltuarius,* and *Strophurus* (often included in *Diplodactylus*)
Eublepharidae	—	25	*Aeluroscalabotes, Coleonyx, Eublepharis, Goniurosaurus, Hemitheconyx,* and *Holodactylus*
Gekkonidae	Gekkoninae	670+	*Afroedura, Afrogecko, Agamura, Ailuronyx, Alsophylax, Aristelligella, Aristelliger, Asaccus, Asiocolotes, Blaesodactylus, Bogertia, Briba, Bunopus, Calodactylodes, Carinatogecko, Chondrodactylus, Christinus, Cnemaspis, Colopus, Cosymbotus, Crossobamon, Cryptactites, Cyrto-dactylus, Dixonius, Ebenavia, Euleptes, Geckoella, Geckolepis, Geckonia, Gehyra, Gekko, Goggia, Gymnodactylus, Haemodracon, Hemidactylus, Hemiphyllodactylus, Heteronotia, Homonota, Homopholis, Lepidodacty-lus, Luperosaurus, Lygodactylus, Matuatua, Microgecko, Microscalabotes, Nactus, Narudasia, Pachydactylus, Palmatogecko, Paragehyra, Paroe-dura, Perochirus, Phelsuma, Phyllodactylus, Phyllopezus, Pristurus, Pseudogekko, Ptenopus, Ptychozoon, Ptyodactylus, Quedenfeldtia, Rhoptropus, Saurodactylus, Stenodactylus, Tarentola, Tenuidactylus, Teratolepis, Teratoscincus, Thecadactylus, Tropiocolotes, Urocotyledon,* and *Uroplatus*
	Sphaerodactylinae	133+	*Coleodactylus, Gonatodes, Lepidoblepharis, Pseudogonatodes,* and *Sphaerodactylus*
Pygopodidae	—	35+	*Aclys, Aprasia, Delma, Lialis, Ophidiocephalus, Paradelma, Pletholax,* and *Pygopus*
TOTAL		973+	

SCLEROGLOSSA: *INCERTAE SEDIS*

FAMILY	SUBFAMILY	NO. SPECIES	GENERA
Dibamidae	—	11	*Anelytropsis* and *Dibamus*
Amphisbaenidae	—	130+	*Amphisbaena, Ancylocranium, Anops, Aulura, Baika, Blanus, Bronia, Cadea, Cercolophia, Chirindia, Cynisca, Dalophia, Geocalamus, Leposter-non, Loveridgea, Mesobaena, Monopeltis,* and *Zygaspis*
Bipedidae	—	3	*Bipes*
Rhineuridae	—	1	*Rhineura*
Trogonophidae	—	6	*Agamodon, Diplometopon, Pachycalamus,* and *Trogonophis*
TOTAL		151+	

SCLEROGLOSSA: AUTARCHOGLOSSA (excluding snakes)

FAMILY	SUBFAMILY	NO. SPECIES	GENERA
SCINCOMORPHA: Lacertoidea			
Xantusiidae	—	20	*Cricosaura, Lepidophyma,* and *Xantusia*
Lacertidae	—	220+	*Acanthodactylus, Adolfus, Algyroides, Australolacerta, Eremias, Gallotia, Gastropholis, Heliobolus, Holaspis, Ichnotropis, Lacerta, Latastia, Meroles, Mesalina, Nucras, Omanosaura, Ommateremias, Ophisops, Pedioplanis, Philochortus, Podarcis, Poromera, Psammodromus, Pseuderemias, Rhabderemias, Takydromus,* and *Tropidosaura*
Gymnophthalmidae	—	160+	*Alopoglossus, Amapasaurus, Anadia, Anotosaura, Arthrosaura, Bachia, Calyptommatus, Cercosaura, Colobodactylus, Colobosaurus, Colobosauroides, Echinosaura, Ecpleopus, Euspondylus, Gymnophthalmus, Heterodactylus, Iphisa, Leposoma, Macropholidus, Micrablepharus, Neusticurus, Nothobachia, Opipeuter, Pantodactylus, Pholidobolus, Placosoma, Prionodactylus, Procellosaurinus, Proctoporus, Psilophthalmus, Ptychoglossus, Riolama, Stenolepis, Teuchocercus, Tretioscincus,* and *Vanzosaura*
Teiidae	Teiinae	110+	*Ameiva, Cnemidophorus, Dicrodon, Kentropyx,* and *Teius*
	Tupinambinae	8	*Callopistes, Crocodilurus, Dracaena,* and *Tupinambis*
SCINCOMORPHA: Scincoidea			
Cordylidae	—	40+	*Chamaesaura, Cordylus, Platysaurus,* and *Pseudocordylus*
Gerrhosauridae	—	30+	*Angolosaurus, Cordylosaurus, Gerrhosaurus, Tetradactylus, Tracheloptychus,* and *Zonosaurus*
Scincidae	Acontinae	17	*Acontias, Acontophiops,* and *Typhlosaurus*
	Feyliniinae	6	*Feylinia*
	Lygosominae	760+	*Ablepharus, Afroblepharus, Anomalopus, Apterygodon, Ateuchosaurus, Bartleia, Bassiana, Brachymeles, Caledoniscincus, Calyptotis, Carinascincus, Carlia, Clairescincus, Coeranoscincus, Coggeria, Cophoscincopus, Corucia, Cryptoblepharus, Ctenotus, Cyclodina, Dasia, Egernia, Emoia, Eremiascincus, Eroticoscincus, Eugonglyus, Eulamprus, Eulepis, Eumecia, Fojia, Geomyersia, Geoscincus, Glaphyromorphus, Gnypetoscincus, Graciliscincus, Haackgreerius, Harrisoniascincus, Hemiergis, Isopachys, Lacertaspis, Lacertoides, Lamprolepis, Lampropholis, Lankascincus, Larutia, Leiolopisma, Leptoseps, Leptosiaphos, Lerista, Lioscincus, Lipinia, Lobulia, Lygisaurus, Lygosoma, Mabuya, Marmorosphax, Menetia, Morethia, Nannoscincus, Niveoscincus, Notoscincus, Ophioscincus, Panaspis, Papuascincus, Paralipinia, Parvoscincus, Phoboscincus, Prasinohaema, Proablepharus, Pseudemoia, Riopa, Saiphos, Saproscincus, Scincella, Sigaloseps, Simiscincus, Sphenomorphus, Tachygyia, Techmarscincus, Tiliqua, Tribolonotus, Tropidophorus, Tropidoscincus, Typhlosaurus,* and *Vietnascincus*

(continued)

FAMILY	SUBFAMILY	NO. SPECIES	GENERA
Scincidae (continued)	Scincinae	210+	*Amphiglossus, Androngo, Barkudia, Chalcides, Chalcidoseps, Cryptoscincus, Davewakeum, Eumeces, Gongylomorphus, Janetaescincus, Macroscincus, Melanoseps, Neoseps, Nessia, Ophiomorus, Pamelaescincus, Paracontias, Proscelotes, Pseudacontias, Pygomeles, Scelotes, Scincopus, Scincus, Scolecoseps, Sepsina, Sepsophis, Sphenops, Typhlacontias,* and *Voeltzkowia*

ANGUIMORPHA

FAMILY	SUBFAMILY	NO. SPECIES	GENERA
Anguidae	Anguinae	17	*Anguis* and *Ophisaurus*
	Anniellinae	2	*Anniella*
	Diploglossinae	40+	*Celestus, Diploglossus, Ophiodes, Sauresia,* and *Wetmorena*
	Gerrhonotinae	42+	*Abronia, Barisia, Elgaria, Gerrhonotus,* and *Mesaspis*
Xenosauridae	—	10+	*Shinisaurus* and *Xenosaurus*

Varanoidea

FAMILY	SUBFAMILY	NO. SPECIES	GENERA
Helodermatidae	—	2	*Heloderma*
Lanthanotidae	—	1	*Lanthanotus*
Varanidae	—	50	*Varanus*
TOTAL		1,745+	

GLOSSARY

ABIOTIC Any aspect of the non-living environment, such as rain, sunlight, or temperature.

ACRODONT Teeth that attach on the surface of the jawbone, not set in sockets and not replaced.

ADAPTATION Conformity between an organism and its environment, or all the ways that a given organism copes with its physical and biotic environments. Adaptations result from natural selection and enhance fertility or survival.

ADAPTIVE RADIATION Diversification into a large group of ecologically diverse species.

AEROBIC Metabolic activities that rely on oxidation to provide energy.

ALATE A winged reproductive stage in social insects such as in ants and termites.

ALLOCHRONIC Not in synchrony, as in alternating production of eggs between the two ovaries.

ALLOMETRY Differential growth rates of body parts with age.

ALLOPATRY The occurrence of two species in different, nonoverlapping geographic areas.

AMNIOTIC EGG A type of egg produced by reptiles, birds, and mammals in which an embryo is enclosed by extra-embryonic membranes. The amnion isolates the embryo from the remainder of the egg during development.

ANAEROBIC Metabolic energy obtained without the use of oxygen.

ANGUIMORPHA A large monophyletic clade that includes anguids, xeno-saurids, beaded lizards *(Heloderma)*, *Lanthanotus*, varanids, mosasaurs, and snakes, and which is the sister group to Scincomorpha, which together comprise the Autarchoglossa.

ARBOREAL Living above ground in trees or shrubs.

AUTARCHOGLOSSA A large monophyletic clade that includes Scincomorpha and Anguimorpha; the sister group to Gekkota.

AUTOTOMY "Self loss": referring to separation of a piece of a lizard's tail.

BRILLE A spectacle, or clear scale enclosing the eye, as in most geckos, xantusiids, some skinks, and all snakes.

BRUMATION A period of inactivity during winter, the ectothermic equivalent of hibernation in endotherms.

BUCCAL Oral cavity of the mouth.

CARPHODACTYLINAE A subfamily of diplodactylid geckos.

CHEMORECEPTION Detection of chemicals.

CHORIOALLANTOIS Combination of two extraembryonic membranes, the chorion and the allantois, that form an advanced placenta in some reptiles.

CLADE A group of descendant species that share a common ancestor.

CLAVATE Spade shaped.

CLOACA The vent at the end of the digestive tract, which also includes orifices from the urinary and reproductive tracts.

CONGENERS Members of the same genus.

CONSPECIFICS Members of the same species.

CREPUSCULAR Active at dusk, dawn, or both.

CRYPSIS (*adj.* CRYPTIC) Anatomical and behavioral traits that camouflage an animal, rendering it nearly indistinguishable from its background.

CYCLOID SCALES Circular-shaped scales.

DEWLAP A fanlike extension of the throat and/or lower neck.

DIAPSID A skull condition with an upper and lower temporal arch forming two temporal openings, as seen in tuatara.

DIMORPHISM Morphological differences between sexes, such as color, size, head size, and ornamentation.

DIPLOID A fertilized egg or somatic body cell that contains both maternal and paternal genetic material (two full sets of chromosomes).

DISTAL A part of an appendage that is away from the body.

DIURNAL Active during the day.

DORSAL The top side (upper surface) of an animal.

ECOMORPH An animal's body plan that can be directly related to its ecology.

ECOTONE An edge community at the boundary of two distinct biomes.

ECTOTHERM An animal that relies on heat sources and sinks in the external environment to gain or lose heat.

ELAPID A member of a family of venomous snakes (Elapidae) that includes cobras and coral snakes.

ENDEMIC A taxon restricted to a specific defined geographical region.

ENDOSYMBIONT A beneficial mutualistic microorganism that lives inside a host, usually in its gut.

ENDOTHERM An animal that generates bodily heat metabolically to control its body temperature.

EPITHELIUM The outer layer of cells, such as those lining the intestine or the surface of the skin.

ESTIVATE To enter a dormant, low metabolic state.

EUKARYOTE A relatively derived and complex organism that has cell nuclei (as opposed to simpler prokaryotes such as bacteria, which have no cell nuclei).

EUTHERIAN Placental mammals.

EXTANT Surviving to the present day.

FACULTATIVE A nonobligatory relationship.

FECUNDITY Number of eggs or offspring produced.

FEMORAL PORES Glands found on the underside of the base of the hindlegs, which produce lipid-based substances used to deposit scent trails.

FENESTRAE Openings in bony structures through which pass softer parts of anatomy such as nerves and muscles.

FITNESS An organism's ability to perpetuate its genes in a population gene pool.

FOSSORIAL Living underground.

FOVEA A focal point in the center of the retina highly innervated with sensitive visual cells.

GAMETE A haploid cell that carries only half of a parental genome (one set of chromosomes), such as a sperm or an egg.

GEKKOTA A large monophyletic clade of primarily nocturnal lizards that includes geckos and Australian pygopodid lizards; the sister group to Autarchoglossa.

GENE POOL All the genes present in a given population at a particular moment in time.

GULAR The throat region of the lower neck.

GUSTATORY Chemical signals that stimulate taste buds.

HAPLOID A sperm cell or an egg cell carrying only half of a parent's genes.

HELIOTHERMIC Deriving bodily heat from the sun, as in basking.

HEMIBACULUM A bone in the hemipenes of some varanid lizards.

HEMIPENES Paired copulatory organs that evert from the base of the tail of a male squamate.

HETEROGAMETY The situation in which sex cells or gametes (sperm and eggs) have different sex chromosomes, as in XY heterogamety.

HETEROZYGOSITY Having two different alleles at a particular locus on homologous chromosomes.

HISTOCOMPATIBILITY The ability to accept a tissue transplant; a result of similarity in antigens between the donor and the recipient. For lizards, the term often refers to skin transplant experiments.

HOLOCRINE GLAND A gland that secretes chemical products.

HOMEOTHERMY Maintenance of a stable internal body temperature.

HYGROSCOPIC Taking up of moisture from the external environment.

HYMENOPTERA The insect order containing ants, bees, and wasps.

HYPERNATREMIA High in sodium.

IGUANIA A large monophyletic clade of relatively primitive lizards, consisting of iguanids, agamids, and chameleons; the sister group to all other lizards (Scleroglossa).

INCERTAE SEDIS The situation in which a taxon is known to have descended from within a particular clade but its exact position in that clade remains undetermined.

INFUNDIBULUM The opening of the fallopian tube (or oviduct) in the abdominal cavity.

INGUINAL Referring to the groin region.

INTEGUMENTARY Referring to the skin.

INTROMISSION Insertion of a male's hemipenis into a female's cloaca.

ISOPTERA The insect order containing termites.

JACOBSON'S ORGAN Paired organs in the roof of the mouth used in analyzing scents; part of the vomeronasal system.

KARYOTYPE The complete set of chromosomes carried by a given individual.

LAMELLA(E) Scale(s) on the underside of a lizard's toe.

LECITHOTROPHY Embryonic development that relies entirely on yolk.

LEK A group of males gathered in a common area to display and court females.

MATROTROPHY Development that relies partially or wholly on maternal provision of energy and material to a developing embryo.

MAXITHERMY Maintenance of a high active body temperature close to the upper thermal limit to provide high performance ability and maximal fitness.

MEGAPODE A bird in the family Megapodiidae, occurring in Australia and southeast Asia. Megapodes bury their eggs in mounds, and the eggs develop without incubation.

MEIOSIS The reduction division whereby homologous chromosomes align and separate into haploid sex cells, or gametes.

MESIC Relatively moist.

MESOKINESIS A skull condition whereby joints allow the muzzle to move with respect to the braincase, rising upward when the mouth is opened and clamping downward when it is closed.

METATAXON A higher taxonomic unit that has not yet received an official name.

MONOPHYLETIC Referring to a natural group whose members have all evolved from a single common ancestor.

MYRMECOPHAGY Ant eating.

NARES Nostrils.

NOCTURNAL Active at night.

OLFACTORY Referring to the sense of smell.

ONTOGENETIC Occurring with age, as a result of the development of an organism.

OSTEODERM Bones embedded within scales.

OVIDUCTAL EGG An egg (shelled or unshelled) in the oviduct.

OVIPAROUS Egg laying.

OVIPOSITION SITE A place where eggs are laid (a nest in most species but a surface for many geckos).

PAPILLAE Tiny projections.

PARIETAL SCALES Scales on top of the head located over the parietal bones.

PARTHENOFORM A unisexual parthenogenetic species.

PARTHENOGENESIS Clonal reproduction without males, resulting in all-female unisexual species.

PATAGLIA Winglike structures.

PEDUNCLE A stalk, such as the thin end of a knob-tailed gecko's tail just anterior to the terminal knob.

PHENOTYPE An organism's external appearance.

PHEROMONE A chemical used in communication.

PHYLOECOLOGY The study of ecological characteristics in a phylogenetic or historical context. Phyloecological studies seek to identify the evolutionary history of ecological traits.

PHYLOGENETIC Having to do with evolutionary history.

PLATYNOTA A clade of lizards that includes helodermatids, lanthanotids, varanids, and mosasaurids (and perhaps snakes).

PLEURODONT A type of dentition in which teeth are set in sockets in the jawbones.

POIKILOTHERMY The condition in which body temperature fluctuates with ambient temperature.

PREFORMED WATER Water acquired from food; e.g., the water in the body of an insect that a lizard eats.

PROPAGULE Anything from an egg that just begins development to a hatchling; a propagule gives rise to an adult animal (or plant).

PTERYGIUM A winglike structure.

QUADRATE A bone that connects the upper jaw with the mandible in squamates.

RIPARIAN Living along the banks of creeks or rivers.

SALTATE To leap; a form of locomotion.

SAXICOLOUS Rock dwelling.

SCANSORIAL Living on surfaces (usually vertical), an ability enhanced in some lizard species by adhesive pads on toes and on tips of tails.

SCINCOMORPHA A large monophyletic clade that includes lacertids, teiids, gymnophthalmids, skinks, and xantusiids; the sister group to Anguimorpha.

SCLEROGLOSSA A large monophyletic clade consisting of Autarchoglossa and Gekkota; the sister group to Iguania.

SETA(E) Hairlike structure(s) on the underside of adhesive toe pads.

SEXUAL DIMORPHISM Sex-specific differences in anatomy or behavior.

SPATULA(E) Tiny ending(s) at the tips of setae.

SPHAERODACTYLINAE A diurnal subfamily of Gekkonidae restricted to the New World.

SQUAMATE A lizard or snake.

STREPTOSTYLY A hanging jaw setup whereby the lower jaw is attached to the quadrate bone, which is free to rotate on the skull at the base of the upper jaw.

SVL Snout-vent length; a linear measurement from the tip of the snout to the cloacal opening.

SYMPATRY The co-occurrence of two (or more) species at the same place.

SYNAPOMORPHY A shared, derived trait.

TAXON (*pl.* TAXA) A named taxonomic unit, such as a species, genus, family, or higher unit of classification. Any clade is a taxon.

TETRAPOD A chordate with four legs (amphibians, reptiles, and mammals) or one derived from an ancestor with four legs (e.g., snakes).

THERMOCONFORMER An animal that does not actively thermoregulate but allows its body temperature to mirror ambient environmental temperature.

THERMOREGULATOR An animal that maintains a narrow range of body temperature in the face of variable external environmental temperatures.

THIGMOTHERMIC Obtaining heat from the substrate via conduction.

TROPHIC LEVEL A functional classification of organisms according to their feeding relationships.

URODAEAL GLAND A gland just inside the cloaca.

VARANOIDEA A clade that includes beaded lizards *(Heloderma), Lanthanotus,* varanids, mosasaurs, and snakes.

VENTRAL Referring to the bottom surface of an animal.

VENTROLATERAL Referring to the bottom sides of an animal.

VITELLOGENESIS The process by which lipids from an organism's body are transferred into production of yolk of developing eggs.

VIVIPAROUS Bearing live young.

VOMEROLFACTION The ability to discriminate chemicals based on a sensitive chemosensory system in the roof of the mouth, which can detect and analyze non-airborne chemicals picked up with the tongue.

VOMERONASAL Having to do with the chemosensory system that perceives vomodors; also called the vomerolfactory system, or Jacobson's organ.

VOMODOR A non-airborne chemical.

XERIC Relatively arid.

ZYGODACTYLY A condition found in chameleons in which two toes are yoked together and act in opposition to three other toes yoked together.

REFERENCES

Akersten, W. A. 1985. Canine function in *Smilodon* (Mammalia: Felidae: Machairodontinae). *Nat. Hist. Mus. Los Angeles County, Contr. Sci.* 356:1–22.

Alberts, A. C. 1989. Ultraviolet visual sensitivity in desert iguanas: Implications for pheromone detection. *Anim. Behav.* 38: 129–137.

———. 1992. Pheromonal self-recognition in the desert iguana. *Copeia* 1992:229–232.

Alberts, A. C., J. A. Phillips, and D. I. Werner. 1993. Sources of intraspecific variability in the protein composition of lizard femoral gland secretions. *Copeia* 1993:775–781.

Anderson, R. A., and W. H. Karasov. 1981. Contrasts in energy intake and expenditure in sit-and-wait and widely foraging lizards. *Oecologica* 49:67–72.

Anderson, R. A., and L. J. Vitt. 1990. Sexual selection versus alternative causes of sexual dimorphism in teiid lizards. *Oecologia* 84:145–157.

Andrews, R. M. 1979. The lizard *Corytophanes cristatus:* An extreme "sit-and-wait" predator. *Biotropica* 11:136–139.

———. 1985. Oviposition frequency of *Anolis carolinensis. Copeia* 1985:259–262.

Andrews, R. M., and T. Mathies. 2000. Natural history of reptilian development: Constraints on the evolution of viviparity. *BioScience* 50:227–238.

Andrews, R. M., and A. S. Rand. 1974. Reproductive effort in anoline lizards. *Ecology* 55:1317–1327.

Arnold, E. N. 1984a. Variation in the cloacal and hemipenial muscles of lizards and its bearing on their relationships. *Symp. Zool. Soc. London* 52:47–85.

———. 1984b. Evolutionary aspects of tail shedding in lizards and their relatives. *J. Nat. Hist.* 18:127–169.

———. 1984c. Ecology of lowland lizards in the eastern United Arab Emirates. *J. Zool., Lond.* 204:1–26.

———. 1988. Caudal autotomy as a defense. In *Biology of the Reptilia,* vol. 16: *Ecology B, Defense and life history,* ed. C. Gans and R. B. Huey, 235–273. New York: Alan R. Liss.

Auffenberg, W. 1978. Social feeding behavior in *Varanus komodoensis.* In *Behavior and neurology of lizards,* ed. N. Greenberg and P. D. MacLean, 301–331. Poolesville, Md.: National Institute of Mental Health.

———. 1981. *The behavioral ecology of the Komodo monitor.* Gainesville: University Press of Florida.

———. 1988. *Gray's monitor lizard.* Gainesville: University Press of Florida.

———. 1994. *The Bengal monitor.* Gainesville: University Press of Florida.

Auffenberg, W., and T. Auffenberg. 1988. Resource partitioning in a community of Philippine skinks (Sauria: Scincidae). *Bull. Florida State Mus. Biol. Sci.* 32:151–219.

Autumn, K., and B. Han. 1989. Mimicry of scorpions by juvenile lizards, *Teratoscincus roborowskii* (Gekkonidae). *Chinese Herpetol. Res.* 2:60–64.

Autumn, K., Y. A. Liang, S. T. Hsieh, W. Zesch, W. P. Chan, T. W. Kenny, R. Fearing, and R. J. Full. 2000. Adhesive force of a single gecko foot-hair. *Nature* 405:681–685.

Avery, R. A. 1970. Utilization of caudal fat by hibernating common lizards, *Lacerta vivipara. Comp. Biochem. Physiol.* 37:119–121.

Avila-Pires, T. C. S. 1995. Lizards of Brazilian Amazonia (Reptilia: Squamata). *Zoologische Verhandelingen* 299:1–706.

Baird, T. A., M. A. Acree, and C. L. Sloan. 1996. Age and gender-related differences in the social behavior and mating success of free-living collared lizards, *Crotaphytus collaris. Copeia* 1996:336–347.

Ballinger, R. E. 1973. Experimental evidence of the tail as a balancing organ in the lizard *Anolis carolinensis. Herpetologica* 29:65–66.

———. 1983. Life-history variations. In *Lizard ecology: Studies of a model organism,* ed. R. B. Huey, E. R. Pianka, and T. W. Schoener, 241–260. Cambridge, Mass.: Harvard University Press.

Ballinger, R. E., and J. D. R. Clark. 1973. Energy content of lizard eggs and the measurement of reproductive effort. *J. Herpetol.* 7:129–132.

Ballinger, R. E., J. A. Lemos-Espinal, S. Sanoja-Sarabia, and N. R. Coady. 1995. Ecological observations of the lizard *Xenosaurus grandis* in Cuautlapán, Veracruz, Mexico. *Biotropica* 27:128–132.

Ballinger, R. E., J. A. Lemos-Espinal, and G. R. Smith. 2000. Reproduction in females of three species of crevice-dwelling lizards (genus *Xenosaurus*) from Mexico. *Stud. Neotrop. Fauna Environ.* 35:179–183.

Ballinger, R. E., J. W. Nietfeldt, and J. J. Krupa. 1979. An experimental analysis of the role of the tail in attaining high running speed in *Cnemidophorus sexlineatus* (Reptilia: Squamata: Lacertilia). *Herpetologica* 35:114–116.

Bartholomew, G. A., and R. C. Lasiewski. 1965. Heating and cooling rates, heart rate, and simulated diving in the Galápagos marine iguana. *Comp. Biochem. Physiol.* 16:573–582.

Barwick, R. E. 1982. The growth and ecology of the gecko *Hoplodactylus duvauceli* at the Brothers Islands. In *New Zealand herpetology,* ed. D. G. Newman, 337–391. New Zealand Wildlife Service, Occ. Publ. 2.

Basso, E. B. 1973. *The Kalapalo Indians of central Brazil.* New York: Holt, Rinehart & Winston.

Bauer, A. M. 1986. Saltation in the pygopodid lizard *Delma tincta. J. Herpetology* 20: 462–463.

———. 1990. Gekkonid lizards as prey of invertebrates and predators of vertebrates. *Herp. Rev.* 21:83–87.

————. 1995. Geckos of the genus *Rhacodactylus*. *Reptiles,* 32–49.

————. 1997. Peritoneal pigmentation and generic allocation in the Chamaeleonidae. *Afr. J. Herpetol.* 46:117–123.

————. 1998. Morphology of the adhesive tail tips of carphodactyline geckos (Reptilia: Diplodactylidae). *J. Morph.* 235: 41–58.

Bauer, A. M., and D. A. Good. 1986. Scaling of scansorial surface area in the genus *Gekko.* In *Studies in herpetology, proceedings of the European Herpetological Meeting, Prague,* ed. Z. Rocek, 363–366. Prague: Charles University.

Bauer, A. M., and A. P. Russell. 1992. Regional integumentary loss as an escape strategy in island gekkonid lizards. *Proc. Sixth Ord. Gen. Meet. S. E. H.* 69–71.

————. 1994. Is autotomy frequency reduced in geckos with "actively functional" tails? *Herp. Nat. Hist.* 2:1–15.

Bauer, A. M., and R. A. Sadlier. 2000. *The herpetofauna of New Caledonia.* N.p.: Society for the Study of Amphibians and Reptiles.

Bauer, A. M., J. Dougherty, and A. P. Russell. 1992. Vocalizations of the New Caledonian giant gecko, *Rhacodactylus leachianus. Amphibia-Reptilia* 13:412–417.

Bauer, A. M., A. P. Russell, and B. D. Edgar. 1990. Utilization of the termite *Hodotermes mossambicus* (Hagen) by gekkonid lizards near Keetmanshoop, South West Africa. *S. Afr. J. Zool.* 24:239–243.

Bauer, A. M., A. P. Russell, and R. E. Shadwick. 1989. Mechanical properties and morphological correlates of fragile skin in gekkonid lizards. *J. Exp. Biol.* 145:79–102.

————. 1990. Skin mechanics and morphology of the gecko *Sphaerodactylus roosevelti. Amer. Zool.* 30:570.

————. 1993. Skin mechanics and morphology of the gecko *Teratoscincus scincus. Amphibia-Reptilia* 14:321–331.

Bauwens, D., and R. Díaz-Uriarte. 1997. Covariation of life history traits in lacertid lizards: A comparative study. *Amer. Natur.* 149:91–111.

Bauwens, D., and R. F. Verheyen. 1985. The timing of reproduction in the lizard *Lacerta vivipara:* Differences between individual females. *J. Herpetol.* 19: 353–364.

Bayless, M. K. 1998. The artrellia: Dragon of the trees. *Reptiles* 6:32–47.

Beck, D. D. 1990. Ecology and behavior of the Gila monster in southwestern Utah. *J. Herpetol.* 24:54–68.

Beck, D. D., and C. H. Lowe. 1991. Ecology of the beaded lizard, *Heloderma horridum,* in a tropical dry forest in Jalisco, Mexico. *J. Herpetol.* 25:395–406.

Beck, D. D., and A. Ramírez-Bautista. 1991. Combat behavior of the beaded lizard, *Heloderma h. horridum,* in Jalisco, Mexico. *J. Herpetol.* 25:481–484.

Beck, D. D., M. R. Dohm, J. T. Garland, A. Ramírez-Bautista, and C. H. Lowe. 1995. Locomotor performance and activity energetics of helodermatid lizards. *Copeia* 1995:577–585.

Beebe, W. 1945. Field notes on the lizards of Kartabo, British Guiana, and Caripito, Venezuela. Part 3. Teiidae, Amphisbaenidae, and Scincidae. *Zoologica* 30:7–32.

Bellairs, A., and G. Underwood. 1951. The origin of snakes. *Biol. Rev. Cambridge Phil. Soc.* 26:193–237.

Benabib, M. 1994. Reproduction and lipid utilization of tropical populations of *Sceloporus variabilis. Herp. Monogr.* 8:160–180.

Benabib, M., K. M. Kjer, and J. W. Sites, Jr. 1997. Mitochondrial DNA sequence-based phylogeny and the evolution of viviparity in the *Sceloporus scalaris* group (Reptilia, Squamata). *Evolution* 51:1262–1275.

Bentley, P. J., and W. F. Blumer. 1962. Uptake of water by the lizard *Moloch horridus. Nature* 194:699–700.

Berry, K. H. 1974. The ecology and social behavior of the chuckwalla, *Sauromalus obesus obesus* Baird. *Univ. Calif. Publ. Zool.* 101:1–44.

Bezy, R. L. 1967. Variation, distribution, and taxonomic status of the Arizona night lizard *(Xantusia arizonae). Copeia* 1967:653–661.

————. 1988. The natural history of the night lizards, family Xantusiidae. In *Proceedings of the Conference on California Herpetology,* ed. H. F. De Lisle, P. R. Brown, B. Kaufman, and B. M. McGurty, 1–12. Los Angeles: Southwestern Herpetologists Society.

————. 1989. Morphological differentiation in unisexual and bisexual xantusiid lizards of the genus *Lepidophyma* in Central America. *Herpetol. Monogr.* 3: 61–80.

Blackburn, D. 1982. Evolutionary origins of viviparity in the Reptilia. I. Sauria. *Amphib.-Rept.* 3:185–205.

————. 1985. Evolutionary origins of viviparity in the Reptilia. II. Serpentes, Amphisbaenia, and Ichthyosauria. *Amphib.-Rept.* 6:259–291.

Blackburn, D. G., and L. J. Vitt. 1992. Reproduction in viviparous South American lizards of the genus *Mabuya.* In *Reproductive biology of South American vertebrates,* ed. W. C. Hamlett, 150–164. New York: Springer-Verlag.

————. 2002. Specializations of the chorioallantoic placenta in the Brazilian scincid lizard, *Mabuya heathi:* A new placental morphotype for reptiles. *J. Morph.* 254: 121–131.

Blackburn, D. G., L. J. Vitt, and C. A. Beuchat. 1984. Eutherian-like reproductive specializations in a viviparous reptile. *Proc. Natl. Acad. Sci. USA* 81:4860–4863.

Blair, W. F. 1960. *The rusty lizard: A population study.* Austin: University of Texas Press.

Bogert, C. M., and R. M. del Campo. 1956. The Gila monster and its allies: The relationships, habits, and behavior of the lizards of the family Helodermatidae. *Bull. Amer. Mus. Nat. Hist.* 109:1–238.

Böhme, W., R. Hutterer, and W. Bings. 1985. Die Stimme der Lacertidae, speziell der Kanareneidechsen (Reptilia: Sauria). *Bonn. Zool. Beitr.* 36:337–354.

Borsuk-Bialynicka, M. 1984. Anguimorphans and related lizards from the late Cretaceous of the Gobi Desert, Mongolia. *Palaeontologia Polonica* 46:5–105.

Borsuk-Bialynicka, M., M. Lubka, and W. Böhme. 1999. A lizard from Baltic amber (Eocene) and the ancestry of the crown group lacertids. *Acta Paleont. Polonica* 44:349–382.

Branch, B. 1988. *Field guide to the snakes and other reptiles of southern Africa.* Sanibel Island, Fla.: Ralph Curtis Books.

Branch, W. R., and M. J. Whiting. 1997. A new *Platysaurus* (Squamata: Cordylidae) from northern Cape Province, South Africa. *Afr. J. Herpetol.* 46:124–136.

Brattstrom, B. H. 1952. The food of the nightlizards, genus *Xantusia. Copeia* 1952:168–172.

————. 1965. Body temperature of reptiles. *Amer. Midl. Nat.* 73:376–422.

Broadley, D. G. 1964. A review of the crag lizards (genus *Pseudocordylus*) of Natal. *Ann. Natal Mus.* 16:99–110.

————. 1968. A revision of the African genus *Typhlosaurus* Wiegmann (Sauria: Scincidae). *Arnoldia* (Rhodesia) 3:1–20.

————. 1971. A review of *Rhampholeon marshalli* Boulenger, with the description of a new subspecies from Mozambique (Sauria: Chamaeleonidae). *Arnoldia* (Rhodesia) 5:1–6.

————. 1974. Reproduction in the genus *Platysaurus* (Sauria: Cordylidae). *Herpetologica* 30:379–380.

————. 1978. A revision of the genus *Platysaurus* A. Smith (Sauria: Cordylidae). *Occ. Pap. Natl. Museums and Monuments of Rhodesia* 6:129–185.

Brown, D. E., and N. B. Carmony. 1991. *Gila monster: Facts and folklore of America's Aztec lizard.* Silver City, N.M.: High-Lonesome Books.

Brown, W. C., and A. C. Alcala. 1957. Viability of lizard eggs exposed to sea water. *Copeia* 1957:39–41.

Bull, C. M. 1988. Mate fidelity in an Australian lizard, *Trachydosaurus rugosus*. *Behav. Ecol. Sociobiol.* 23:45–49.

———. 1994. Population dynamics and pair fidelity in sleepy lizards. In *Lizard ecology: Historical and experimental perspectives,* ed. L. J. Vitt and E. R. Pianka, 159–174. Princeton: Princeton University Press.

Bull, C. M., and B. C. Bagurst. 1998. Home range overlap of mothers and their offspring in the sleepy lizard, *Tiliqua rugosa*. *Behav. Ecol. Sociobiol.* 42:357–362.

Bull, C. M., and Y. Pamula. 1996. Sexually dimorphic head sizes and reproductive success in the sleepy lizard, *Tiliqua rugosa*. *J. Zool. London* 240:511–521.

———. 1998. Enhanced vigilance in monogamous pairs of the lizard *Tiliqua rugosa*. *Behav. Ecol.* 9:452–455.

Bull, C. M., S. J. B. Cooper, and B. C. Baghurst. 1998. Social monogamy and extra-pair fertilization in an Australian lizard, *Tiliqua rugosa*. *Behav. Ecol. Sociobiol.* 44:63–72.

Bull, J. J. 1980. Sex determination in reptiles. *Quart. Rev. Biol.* 55:3–21.

Burghardt, G. M., and A. S. Rand, eds. 1982. *Iguanas of the world: Their behavior, ecology, and conservation.* Park Ridge, N.J.: Noyes.

Burghardt, G. M., H. W. Greene, and A. S. Rand. 1977. Social behavior in hatchling green iguanas: Life at a reptile rookery. *Science* 195:689–691.

Bursey, C. R., S. R. Goldberg, and D. N. Woolery. 1996. *Oochoristica piankai* sp. n. (Cestoda: Linstowiidae) and other helminths of *Moloch horridus* (Sauria: Agamidae) from Australia. *J. Helminthol. Soc. Wash.* 63: 215–221.

Bustard, H. R. 1967a. The comparative behavior of chameleons: Fight behavior in *Chamaeleo gracilis* Hallowell. *Herpetologica* 23:44–50.

———. 1967b. Gekkonid lizards adapt fat storage to desert environments. *Science* 158:1197–1198.

Cadle, J. E., and P. Chuna M. 1995. A new lizard of the genus *Macropholidus* (Teiidae) from a relictual humid forest of northwestern Peru, and notes on *Macropholidus ruthveni* Noble. *Breviora* 1995:1–39.

Caldwell, J. P. 1996. The evolution of myrmecophagy and its correlates in poison frogs (family Dendrobatidae). *J. Zool., Lond.* 240:75–101.

Caldwell, M. W. 1996. Ichthyosauria: A preliminary phylogenetic analysis of diapsid affinities. *N. Jb. Geol. Paläont. Abh.* 200: 361–386.

Caldwell, M. W., and M. S. Y. Lee. 1997. A snake with legs from the marine Cretaceous of the Middle East. *Nature* 386: 705–709.

Campbell, J. A., and D. R. Frost. 1993. Anguid lizards of the genus *Abronia*: Revisionary notes, descriptions of four new species, a phylogenetic analysis, and key. *Bull. Am. Mus. Nat. Hist.* 216:1–121.

Campbell, J. A., and W. W. Lamar. 1989. *The venomous reptiles of Latin America.* Ithaca, N.Y.: Comstock.

Carothers, J. H. 1986. An experimental confirmation of morphological adaptation: Toe fringes in the sand-dwelling lizard *Uma scoparia*. *Evolution* 40:871–874.

Carpenter, C. C. 1966. The marine iguana of the Galápagos Islands, its behavior and ecology. *Proc. California Acad. Sci.* 34: 329–376.

———. 1982. The aggressive displays of iguanine lizards. In *Iguanas of the world: Their behavior, ecology, and conservation,* ed. G. M. Burghardt and A. S. Rand, 215–231. Park Ridge, N.J.: Noyes.

Carretero, M. A., and G. A. Llorente. 1993. Feeding of two sympatric lacertids in a sandy coastal area (Ebro Delta, Spain). In *Lacertids of the Mediterranean region: A biological approach,* ed. E. D. Valakos, W. Böhme, V. Pérez-Mellado, and P. Maragou, 155–172. Athens, Greece: Hellenic Zoological Society.

Carroll, R. L. 1988. Late Paleozoic and early Mesozoic lepidosauromorphs and their relation to lizard ancestry. In *Phylogenetic relationships of the lizard families,* ed. R. Estes and G. Pregill, 99–118. Stanford: Stanford University Press.

Carter, D. B. 1999. Nesting and evidence of parental care by the lace monitor, *Varanus varius*. In *Advances in monitor research II,* ed. H.-G. Horn and W. Böhme. *Mertensiella* 11:137–147.

Casas-Andreu, G., and M. A. Currola-Hidalgo. 1993. Comparative ecology of two species of *Cnemidophorus* in central Jalisco, Mexico. In *Biology of whiptail lizards (genus* Cnemidophorus*),* ed. J. W. Wright and L. J. Vitt, 133–150. Norman: Oklahoma Museum of Natural History.

Case, T. J. 1976. Body size differences between populations of the chuckwalla, *Sauromalus obesus*. *Ecology* 57:313–323.

———. 1979. Character displacement and coevolution in some *Cnemidophorus* lizards. *Fortschr. Zool.* 25:235–282.

———. 1983. Sympatry and size similarity in *Cnemidophorus*. Chapter 14 in *Lizard ecology: Studies of a model organism,* ed. R. B. Huey, E. R. Pianka, and T. W. Schoener, 297–325. Cambridge, Mass.: Harvard University Press.

Cei, J. M. 1986. *Reptiles del centro, centro-oeste y sur de la Argentina. Herpetofauna de las zonas áridas y semiáridas.* Monografie 4:1–527. Torino, Italy: Museo Regionale di Scienze Naturali.

———. 1993. *Reptiles del noroeste, nordeste y este de la Argentina.* Monografie 14:1–946. Torino, Italy: Museo Regionale di Scienze Naturali.

Censky, E. J. 1996. The evolution of sexual size dimorphism in the teiid lizard *Ameiva plei*: A test of alternative hypotheses. In *Contributions to West Indian herpetology: A tribute to Albert Schwartz,* ed. R. Powell and R. W. Henderson, 277–289. Ithaca, N.Y.: Society for the Study of Amphibians and Reptiles.

Censky, E. J., and D. R. Paulson. 1992. Revision of the *Ameiva* (Reptilia: Teiidae) of the Anguilla Bank, West Indies. *Ann. Carnegie Mus.* 61:177–195.

Chan-Ard, T., W. Grossman, A. Gumprecht, and K.-D. Schulz. 1999. *Amphibians and reptiles of peninsular Malaysia and Thailand.* Wuerselen, Ger.: Bushmaster Productions.

Chiszar, D., T. Gingery, B. Gingery, and H. M. Smith. 1999. *Phymaturus patagonicus* (Argentine chuckwalla): Facultative parthenogenesis. *Herp. Rev.* 30:98.

Christain, D. P. 1975. Vulnerability of meadow voles, *Microtus pennsylvanicus,* to predation by feral cats. *Am. Midl. Nat.* 93: 498–502.

Christiansen, J. L. 1971. Reproduction of *Cnemidophorus inornatus* and *Cnemidophorus neomexicanus* (Sauria, Teiidae) in northern New Mexico. *Amer. Mus. Novit.* 2442:1–48.

Cifelli, R. L., and R. L. Nydam. 1995. Primitive, helodermatid-like platynotan from the early Cretaceous of Utah. *Herpetologica* 51:286–291.

Ciofi, C. 1999. The Komodo dragon. *Sci. Am.* 280:84–91.

Clark, D. R., Jr. 1971. The strategy of tail-autotomy in the ground skink, *Lygosoma laterale*. *J. Exp. Zool.* 176:295–302.

Clarke, R. F. 1965. An ethological study of the iguanid lizard genera *Callisaurus, Cophosaurus,* and *Holbrookia. Emporia State Res. Stud.* 13:1–66.

Clutton-Brock, T. H., and P. H. Harvey, eds. 1978. *Readings in sociobiology.* San Francisco: Freeman.

Cogger, H. G. 1992. *Reptiles and amphibians of Australia.* Ithaca, N.Y.: Cornell/ Comstock.

Colbert, E. H. 1967. Adaptations for gliding in the lizard *Draco. Amer. Mus. Novit.* 2283:1–20.

Colburn, T., D. Dumanoski, and J. P. Myers. 1996. *Our stolen future.* Dutton.

Cole, C. J. 1984. Unisexual lizards. *Sci. Amer.* 250:94–100.

Cole, C. J., and C. R. Townsend. 1990. Parthenogenetic lizards as vertebrate systems. *J. Exp. Zool.* 4:174–176.

Cole, C. J., H. C. Desauer, C. R. Townsend, and M. G. Arnold. 1990. Unisexual lizards of the genus *Gymnophthalmus* (Reptilia: Teiidae) in the neotropics: Genetics, origin, and systematics. *Amer. Mus. Novit.* 2994:1–29.

Colli, G. R., and D. S. Zamboni. 1999. Ecology of the worm-lizard *Amphisbaena alba* in the cerrado of central Brazil. *Copeia* 1999:733–742.

Colli, G. R., A. K. Péres, and H. J. Cunha. 1998. A new species of *Tupinambis* (Squamata: Teiidae) from central Brazil, with an analysis of morphological and genetic variation in the genus. *Herpetologica* 54:477–492.

Colli, G. R., R. P. Bastos, and A. F. B. Araujo. 2002. The character and dynamics of the cerrado herpetofauna. In *The cerrados of Brazil*, ed. P. S. Oliveira and R. J. Marquis, 223–241. New York: Columbia University Press.

Coman, B. J., and H. Brunner. 1972. Food habits of the feral house cat in Victoria. *J. Wildl. Mgmt.* 36:848–853.

Congdon, J. D., and A. E. Dunham. 1994. Contributions of long-term life history studies to conservation biology. In *Principles of conservation biology,* ed. G. Meffe and R. Carroll, 181–182. Sunderland, Mass.: Sinauer Associates.

Congdon, J. D., A. E. Dunham, and R. C. van Loben Sels. 1994. Demographics of common snapping turtles *(Chelydra serpentina):* Implications for conservation and management of long-lived organisms. *Amer. Zool.* 34:397–408.

Congdon, J. D., L. J. Vitt, and W. W. King. 1974. Geckos: Adaptive significance and energetics of tail autotomy. *Science* 184:1379–1380.

Congdon, J. D., L. J. Vitt, R. C. van Loben Sels, and R. D. Ohmart. 1982. The ecological significance of water flux rates in arboreal desert lizards of the genus *Urosaurus. Physiol. Zool.* 55:317–322.

Cooper, W. E., Jr. 1981. Head-first swallowing of large prey by a scincid lizard, *Eumeces laticeps. J. Herpetol.* 15:371–373.

———. 1984. Female secondary sexual coloration and sex recognition in the keeled earless lizard, *Holbrookia propinqua. Anim. Behav.* 32:1142–1150.

———. 1985. Female residency and courtship intensity in a territorial lizard, *Holbrookia propinqua. Amphibia-Reptilia* 6:63–69.

———. 1986. Chromatic components of female secondary sexual coloration: Influence on social behavior of male keeled earless lizards *(Holbrookia propinqua). Copeia* 1986:980–986.

———. 1989. Prey odor discrimination in the varanoid lizards *Heloderma suspectum* and *Varanus exanthematicus. Ethology* 81:250–258.

———. 1990. Chemical detection of predators by a lizard, the broad-headed skink *(Eumeces laticeps). J. Exp. Zool.* 256:162–167.

———. 1992. Prey discrimination and post-strike elevation in tongue flicking by a cordylid lizard, *Gerrhosaurus nigrolineatus. Copeia* 1992:146–154.

———. 1996. Chemosensory recognition of familiar and unfamiliar conspecifics by the scincid lizard *Eumeces laticeps. Ethology* 102:1–11.

———. 1997a. Pheromonal discrimination of sex by male and female leopard geckos *(Eublepharis macularius). J. Chem. Ecol.* 23:2967–2977.

———. 1997b. Independent evolution of squamate olfaction and vomerolfaction and correlated evolution of vomerolfaction and lingual structure. *Amphib.-Rept.* 18:85–105.

———. 1998a. Reactive and anticipatory display to deflect predatory attack to an autotomous lizard tail. *Can. J. Zool.* 76:1507–1510.

———. 1998b. Prey chemical discrimination indicated by tongue-flicking in the eublepharid gecko *Coleonyx variegatus. J. Exp. Zool.* 281:21–25.

———. 1999. Tradeoffs between courtship, fighting, and antipredatory behavior by a lizard, *Eumeces laticeps. Behav. Ecol. Sociobiol.* 47:54–59.

———. 2000. Pursuit deterrence in lizards. *Saudi. J. Bio. Sci.* 7:15–28.

Cooper, W. E., Jr., and A. C. Alberts. 1990. Responses to chemical food stimuli by an herbivorous actively foraging lizard, *Dipsosaurus dorsalis. Herpetologica* 46:259–266.

Cooper, W. E., Jr., and G. M. Burghardt. 1990. Vomerolfaction and vomodor. *J. Chem. Ecol.* 16:103–105.

Cooper, W. E., Jr., and G. W. Ferguson. 1973. Estrogenic priming of color change induced by progesterone in the collared lizard, *Crotaphytus collaris. Herpetologica* 29:107–110.

Cooper, W. E., Jr., and N. Greenberg. 1992. Reptilian coloration and behavior. In *Biology of the Reptilia,* vol. 18: *Physiology E, Hormones, brain, and behavior,* ed. D. Crews and C. Gans, 298–422. Chicago: University of Chicago Press.

Cooper, W. E., Jr., and S. E. Trauth. 1992. Discrimination of conspecific male and female cloacal chemical stimuli by male and possession of a probable pheromone gland by females in a cordylid lizard, *Gerrhosaurus nigrolineatus. Herpetologica* 48:229–236.

Cooper, W. E., Jr., and J. H. van Wyk. 1994. Absence of prey chemical discrimination by tongue-flicking in an ambush-foraging lizard having actively foraging ancestors. *Ethology* 97:317–328.

Cooper, W. E., Jr., and L. J. Vitt. 1984a. Conspecific odor detection by the male broad-headed skink, *Eumeces laticeps:* Effects of sex and site of odor source and of male reproductive condition. *J. Exp. Zool.* 230:199–209.

———. 1984b. Detection of conspecific odors by the female broad-headed skink, *Eumeces laticeps. J. Exp. Zool.* 229:49–54.

———. 1985. Blue tails and autotomy: Enhancement of predation avoidance in juvenile skinks. *Zeitsch. Tierpsych.* 70:265–276.

———. 1986a. Tracking of female conspecific odor trails by male broad-headed skinks *(Eumeces laticeps). Ethology* 71:242–248.

———. 1986b. Interspecific odour discriminations among syntopic congeners in scincid lizards (genus *Eumeces). Behaviour* 97:1–9.

———. 1986c. Lizard pheromones: Behavioral responses and adaptive significance in skinks of the genus *Eumeces.* In *Chemical signals in vertebrates,* vol. 4, ed. D. Duvall, D. Müller-Schwarze, and R. M. Silverstein, 323–340. New York: Plenum Press.

———. 1986d. Interspecific odour discrimination by a lizard *(Eumeces laticeps). Anim. Behav.* 34:367–376.

———. 1987a. Intraspecific and interspecific aggression in lizards of the scincid genus *Eumeces:* Chemical detection of conspecific sexual competitors. *Herpetologica* 43:7–14.

———. 1987b. Ethological isolation, sexual behavior, and pheromones in the *Fasciatus* species group of the lizard genus *Eumeces. Ethology* 75:328–336.

———. 1987c. Deferred agonistic behavior in a long-lived scincid lizard, *Eumeces laticeps. Oecologia* 72:321–326.

———. 1989. Prey odor discrimination by the broad-headed skink *(Eumeces laticeps). J. Exp. Zool.* 249:11–16.

———. 1991. Influence of detectability and ability to escape on natural selection of conspicuous autotomous defenses. *Can. J. Zool.* 69:757–764.

———. 1993. Female mate choice of large male broad-headed skinks. *Anim. Behav.* 45:683–693.

———. 1997. Maximizing male reproductive success in the broad-headed skink (*Eumeces laticeps*): Preliminary evidence for mate guarding, size-assortative pairing, and opportunistic extra-pair mating. *Amphib.-Rept.* 18:59–73.

Cooper, W. E., Jr., C. S. Adams, and J. L. Dobie. 1983. Female color change in the keeled earless lizard, *Holbrookia propinqua:* Relationship to the reproductive cycle. *Southw. Natur.* 28:275–280.

Cooper, W. E., Jr., C. S. Deperno, and J. Arnett. 1994. Prolonged poststrike elevation in tongue-flicking rate with rapid onset in Gila monster, *Heloderma suspectum:* Relation to diet and foraging and implications for evolution of chemosensory searching. *J. Chem. Ecol.* 20:2867–2881.

Cooper, W. E., Jr., J. A. Lemos-Espinal, and G. R. Smith. 1998. Presence and effect of defensiveness or context on detectability of prey chemical discrimination in the lizard *Xenosaurus platyceps. Herpetologica* 54:409–413.

Cooper, W. E., Jr., P. López, and A. Salvador. 1994. Pheromone detection by an amphisbaenian. *Anim. Behav.* 47:1401–1411.

Cooper, W. E., Jr., M. T. Mendonca, and L. J. Vitt. 1987. Induction of orange head coloration and activation of courtship and aggression by testosterone in the male broad-headed skink (*Eumeces laticeps*). *J. Herpetol.* 21:96–101.

Cooper, W. E., Jr., J. H. van Wyk, and P. Le F. N. Mouton. 1996. Pheromonal detection and sex discrimination of conspecific substrate deposits by the rock-dwelling cordylid lizard *Cordylus cordylus. Copeia* 1996:839–845.

Cooper, W. E., Jr., M. J. Whiting, and J. H. van Wyk. 1997. Foraging modes of cordyliform lizards. *S. Afr. J. Zool.* 32:9–13.

Cooper, W. E., Jr., J. H. van Wyk, P. Le F. N. Mouton, A. M. Al-Johany, J. A. Lemos-Espinal, M. Paulissen, and M. Flowers. 2000. Lizard antipredatory behaviors preventing extraction from crevices. *Herpetologica* 56:394–401.

Cowles, R. B. 1930. The life history of *Varanus niloticus. J. Entomol. Zool.* 22:1–31.

Cowles, R. B., and C. M. Bogert. 1944. A preliminary study of the thermal requirements of desert reptiles. *Bull. Amer. Mus. Nat. Hist.* 83:261–296.

Crews, D., J. M. Bergeron, J. J. Bull, D. Flores, A. Tousignant, J. K. Skipper, and T. Wibbles. 1994. Temperature-dependent sex determination in reptiles:

Proximate mechanisms, ultimate outcomes, and practical applications. *Develop. Genetics* 15:297–312.

Cuellar, O. 1966a. Delayed fertilization in the lizard *Uta stansburiana. Copeia* 1966:549–552.

———. 1966b. Oviducal anatomy and sperm storage structures in lizards. *J. Morph.* 119:7–20.

———. 1974. On the origin of parthenogenesis in vertebrates: The cytogenetic factors. *Amer. Natur.* 108:625–648.

———. 1976. Intraclonal histocompatibility in a parthenogenetic lizard: Evidence of genetic homogeneity. *Science* 193:150–153.

———. 1984. Histocompatibility in Hawaiian and Polynesian populations of the parthenogenetic gecko *Lepidodactylus lugubris. Evolution* 38:176–185.

Cundall, D., and H. W. Greene. Feeding in snakes. In *Feeding,* ed. K. Schwenk, 293–333. San Diego: Academic Press.

Cunningham, J. D. 1956. Food habits of the San Diego alligator lizard. *Herpetologica* 12:225–230.

Cusumano, M., and R. Powell. 1991. A note on the diet of *Amphisbaena gonavensis* in the Dominican Republic. *Amphib.-Rept.* 12:350–352.

Daniel, J. C. 1983. *The book of Indian reptiles.* N.p.: Bombay Natural History Society, Oxford University Press.

Darevsky, I. S. 1958. Natural parthenogenesis in certain subspecies of rock lizard *Lacerta saxicola* Eversmann. *Doklady Akad. Nauk SSSR, Biol. Sci. Sect.* 122:877–879.

———. 1967. *Rock lizards of the Caucasus.* New Delhi: Indian National Scientific Documentation Centre.

Dearing, M. D., and J. J. Schall. 1992. Testing models of optimal diet assembly by the generalist herbivorous lizard *Cnemidophorus murinus. Ecology* 73:845–858.

deBraga, M., and O. Rieppel. 1997. Reptile phylogeny and the interrelationships of turtles. *Zool. J. Linn. Soc.* 120:281–354.

DeLisle, H. F. 1996. *The natural history of monitor lizards.* Malabar, Fla.: Krieger.

Denney, R. N. 1974. The impact of uncontrolled dogs on wildlife and livestock. *Trans. Am. Wildl. Nat. Res. Conf.* 39:257–291.

Deran, S. M., J. C. O'Reilly, and T. Theimer. 1994. Mechanism of defensive inflation in the chuckwalla, *Sauromalus obesus. J. Exp. Zool.* 270:451–459.

Derr, M. 2000. In Caribbean, endangered iguanas get their day. *New York Times,* 10 Oct., D5.

Descola, P. 1994. *In the society of nature: Native ecology in Amazonia.* New York: Cambridge University Press.

Dial, B. E. 1981. Function and energetics of autotomized tail thrashing in *Lygosoma laterale* (Sauria: Scincidae). *Amer. Zool.* 21:1001.

———. 1986. Tail display in two species of iguanid lizards: A test of the "predator signal" hypothesis. *Amer. Natur.* 127:103–111.

Dial, B. E., and L. C. Fitzpatrick. 1981. The energetic costs of tail autotomy to reproduction in the lizard *Coleonyx brevis* (Sauria: Gekkonidae). *Oecologia* 51:310–317.

———. 1983. Lizard tail autotomy: Function and energetics of post-autotomy tail movement in *Scincella lateralis. Science* 219:391–393.

———. 1984. Predator escape success in tailed versus tailless *Scincella lateralis* (Sauria: Scincidae). *Anim. Behav.* 32:

Dial, B., and L. L. Grismer. 1992. A phylogenetic analysis of physiological-ecological character evolution in the lizard genus *Coleonyx* and its implications for historical biogeographic reconstruction. *Syst. Biol.* 41:178–195.

———. 1994. Phylogeny and physiology: Evolution of lizards in the genus *Coleonyx.* In *Conference on the herpetology of North American Deserts,* ed. H. DeLisle, 239–254. Los Angeles: Southwestern Herpetological Society.

Diamond, J. M. 1987. Natural selection: Did Komodo dragons evolve to eat pygmy elephants? *Nature* 326:832.

Díaz-Uriarte, R. 1999. Anti-predator behavior changes following an aggressive encounter in the lizard *Tropidurus hispidus. Proc. R. Soc. Lond. B* 266:2457–5464.

Dischner, H. 1958. Zur Wirkungsweise der Zunge bei Chamäleons. *Natur und Volk* 9:320–324.

Distel, H., and J. Veazey. 1982. The behavioral inventory of the green iguana (*Iguana iguana*). In *Iguanas of the world: Their behavior, ecology, and conservation* ed. G. M. Burghardt and A. S. Rand, 252–270. Park Ridge, N.J.: Noyes.

Dixon, J. R., and P. Soini. 1986. *The reptiles of the upper Amazon Basin, Iquitos region, Peru.* Milwaukee: Milwaukee Public Museum.

Donadio, O. E., and J. M. Gallardo. 1984. Biología y conservación de las especies del género *Tupinambis* (Squamata, Sauria, Teiidae) en la république Argentina. *Rev. Mus. Arg. Cienc. Nat. "Bernardo Rivadavia" Inst. Invest. Cienc. Nat.* 13:117–127.

Donaldson, W. L., A. H. Price, and J. Morse. 1994. The current status and future prospects of the Texas horned lizard (*Phrynosoma cornutum*) in Texas. *Tex. J. Sci.* 46:97–113.

Donoso-Barros, R. 1966. *Reptiles de Chile.* Santiago: Ediciones de la Universidad de Chile.

Doughty, P., and R. Shine. 1995. Life in two dimensions: Natural history of the southern leaf-tailed gecko, *Phyllurus platurus. Herpetologica* 51:193–201.

Duellman, W. E., and A. S. Duellman. 1959. Variation, distribution, and ecology of the iguanid lizard *Enyaliosaurus clarki* of Michoacan, Mexico. *Occ. Pap. Mus. Zool. Univ. Michigan* No. 598:1–10.

Duellman, W. E., and E. R. Pianka. 1990. Biogeography of nocturnal insectivores: Historical events and ecological filters. *Ann. Rev. Ecol. Syst.* 21:57–68.

Dugan, B., and T. V. Wiewandt. 1982. Socioeconomic determinants of mating strategies in iguanine lizards. In *Iguanas of the world: Their behavior, ecology, and conservation,* ed. G. M. Burghardt and A. S. Rand, 303–319. Park Ridge, N.J.: Noyes.

Dunham, A. E. 1980. An experimental study of interspecific competition between the iguanid lizards *Sceloporus merriami* and *Urosaurus ornatus. Ecol. Monogr.* 50:309–330.

Dunham, A. E., and D. B. Miles. 1985. Patterns of covariation in life history traits of squamate reptiles: The effects of size and phylogeny reconsidered. *Amer. Natur.* 126: 231–257.

Dunham, A. E., D. B. Miles, and D. N. Reznick. 1988. Life history patterns in squamate reptiles. In *Biology of the Reptilia,* vol. 16: *Ecology B, Defense and life history,* ed. C. Gans and R. B. Huey, 441–522. New York: Alan R. Liss.

Durtsche, R. D. 1992. Feeding time strategies of the fringe-toed lizard, *Uma inornata,* during breeding and non-breeding seasons. *Oecologia* 89:85–89.

———. 1995. Foraging ecology of the fringe-toed lizard, *Uma inornata,* during periods of high and low food abundance. *Copeia* 1995:915–926.

———. 2000. Ontogenetic plasticity of food habits in the Mexican spiny-tailed iguana, *Ctenosaura pectinata. Oecologia* 124:185–195.

Echelle, A. A., A. F., Echelle, and H. S. Fitch. 1971. A comparative analysis of aggressive display in nine species of Costa Rican *Anolis. Herpetologica* 27:271–288.

Edwards, C. M., S. A. Stanley, R. Davis, A. E. Brynes, G. S. Frost, L. J. Seal, M. A. Ghafel, and S. R. Bloom. 2001. Exendin-4 reduces fasting and postprandial glucose and decreases energy intake in healthy volunteers. *Amer. J. Physio. Endocrinol. Metab.* 281:155–161.

Ehmann, H. 1979. The rediscovery of the bronzeback legless lizard. *Wildlife in Australia* 16:13–14.

Ehmann, H., and D. Metcalf. 1978. The rediscovery of *Ophidiocephalus taeniatus* Lucas and Frost (Pygopodidae, Lacertilia) the bronzeback. *Herpetofauna* 9:8–10.

Eisner, T., D. Alsop, K. Hicks, and J. Meinwald. 1978. Defensive secretions of millipedes. In *Handbook of experimental pharmacology,* ed. G. V. R. Born, O. Eichler, A. Farah, H. Herken, and A. D. Welch, 41–72. Berlin: Springer-Verlag.

Emlen, J. N. 1966. The role of time and energy in food preference. *Amer. Natur.* 100:611–617.

Endo, R., and T. Shikama. 1942. Mesozoic reptilian fauna in the Jehol mountain land, Manchoukuo. *Bull. Cent. Nat. Mus. Manchoukuo* 3:1–12.

Estes, R. 1983. The fossil record and early distribution of lizards. In *Advances in herpetology and evolutionary biology: Essays in honor of Ernest E. Williams,* ed. A. G. J. Rhodin and K. Miyata, 365–398. Cambridge, Mass.: Museum of Comparative Zoology, Harvard University.

Estes, R., and G. Pregill, eds. 1988. *Phylogenetic relationships of the lizard families: Essays commemorating Charles L. Camp.* Stanford: Stanford University Press.

Estes, R., K. de Queiroz, and J. Gauthier. 1988. Phylogenetic relationships within Squamata. In *Phylogenetic relationships of the lizard families,* ed. R. Estes and G. Pregill, 119–281. Stanford: Stanford University Press.

Etheridge, R. 1967. Lizard caudal vertebrae. *Copeia* 1967:699–721.

Etheridge, R., and K. de Queiroz. 1988. A phylogeny of the Iguanidae. In *Phylogenetic relationships of the lizard families: Essays commemorating Charles L. Camp,* ed. R. Estes and G. Pregill, 283–367. Stanford: Stanford University Press.

Etheridge, R., and R. E. Espinoza. 2000. *Taxonomy of the Liolaeminae (Squamata: Iguania: Tropiduridae) and a semi-annotated bibliography.* Smithsonian Herpeteological Service, No. 126.

Evans, S. E. 1981. Caudal autotomy in a lower Jurassic eosuchian. *Copeia* 1981: 883–884.

Farlow, J. O., and E. R . Pianka. 2000. Body form and trackway pattern in Australian desert monitors (Squamata: Varanidae): Comparing zoological and ichnological diversity. *Palaios* 15:235–247.

Ferguson, G. W. 1973. Character displacement of the push-up displays of two partially sympatric species of spiny lizards: *Sceloporus* (Sauria: Iguanidae). *Herpetologica* 29:281–284.

Ferguson, G. W., and C. H. Bohlen. 1972. The regulation of prairie swift (lizard) populations: A progress report. In *Proceedings of the Midwestern Prairie Conference,* no. 3, 69–73. Manhattan: Kansas State University.

Ferguson, G. W., C. H. Bohlen, and P. Wooley. 1980. *Sceloporus undulatus:* Comparative life history and regulation of a Kansas population. *Ecology* 61:313–322.

Fitch, H. S. 1935. Natural history of the alligator lizards. *Trans. Acad. Sci. St. Louis* 29:1–38.

———. 1954. Life history and ecology of the five-lined skink *Eumeces fasciatus. Mus. Nat. Hist., Univ. Kansas, Misc. Publ.* 8:1–156.

———. 1970. Reproductive cycles in lizards and snakes. *Misc. Publ. Univ. Kansas Mus. Nat. Hist.* 52:1–247.

———. 1989. A field study of the slender glass lizard, *Ophisaurus attenuatus,* in northeastern Kansas. *Occ. Pap. Mus. Nat. Hist. Univ. Kansas* 125:1–50.

Fitzgerald, L. A. 1994a. The interplay between life history and environmental stochasticity: Implications for the management of exploited lizard populations. *Amer. Zool.* 34:371–381.

———. 1994b. *Tupinambis* lizards and people: A sustainable use approach to conservation and development. *Conserv. Biol.* 8:12–16.

Fitzgerald, L. A., J. M. Chani, and O. E. Donadío. 1991. *Tupinambis* lizards in Argentina: Implementing management of a traditionally exploited resource. In *Neotropical wildlife: Use and conservation,* ed. J. Robinson and K. Redford, 303–317. Chicago: University of Chicago Press.

Fitzgerald, L. A., F. B. Cruz, and G. Perotti. 1993. The reproductive cycle and the size at maturity of *Tupinambis rufescens* (Sauria: Teiidae) in the dry chaco of Argentina. *J. Herpetol.* 27:70–78.

Fitzgerald, L.A., G. Porini, and V. Lichtschein. 1994. El manejo de *Tupinambis* en Argentina: Historia, estado actual y perspectivas futuras. *Interciencia* 19.

FitzSimons, V. F. M. 1943. The lizards of South Africa. *Mem. Transvaal Mus.* 1:1–528.

Fleishman, L. J. 1992. The influence of the sensory system and the environment on motion patterns in the visual displays of anoline lizards and other vertebrates. *Amer. Natur.* 139:S36–S61.

Fleishman, L. J., E. R. Loew, and M. Leal. 1993. Ultraviolet vision in lizards. *Nature* 365:397.

Fleishman, L. J., W. J. McClintock, R. B. D'Eath, D. H. Brainard, and J. A. Endler. 1998. Colour reception and the use of video playback experiments in animal behavior. *Anim. Behav.* 56:1035–1040.

Fleming, A. F. 1993. The female reproductive cycle of the lizard *Pseudocordylus m. melanotus* (Sauria: Cordylidae). *J. Herpetol.* 27:103–107.

Fleming, A. F., and J. H. van Wyk. 1992. The female reproductive cycle of the lizard *Cordylus p. polyzonus* (Sauria: Cordylidae) in the southwestern Cape Province, South Africa. *J. Herpetol.* 26:121–127.

Flores-Villela, O. 1993. *Herpetofauna Mexicana.* Carnegie Museum of Natural History Special Publication No. 17. Pittsburgh: Carnegie Museum of Natural History.

Fox, S. F. 1978. Natural selection on behavioral phenotypes of the lizard *Uta stansburiana. Ecology* 59:834–847.

———. 1982. Fitness, home-range quality, and aggression in *Uta stansburiana.* In *Lizard ecology,* ed. R. B. Huey, E. R. Pianka, and T. W. Schoener, 149–168. Cambridge, Mass.: Harvard University Press.

Fox, S. F., and T. A. Baird. 1992. The dear enemy phenomenon in the collared lizard, *Crotaphytus collaris,* with a cautionary note on experimental methodology. *Anim. Behav.* 44:780–782.

Fox, S. F., and M. A. Rostker. 1982. Social cost of tail loss in *Uta stansburiana. Science* 218:692–693.

Fox, S. F., N. A. Heger, and L. S. Delay. 1990. Social cost of tail loss in *Uta stansburiana:* Lizard tails as status-signalling badges. *Anim. Behav.* 39:549–554.

Frankenberg, E. 1974. Vocalization of males of three geographical forms of *Ptyodactylus* from Israel (Reptilia: Sauria: Gekkoninae). *J. Herpetol.* 8:59–70.

———. 1975. Distress calls of gekkonid lizards from Israel and Sinai. *Isr. J. Zool.* 24:43–53.

———. 1982. Vocal behavior of the Mediterranean house gecko, *Hemidactylus turcicus. Copeia* 1982:770–775.

Fraser, S. P., and G. C. Grigg. 1984. Control of thermal conductance is insignificant to thermoregulation in small reptiles. *Physiol. Zool.* 57:392–400.

Frazzetta, T. 1983. Adaptation and function of cranial kinesis in reptiles: A time-motion analysis of feeding in alligator lizards. In *Advances in herpetology and evolutionary biology: Essays in honor of Ernest E. Williams,* ed. G. J. Rhodin and K. Miyata, 222–244. Cambridge, Mass.: Museum of Comparative Zoology, Harvard University.

Frey, E., H. D. Sues, and W. Munk. 1997. Gliding mechanism in the late Permian reptile *Coelurosaurus. Science* 275:1450–1452.

Fritts, T. H. 1966. Notes on the reproduction of *Xenosaurus grandis* (Squamata: Xenosauridae). *Copeia* 1966:598.

Frost, D. R. 1992. Phylogenetic analysis and taxonomy of the *Tropidurus* group of lizards (Iguania: Tropiduridae). *Amer. Mus. Novit.* 3033:1–68.

Frost, D. R., and R. Etheridge. 1989. A phylogenetic analysis and taxonomy of iguanian lizards. *Misc. Publ. Mus. Nat. Hist. Univ. Kansas* 81:1–65.

Frost, D. R., R. Etheridge, D. James, and T. A. Titus. 2001. Total evidence, sequence alignment, evolution of polychrotid lizards, and a reclassification of the Iguania (Squamata: Iguania). *Amer. Mus. Novitates* 3343:1–38.

Frost, D. R., M. T. Rodrigues, T. Grant, and T. A. Titus. 2001. Phylogenetics of the lizard genus *Tropidurus* (Squamata: Tropiduridae: Tropidurinae): Direct optimization, descriptive efficiency, and sensitivity analysis of congruence between molecular data and morphology. *Mol. Phyl. Evol.* 21:352–371.

Fuentes, E. R. 1976. Ecological convergence of lizard communities in Chile and California. *Ecology* 57:3–17.

Fuentes, E. R., and F. M. Jaksić. 1980. Ecological species replacement of *Liolaemus* lizards along a habitat gradient. *Oecologia* 46:45–48.

Gallagher, D. S., Jr., and J. R. Dixon. 1991. Taxonomic revision of the South American lizard genus *Kentropyx* Spix (Sauria: Teiidae). *Boll. Mus. Regionale Sci. Nat. Torino* 10:125–171.

Gans, C. 1961. The feeding mechanism of snakes and its possible evolution. *Amer. Zool.* 1:217–227.

———. 1971. Studies on amphisbaenians (Amphisbaenia: Reptilia). 4. A review of the amphisbaenid genus *Leposternon. Bull. Am. Mus. Nat. Hist.* 144:379–464.

———. 1974. *Biomechanics: An approach to vertebrate biology.* Ann Arbor: University of Michigan Press.

———. 1975. Tetrapod limblessness: Evolution and functional corollaries. *Amer. Zool.* 15:455–467.

———. 1978. The characteristics and affinities of the Amphisbaenia. *Trans. Zool. Soc. Lond.* 34:347–416.

———. 1983. Is *Sphenodon punctatus* a maladapted relic? In *Advances in herpetology and evolutionary biology: Essays in honor of Ernest E. Williams,* ed. A. G. J. Rhodin and K. Miyata, 613–620. Cambridge, Mass.: Museum of Comparative Zoology, Harvard University.

———. 1990. Patterns in amphisbaenian biogeography: A preliminary analysis. In *Vertebrates in the tropics,* ed. G. Peters and R. Hutterer, 133–143. Bonn: Museum Alexander Koenig.

Gans, C., T. Krakauer, and C. V. Paganelli.

1968. Water loss in snakes, interspecific and intraspecific variability. *Comp. Biochem. Physiol.* 27A:747–761.

Gao, K. 1994. First discovery of late Cretaceous cordylids (Squamata) from Madagascar. *J. Vert. Paleont.* 14, suppl. 3 (Abstracts of papers, 54th annual meeting, Society of Vertebrate Paleontology, Seattle): 26A.

Gao, K., and R. C. Fox. 1991. New teiid lizards from the Upper Cretaceous Oldham Formation (Judithian) of southeastern Alberta, Canada, with a review of the Cretaceous record of teiids. *Ann. Carnegie Mus.* 60:145–162.

———. 1996. Taxonomy and evolution of late Cretaceous lizards (Reptilia: Squamata) from western Canada. *Bull. Carnegie Mus. Nat. Hist.* 33:1–107.

Gardner, J. D., and R. L. Cifelli. 1999. A primitive snake from the Cretaceous of Utah. *Special Papers in Paleontology* 60:87–100.

Garland, T., R. B. Huey, and A. F. Bennett. 1991. Phylogeny and coadaptation of thermal physiology in lizards: A reanalysis. *Evolution* 45:1969–1975.

Gauthier, J., D. Cannatella, K. de Queiroz, A. G. Kluge, and T. Rowe. 1989. Tetrapod phylogeny. In *The hierarchy of life,* ed. B. Fernholm, K. Bremer, and H. Jörnwall, 337–353. Amsterdam: Elsevier Science.

Gibbons, J. R. H., and I. F. Watkins. 1982. Behavior, ecology, and conservation of south Pacific banded iguanas, *Brachylophus,* including a newly discovered species. In *Iguanas of the world: Their behavior, ecology, and conservation,* ed. G. M. Burghardt and A. S. Rand, 418–441. Park Ridge, N.J.: Noyes.

Gibbons, J. W., D. E. Scott, T. J. Ryan, K. A. Buhlmann, T. D. Tuberville, B. S. Metts, J. L. Greene, T. Mills, Y. Leiden, S. Poppy, and C. T. Winne. 2000. The global decline of reptiles, déjà-vu amphibians. *BioScience* 50:653–666.

Gier, P. J. 1997. Iguanid mating systems: Ecological causes and sexual selection consequences. Ph.D. diss., University of Oklahoma.

Gil, M. J., V. Pérez-Mellado, and F. Guerrero. 1993. Trophic ecology of *Acanthodactylus erythrurus* in central Iberian peninsula. Is there a dietary shift? In *Lacertids of the Mediterranean region: A biological approach,* ed. E. D. Valakos, W. Böhme, V. Pérez-Mellado, and P. Maragou, 199–211. Athens, Greece: Hellenic Zoological Society.

Gillingham, J. C., C. Carmichael, and T. Miller. 1995. Social behavior of the tuatara, *Sphenodon punctatus. Herpetol. Monogr.* 9:5–16.

Glasheen, J. W., and T. A. McMahon. 1991. A hydrodynamic model of locomotion in the basilisk lizard. *Nature* 380:340–342.

Glaw, F., and M. Vences. 1992. *A field guide to the amphibians and reptiles of Madagascar.* Leverkusen, Ger.: Moos-Druck.

Goeldi, E. A. 1897. Die Eier von 13 brasilianischen Reptilien, nebst Bemerkungen über Lebens- und Fortpflanzungsweise letzterer. *Zool. Jahrb.* 10:640–676.

Goldberg, S. R., and R. L. Bezy. 1974. Reproduction in the island night lizard, *Xantusia riversiana. Herpetologica* 30:350–360.

Goulding, M. 1989. *Amazon, the flooded forest.* London: BBC Books.

Gradstein, S. R., and C. Equihua. 1995. An epizoic bryophyte and algae growing on the lizard *Corytophanes cristatus* in Mexican rain forest. *Biotropica* 27: 265–268.

Graves, B. M., and M. Halpern. 1991. Discrimination of self from conspecific chemical cues in *Tiliqua scincoides* (Sauria: Scincidae). *J. Herpetol.* 25:125–126.

Green, B. 1969. Water and electrolyte balance in the sand goanna, *Varanus gouldii* (Gray). Ph.D. thesis, University of Adelaide, Australia.

Greenberg, G., and G. K. Noble. 1944. Social behavior of the American chameleon (*Anolis carolinensis* Voigt). *Physiol. Zool.* 17:392–439.

Greene, H. W. 1973. Defensive tail display by snakes and amphisbaenians. *J. Herpetol.* 7:143–161.

———. 1983. Dietary correlates of the origin and radiation of snakes. *Amer. Zool.* 23:431–441.

———. 1986. Diet and arboreality in the emerald monitor, *Varanus prasinus,* with comments in the study of adaptation. *Fieldiana* 31:1–12.

———. 1997. *Snakes: The evolution of mystery in nature.* Berkeley: University of California Press.

Greene, H. W., and D. Cundall. 2000. Perspectives: Evolutionary biology—limbless tetrapods and snakes with limbs. *Science* 287:1939–1941.

Greer, A. E. 1970. A subfamilial classification of scincid lizards. *Bull. Mus. Comp. Zool.* 139:151–183.

———. 1974. The generic relationships of the scincid lizard genus *Leiolopisma* and its relatives. *Austr. J. Zool.,* suppl. ser., 31:1–67.

———. 1979. A phylogenetic subdivision of Australian skinks. *Rec. Austr. Mus.* 32: 339–371.

———. 1985a. Facial tongue-wiping in Xantusiid lizards: Its systematic implications. *J. Herpetol.* 19:174–175.

———. 1985b. The relationships of the lizard genera *Anelytropsis* and *Dibamus. J. Herpetol.* 19:116–156.

———. 1989. *The biology and evolution of Australian lizards.* Chipping Norton, N.S.W., Austr.: Surrey Beatty & Sons.

———. 1991. Behavioural mimicry in the autotomized tail of a pygopodid lizard. *West. Austr. Natur.* 18:161–163.

Greer, A. E., and F. Parker. 1974. The *fasciatus* species group of *Sphenomorphus* (Lacertilia: Scincidae): Notes on eight previously described species and descriptions of three new species. *Papua New Guinea Sci. Soc. Proc.* 25:31–61.

Grinnell, J. 1924. Geography and evolution. *Ecology* 5:225–229.

Grismer, L. L. 1988. Phylogeny, taxonomy, classification, and biogeography of eublepharid geckos. In *Phylogenetic relationships of the lizard families: Essays Commemorating Charles L. Camp,* ed. R. Estes and G. Pregill, 369–469. Stanford: Stanford University Press.

Guillette, L. J., Jr. 1995. Endocrine disrupting environmental contaminants and developmental abnormalities in embryos. *Human Ecol. Risk Asses.* 1:25–36.

Guillette, J. L., Jr., and F. R. Méndez-de la Cruz. 1993. The reproductive cycle of the viviparous Mexican lizard *Sceloporus torquatus. J. Herpetol.* 27:168–174.

Gutzke, W. H. N., and G. C. Packard. 1986. Sensitive periods for the influence of the hydric environment on eggs and hatchlings of painted turtles (*Chrysemys picta*). *Physiol. Zool.* 59:337–343.

———. 1987. Influence of the hydric and thermal environments on eggs and hatchlings of bull snakes, *Pituophis melanoleucus. Physiol. Zool.* 60:9–17.

Guyer, C., and J. M. Savage. 1986. Cladistic relationships among anoles (Sauria: Iguanidae). *Syst. Zool.* 35:509–531.

———. 1992. Anole systematics revisited. *Syst. Biol.* 41:89–110.

Haacke, W. D. 1969. The call of the barking geckos (Gekkonidae: Reptilia). *Sci. Pap. Namib Desert Res. Sta.* 46:83–93.

Hamilton, W. J., III. 1973. *Life's color code.* New York: McGraw-Hill.

Harkness, L. 1977. Chameleons use accommodation cues to judge distance. *Nature* 267:346–349.

Harris, D. M. 1994. Review of the teiid lizard genus *Ptychoglossus. Herp. Monog.* 8: 226–275.

Harris, V. A. 1964. *The life of the rainbow lizard.* London: Hutchinson.

Harrisson, B. 1961. *Lanthanotus borneensis*—habits and observations. *Sarawak Mus. J.* 10:286–292.

Harrisson, H. 1961. The earless monitor lizard, *Lanthanotus borneensis. Discovery,* July 1961, 290–293.

———. 1963. Earless monitor lizards in Borneo. *Nature* 198:407–408.

———. 1966. A record-size *Lanthanotus* alive (1966): Casual notes. *Sarawak Mus. J.* 14:323–333.

Hasegawa, M., and Y. Taniguchi. 1994. Visual avoidance of a conspicuously colored carabid beetle *Dischissus mirandus* by the lizard *Eumeces okadae. J. Ethology* 12:9–14.

Hasson, O., R. Hibbard, and G. Ceballos. 1989. The pursuit deterrent function of tail-wagging in the zebra-tailed lizard (*Callisaurus draconoides*). *Can. J. Zool.* 67:1203–1209.

Heatwole, H., and J. E. N. Vernon. 1977. Vital limit and evaporative water loss in lizards (Reptilia: Lacertilia): A critique and new idea. *J. Herpetol.* 11:341–348.

Hemley, G. 1984. World trade in tegu skins. *Traffic Bull.* 5:60–62.

Hews, D. K. 1990. Examining hypotheses generated by field measures of sexual selection on male lizards, *Uta palmeri. Evolution* 44:1956–1966.

———. 1993. Food resources affect female distribution and male mating opportunities in the iguanian lizard *Uta palmeri. Anim. Behav.* 46:279–291.

Hews, D. K., R. Knapp, and M. C. Moore. 1994. Early exposure to androgens affects adult expression of alternative male types in tree lizards. *Horm. Behav.* 28:96–115.

Hews, D. K., C. W. Thompson, I. T. Moore, and M. C. Moore. 1997. Population frequencies of alternative phenotypes in tree lizards: Geographic variation and common-garden rearing studies. *Behav. Ecol. Sociobiol.* 41:371–380.

Hicks, R. A., and R. L. Trivers. 1983. The social behavior of *Anolis valecienni.* In *Advances in herpetology and evolutionary biology: Essays in honor of Ernest E. Williams,* ed. A. G. J. Rhodin and K. Miyata, 570–595. Cambridge, Mass.: Museum of Comparative Zoology, Harvard University.

Hillenius, D. 1986. The relationship of *Brookesia, Rhampholeon,* and *Chamaeleo* (Chamaeleonidae, Reptilia). *Bijd. tot de Dierk.* 58:7–11.

Hiller, U. 1968. Untersuchungen zum Feinbau und zur Funktion der Haftborsten von Reptilien. *Z. Morph. Tiere* 62:307–362.

———. 1976. Comparative studies on the functional morphology of two gekkonid lizards. *J. Bombay Nat. Hist. Soc.* 73: 278–282.

Hillis, D. M. 1985. Evolutionary genetics of the Andean lizard genus *Pholidobolus* (Sauria: Gymnophthalmidae): Phylogeny, biogeography, and a comparison of tree construction techniques. *Syst. Biol.* 34: 109–126.

Hillis, D. M., and J. E. Simmons. 1986. Dynamic change of a zone of parapatry between two species of *Pholidobolus* (Sauria: Gymnophthalmidae). *J. Herpetol.* 20:85–87.

Hillis, D. M., D. Moritz, and B. K. Mable, eds. 1996. *Molecular systematics.* 2nd ed. Sinauer.

Hoddenbach, G. A., and F. B. Turner. 1968. Clutch size of the lizard *Uta stansburiana* in southern Nevada. *Amer. Midl. Natur.* 80:262–265.

Hofmann, E. G. 2000. The Chinese crocodile lizard *(Shinisaurus crocodilurus). Reptiles* 8:60–71.

Hollingsworth, B. D. 1998. The systematics of chuckwallas *(Sauromalus)* with a phylogenetic analysis of other iguanid lizards. *Herp. Monogr.* 12:38–191.

Holt, R. D. 1996. Demographic constraints in evolution: Towards unifying the evolutionary theories of senescence and niche conservatism. *Evol. Ecol.* 10:1–11.

Holte, A. E., and M. A. Houck. 2000. Juvenile greater roadrunner (Cuculidae) killed by choking on a Texas horned lizard *(Phrynosoma cornutum). Southw. Natur.* 45:74–76.

Honders, J. 1975. *The world of reptiles and amphibians.* New York: Peebles Press International.

Howland, J. M., L. J. Vitt, and P. T. Lopez. 1990. Life on the edge: The ecology and life history of the tropidurine iguanid lizard *Uranoscodon superciliosum. Can. J. Zool.* 68:1366–1373.

Huey, R. B. 1977. Egg retention in some high-altitude *Anolis* lizards. *Copeia* 1977: 373–375.

Huey, R. B., and E. R. Pianka. 1974. Ecological character displacement in a lizard. *Amer. Zool.* 14:1127–1136.

———. 1977a. Natural selection for juvenile lizards mimicking noxious beetles. *Science* 195:201–203.

———. 1977b. Patterns of niche overlap among broadly sympatric versus narrowly sympatric Kalahari lizards (Scincidae: *Mabuya). Ecology* 58:119–128.

———. 1977c. Seasonal variation in thermoregulatory behavior and body temperature of diurnal Kalahari lizards. *Ecology* 58:1066–1075. (With an appendix by J. A. Hoffman.)

———. 1981. Ecological consequences of foraging mode. *Ecology* 62:991–999.

Huey, R. B., and M. Slatkin. 1976. Cost and benefits of lizard thermoregulation. *Quart. Rev. Biol.* 51:363–384.

Huey, R. B., H. John-Alder, and K. A. Nagy. 1984. Locomotor capacity and foraging

behavior of Kalahari lizards. *Anim. Behav.* 32:41–50.

Huey, R. B., E. R. Pianka, and T. W. Schoener, eds. 1983. *Lizard ecology: Studies of a model organism.* Cambridge, Mass.: Harvard University Press.

Huey, R. B., P. H. Niewiarowski, J. Kaufmann, and J. C. Herron. 1989. Thermal biology of nocturnal ectotherms: Is sprint performance of geckos maximal at low body temperatures? *Phys. Zool.* 62:488–504.

Huey, R. B., E. R. Pianka, and L. J. Vitt. 2001. How often do lizards "run on empty"? *Ecology* 82:1–7.

Huey, R. B., E. R. Pianka, M. E. Egan, and L. W. Coons. 1974. Ecological shifts in sympatry: Kalahari fossorial lizards *(Typhlosaurus). Ecology* 55:304–316.

Hunt, L. E. 1983. A nomenclatural rearrangement of the genus *Anniella* (Sauria: Anniellidae). *Copeia* 1983:79–89.

Hutchison, V. H., and R. K. Dupre. 1992. "Thermoregulation." In *Environmental physiology of the amphibians,* ed. M. E. Feder and W. W. Burggren, 206–249. Chicago: University of Chicago Press.

Hutchinson, M. N. 1993. Family Scincidae. Chapter 31 in *Fauna of Australia,* vol. 2: *Amphibia Reptilia Aves,* ed. G. J. B. Ross, 261–279. Canberra: Australian Biological and Environmental Survey.

Hutchinson, M. N., and S. C. Donnellan. 1993. Chapter 26 in *Fauna of Australia,* vol. 2: *Amphibia Reptilia Aves,* ed. G. J. B. Ross, 210–220. Canberra: Australian Biological and Environmental Survey.

Hutchinson, M. N., S. C. Donnellan, P. R. Baverstock, M. Krieg, S. Simms, and S. Burgin. 1990. Immunological relationships and generic revision of the Australian lizards assigned to the genus *Leiolopisma* (Scincidae: Lygosominae). *Austr. J. Zool.* 38:535–554.

Inger, R. F. 1983. Morphological and ecological variation in the flying lizards (genus *Draco). Fieldiana Zool.* 18:1–35.

Irschick, D. J., C. C. Austin, K. Petren, R. N. Fisher, J. B. Losos, and O. Ellers. 1996. A comparative analysis of clinging ability among pad-bearing lizards. *Biol. J. Linn. Soc.* 59:21–35.

Iverson, J. B. 1978. The impact of feral cats and dogs on populations of the West Indian rock iguana, *Cyclura carinata. Biol. Conserv.* 14:63–73.

———. 1979. Behavior and ecology of the rock iguana *Cyclura carinata. Bull. Florida State Mus. Biol. Sci.* 24:175–358.

———. 1982. Adaptations to herbivory in iguanine lizards. Chapter 4 in *Iguanas of the world,* ed. G. M. Burghardt and

A. S. Rand, 60–76. Park Ridge, N.J.: Noyes.

Iverson, J. B., and M. R. Mamula. 1989. Natural growth in the Bahaman iguana *Cyclura cyclura. Copeia* 1989:502–505.

Ivlev, V. S. 1961. *Experimental feeding ecology of fishes.* New Haven: Yale University Press.

Jackson, J. F. 1978. Differentiation in the genera *Enyalius* and *Strobilurus* (Iguanidae): Implications for Pleistocene climatic changes in eastern Brazil. *Arq. Zool., São Paulo* 30:1–79.

Jaksić, F. M., and S. D. Busack. 1984. Apparent inadequacy of tail-loss figures as estimates of predation upon lizards. *Amphibia-Reptilia* 5:177–179.

Jaksić, F. M., and H. Núñez. 1979. Escaping behavior and morphological correlates in two *Liolaemus* species of central Chile (Lacertilia: Iguanidae). *Oecologia* 42: 119–122.

Jaksić, F. M., and K. Schwenk. 1983. Natural history observations on *Liolaemus magellanicus,* the southernmost lizard in the world. *Herpetologica* 39:457–461.

Jaksić, F. M., H. Núñez, and F. P. Ojeda. 1980. Body proportions, microhabitat selection, and adaptive radiation of *Liolaemus* lizards in central Chile. *Oecologia* 45:178–181.

Jaksić, F. M., H. W. Greene, K. Schwenk, and R. L. Seib. 1982. Predation upon reptiles in Mediterranean habitats of Chile, Spain, and California: A comparative analysis. *Oecologia* 53:152–159.

James, C. D. 1991. Temporal variation in diets and trophic partitioning by coexisting lizards (*Ctenotus:* Scincidae) in central Australia. *Oecologia* 85:553–561.

———. 1996. Ecology of the pygmy goanna *(Varanus brevicauda)* in spinifex grasslands of central Australia. *Austr. J. Zool.* 44:177–192.

James, C. D., and R. Shine. 1985. The seasonal timing of reproduction: A tropical-temperate comparison in Australian lizards. *Oecologia* 67:464–474.

———. 1988. Life-history strategies of Australian lizards: A comparison between the tropics and the temperate zone. *Oecologia* 75:307–316.

James, C. D., J. B. Losos, and D. R. King. 1992. Reproductive biology and diets of goannas (Reptilia: Varanidae) from Australia. *J. Herpetology* 26:128–136.

Janzen, F. J., and G. Paukstis. 1991. Environmental sex determination in reptiles: Ecology, evolution, and experimental design. *Quart. Rev. Biol.* 66:149–179.

Jennings, W. B., and G. G. Thompson. 1999. Territorial behavior in the Australian scincid lizard *Ctenotus fallens. Herpetologica* 55:352–361.

Jenssen, T. A. 1983. Display behavior of two Haitian lizards, *Anolis cybotes* and *Anolis distichus.* In *Advances in herpetology and evolutionary biology: Essays in honor of Ernest E. Williams,* ed. A. G. J. Rhodin and K. Miyata, 552–569. Cambridge, Mass.: Museum of Comparative Zoology, Harvard University.

Jones, R. E., K. T. Fitzgerald, D. Duvall, and D. Banker. 1979. On the mechanisms of alternating and simultaneous ovulation in lizards. *Herpetologica* 35:132–139.

Kequin, G., and M. A. Norrell. 2000. Taxonomic composition and systematics of late Cretaceous lizard assemblages from Ikhaa Tolgod and adjacent localities, Mongolian Gobi Desert. *Bull. Amer. Mus. Nat. Hist.* 249:1–118.

King, D., and B. Green. 1999. *Goanna: The biology of varanid lizards.* 2d ed. Kensington, Australia: New South Wales University Press.

King, M. 1977. Reproduction in the Australian gecko *Phyllodactylus marmoratus* (Gray). *Herpetologica* 33:7–13.

King, M., and D. Hayman. 1978. Seasonal variation of chiasma frequency in *Phyllodactylus marmoratus* (Gray) (Gekkonidae—Reptilia). *Chromosoma* 69:131–154.

King, M., and P. Horner 1993. Family Gekkonidae. In *Fauna of Australia.* Vol. 2: *Amphibia Reptilia Aves,* ed. G. J. B. Ross, 221–233. Canberra: Australian Biological and Environmental Survey.

Kirmse, W. 1988. Foveal and ambient visuomotor control in chameleons (Squamata): Experimental results and comparative review. *Zool. J. Physiol.* 92:341–350.

Klauber, L. M. 1931. A new species of *Xantusia* from Arizona, with a synopsis of the genus. *Trans. San Diego Soc. Nat. Hist.* 7: 1–16.

Klaver, C., and W. Böhme. 1986. Phylogeny and classification of the Chamaeleonidae (Sauria) with special reference to hemipenis morphology. *Bonner Zooloigische Monographien* 22: 5–60.

Kluge, A. G. 1967. Higher taxonomic categories of gekkonid lizards and their evolution. *Bull. Amer. Mus. Nat. Hist.* 135:1–59.

———. 1974. A taxonomic revision of the lizard family Pygopodidae. *Misc. Publ. Mus. Zool. Univ. Mich.* 147:1–221.

———. 1976. Phylogenetic relationships in the lizard family Pygopodidae. *Misc. Publ. Mus. Zool. Univ. Mich.* 152:1–72.

———. 1987. Cladistic relationships in the Gekkonoidea (Squamata, Sauria). *Misc. Publ. Mus. Zool. Univ. Mich.* 173:1–54.

———. 1995. Cladistic relationships of sphaerodactyline lizards. *Am. Mus. Novit.* No. 3139:1–23.

Land, M. F. 1995. Fast-focus telephoto eye. *Nature* 373:658–659.

Lang, M. 1991. Generic relationships within Cordyliformes (Reptilia: Squamata). *Bull. Inst. R. Sci. Nat. Belg. Biol.* 61:121–188.

Lawlor, L. R. 1980a. Overlap, similarity, and competition coefficients. *Ecology* 61:245–251.

———. 1980b. Structure and stability in natural and randomly constructed competitive communities. *Amer. Natur.* 116: 394–408.

Leal, M., and J. A. Rodriguez-Robles. 1977. Signalling displays during predator-prey interactions in a Puerto Rican anole, *Anolis cristatellus. Anim. Behav.* 5:1147–1154.

Lee, J. C. 1974. The diel activity cycle of the lizard, *Xantusia henshawi. Copeia* 1974: 934–940.

———. 1975. The autecology of *Xantusia henshawi henshawi* (Sauria: Xantusiidae). *Transactions of the San Diego Society of Natural History* 17:259–278.

———. 1996. *The amphibians and reptiles of the Yucatán Penninsula.* Ithaca, N.Y.: Cornell University Press.

Lee, M. S. Y. 1997. The phylogeny of varanoid lizards and the affinities of snakes. *Phil. Trans. Roy. Soc. Lond. B* 352:53–91.

———. 1998. Convergent evolution and character correlation in burrowing reptiles: Towards a resolution of squamate relationships. *Biol. J. Linn. Soc.* 65:369–453.

Lee, M. S. Y., and M. W. Caldwell. 1998. Anatomy and relationships of *Pachyrachis problematicus,* a primitive snake with hindlimbs. *Phil. Trans. R. Soc. Lond. B* 353:1521–1552.

Lemos-Espinal, J. A., G. R. Smith, and R. E. Ballinger. 1996. Natural history of the Mexican knob-scaled lizard, *Xenosaurus rectocollaris. Herp. Nat. Hist.* 4:151–154.

———. 1997a. Natural history of *Xenosaurus platyceps,* a crevice-dwelling lizard from Tamaulipas, Mexico. *Herp. Nat. Hist.* 5:181–186.

———. 1997b. Neonate-female associations in *Xenosaurus newmanorum:* A case of parental care in a lizard? *Herp. Rev.* 28: 22–23.

———. 1998. Thermal ecology of the crevice-dwelling lizard *Xenosaurus newmanorum. J. Herpetol.* 32:141–144.

Lieb, C. S. 1985. Systematics and distributions of the skinks allied to *Eumeces tetragrammus* (Sauria: Scincidae). *Contr. Sci. Nat. Hist. Mus. Los Angeles County* 357:1–19.

López, P., and A. Salvador. 1992. The role of chemosensory cues in discrimination of prey odors by the amphisbaenian *Blanus cinereus. J. Chem. Ecol.* 18:87–93.

López, P., A. Salvador, and W. E. Cooper, Jr.

1997. Discrimination of self from other males by chemosensory cues in the Amphisbaenia (*Blanus cinereus*). *J. Comp. Psych.* 111:105–109.

Losos, J. B. 1985. An experimental demonstration of the species-recognition role of *Anolis* dewlap color. *Copeia* 1985:905–910.

———. 1992. The evolution of convergent structure in Caribbean *Anolis* communities. *Syst. Biol.* 41:403–420.

Losos, J. B., and H. W. Greene. 1988. Ecological and evolutionary implications of diet in monitor lizards. *Biol. J. Linn. Soc.* 35:379–407.

Lowe, C. H. 1993. Introduction. In *The biology of whiptail lizards (genus* Cnemidophorus*),* ed. J. W. Wright and L. J. Vitt, 1–25. Norman: Oklahoma Museum of Natural History.

Lowe, C. H., and J. W. Wright. 1966. Evolution of parthenogenetic species of *Cnemidophorus,* whiptail lizards, in western North America. *J. Arizona Acad. Sci.* 4:81–87.

Lowe, C. J., C. R. Schwalbe, and T. B. Johnson. 1986. *The venomous reptiles of Arizona.* Phoenix: Arizona Game and Fish Department.

Lucas, A. H. S., and C. Frost. 1897. Descriptions of two new species of lizards from central Australia. *Proc. Roy. Soc. Victoria* 9:54–56.

Luke, C. 1986. Convergent evolution of lizard toe fringes. *Biol. J. Linn. Soc.* 27:1–16.

Lumholtz, C. 1902. *Unknown Mexico: A record of five years' exploration among the tribes of the western Sierra Madre; in the Tierra Caliente of Tepic and Jalisco; and among the Tarascos of Michoacan.* New York: Scribner.

MacArthur, R. H., and E. R. Pianka. 1966. On optimal use of a patchy environment. *Amer. Natur.* 100:603–609.

MacArthur, R. H., and E. O. Wilson. 1967. *The theory of island biogeography.* Princeton: Princeton University Press.

Macey, J. R., A. Larson, N. B. Ananjeva, and T. J. Papenfuss. 1997. Evolutionary shifts in three major structural features of the mitochondrial genome among iguanian lizards. *J. Mol. Evol.* 44:660–674.

Macey, R. J., J. A. Schulte, III, A. Larson, B. S. Tuniyev, N. Orlov, and T. J. Papenfuss. 1999. Molecular phylogenetics, tRNA evolution, and historical biogeography in anguid lizards and related taxonomic families. *Mol. Phyl. Evol.* 12:250–272.

Magnusson, W. E., L. J. Paiva, R. M. Rocha, C. R. Franke, L. A. Kasper, and A. P. Lima. 1985. The correlates of foraging mode in a community of Brazilian lizards. *Herpetologica* 41:324–332.

Main, A. R., and C. M. Bull. 1996. Mother-offspring recognition in two Australian lizards, *Tiliqua rugosa* and *Egernia stokesii*. *Anim. Behav.* 52:193–200.

Maiorana, V. C. 1977. Tail autotomy, functional conflicts, and their resolution by a salamander. *Nature* 265:533–535.

Manaster, J. 1997. *Horned lizards.* Austin: University of Texas Press.

Manzani, P. R., and A. S. Abe. 1997. A new species of *Tupinambis* Daudin, 1802 (Squamata: Teiidae), from central Brazil. *Bol. Mus. Nac.,* n.s. (Rio de Janeiro) 382:1–10.

Marcellini, D. 1974. Acoustic behavior of the gekkonid lizard *Hemidactylus frenatus*. *Herpetologica* 30:44–52.

———. 1977. Acoustic and visual displays behavior of gekkonid lizards. *Amer. Zool.* 17:251–260.

Markezich, A. L., C. J. Cole, and H. C. Dessauer. 1997. The blue and green whiptail lizards (Squamata: Teiidae: *Cnemidophorus*) of the Peninsula de Paraguana, Venezuela: Systematics, ecology, descriptions of two new taxa, and relationships to whiptails of the Guianas. *Amer. Mus. Novit.* 3207:1–60.

Marquet, P. A., J. C. Ortiz, F. Bozinovic, and F. M. Jaksíc. 1989. Ecological aspects of thermoregulation at high altitudes: The case of Andean *Liolaemus* lizards in northern Chile. *Oecologia* 81:16–20.

Martin, J. 1992. *Masters of disguise: A natural history of chameleons.* New York: Facts on File.

Martin, W. E., Jr., M. J. Whiting, and J. H. van Wyk. 1997. Foraging modes of cordyliform lizards. *S. Afr. Zool.* 32:9–13.

Martins, E. P. 1994. Phylogenetic perspectives on the evolution of lizard territoriality. In *Lizard ecology: Historical and experimental perspectives,* ed. L. J. Vitt and E. R. Pianka, 118–144. Princeton: Princeton University Press.

Mason, R. T. 1992. Reptilian pheromones. In *Biology of the Reptilia,* vol. 18: *Physiology E, Hormones, brain, and behavior,* ed. C. Gans and D. Crews, 114–228. Chicago: University of Chicago Press.

Mathies, T., and R. M. Andrews. 1999. Determinants of embryonic stage at oviposition in the lizard *Urosaurus ornatus*. *Physiol. Biochem. Zool.* 72:645–655.

Mautz, W. J. 1979. The metabolism of reclusive lizards, the Xantusiidae. *Copeia* 1979:577–584.

———. 1982. Patterns of evaporative water loss. In *Biology of the Reptilia,* vol. 12: *Physiology C, Physiological ecology,* ed. C. Gans and F. H. Pough, 443–481. New York: Academic Press.

Mautz, W. J., and T. J. Case. 1974. A diurnal activity cycle in the granite night lizard, *Xantusia henshawi*. *Copeia* 1974:243–251.

Mautz, W. J., and W. Lopez-Forment. 1978. Observations on the diet and activity of *Lepidophyma smithii* (Sauria: Xantusiidae). *Herpetologica* 34:311–313.

Mautz, W. J., and K. A. Nagy. 1987. Ontogenetic changes in diet, field metabolic rate, and water flux in the herbivorous lizard *Dipsosaurus dorsalis*. *Physiol. Zool.* 60:640–658.

Maybury-Lewis, D. 1967. *Akwẽ Shavante society.* Oxford: Clarendon Press.

Mayr, E. 1961. Cause and effect in biology. *Science* 134:1501–1506.

McCoy, M. 2000. *Reptiles of the Solomon Islands.* Kuranda, Australia: ZooGraphics.

McDowell, S., and C. Bogert. 1954. The systematic position of *Lanthanotus* and the affinities of the anguimorphan lizards. *Bull. Amer. Mus. Nat. Hist.* 5:1–142.

McGuire, J. A. 1996. Phylogenetic systematics of crotaphytid lizards (Reptilia: Iguania: Crotaphytidae). *Bull. Carnegie Mus. Nat. Hist.* 32:1–143.

McIlhenny, E. A. 1937. Notes on the five-lined skink. *Copeia* 1937:232–233.

McLaughlin, R. L. 1989. Search modes of birds and lizards: Evidence for alternative movement patterns. *Amer. Natur.* 133:654–670.

McNab, B. K., and W. Auffenberg. 1976. The effect of large body size on the temperature regulation of the Komodo dragon, *Varanus komodoensis*. *Comp. Biochem. Physiol.* 55A:345–350.

Medel, R. G., J. E. Jimenez, S. F. Fox, and F. M. Jaksíc. 1988. Experimental evidence that high population frequencies of lizard tail autotomy indicate innefficient predation. *Oikos* 53:321–324.

Medica, P. A., F. B. Turner, and D. D. Smith. 1973. Hormonal induction of color change in female leopard lizards, *Crotaphytus wislizenii*. *Copeia* 1973:658–661.

Méndez-de la Cruz, F. R., M. Villagrán-Santa Cruz, and R. M. Andrews. 1998. Evolution of viviparity in the lizard genus *Sceloporus*. *Herpetologica* 54:521–532.

Middendorf, G. A., and W. C. Sherbrooke. 1992. Canid elicitation of blood-squirting in a horned lizard *(Phrynosoma cornutum)*. *Copeia* 1992:519–527.

Miller, C. M. 1944. Ecologic relations and adaptations of the limbless lizards of the genus *Anniella*. *Ecol. Monogr.* 14:273–289.

Miller, K., G. C. Packard, and M. J. Packard. 1987. Hydric conditions during incubation influence locomotor performance of hatchling snapping turtles. *J. Exp. Biol.* 127:401–412.

Milstead, W. W., ed. 1967. *Lizard ecology: A symposium.* Columbia: University of Missouri Press.

Milton, D. 1980. An example of communal egg laying in *Oedura tryoni* (De Vis). *Herpetofauna* 11:19–23.

Minnich, J. E. 1982. The use of water. In *Biology of the Reptilia,* vol. 12: *Physiology C, Physiological ecology,* ed. C. Gans and F. H. Pough, 325–395. New York: Academic Press.

Minton, S. R., Jr. 1958. Observations on amphibians and reptiles of the Big Bend region of Texas. *Southwest. Nat.* 3:28–54.

Mittermeier, R. A. 1972. Jamaica's endangered species. *Oryyx* 12:258–262.

Molnar, R. E. 1985. The history of lepidosaurs in Australia. In *Biology of Australasian frogs and reptiles,* ed. G. Grigg, R. Shine, and H. Ehmann, 155–158. Chipping Norton, N.S.W., Austr.: Royal Zoological Society of New South Wales.

Montanucci, R. R. 1973. Systematics and evolution of the Andean lizard genus *Pholidobolus* (Sauria: Teiidae). *Misc. Publ. Mus. Nat. Hist. Univ. Kansas* 59:1–52.

———. 1989. The relationship of morphology to diet in the horned lizard genus *Phrynosoma*. *Herpetologica* 45:208–216.

Moraes, R. C. 1993. Ecologia das especies de *Calyptommatus* (Sauria, Gymnophthalmidae) e partilha de recursos com outros dois microteiídeos. Ph.D. diss., Universidade de São Paulo.

Mori, A. 1990. Tail vibration of the Japanese grass lizard *Takydromus tachydromoides* as a tactic against a snake predator. *J. Ethol.* 8:81–88.

Mori, A., and T. Hikida. 1993. Natural history observations of the flying lizard, *Draco volans sumatranus* (Agamidae, Squamata), from Sarawak, Malaysia. *Raffles Bulletin of Zoology* 41:83–94.

———. 1994. Field observations on the social behavior of the flying lizard, *Draco volans sumatranus,* in Borneo. *Copeia* 1994:124–130.

Mount, R. H. 1963. The natural history of the red-tailed skink, *Eumeces egregius* (Baird). *Am. Midl. Nat.* 70:356–385.

Mouton, P. Le F. N. 2000. Seasonal variation in stomach contents and diet composition in the large girdled lizard, *Cordylus giganteus* (Reptilia: Cordylidae) in the Highveld grasslands of the northeastern Free State, South Africa. *African Zool.* 35:9–27.

Mouton, P. Le F. N., H. Geertsema, and L. Visagie. 2000. Foraging mode of a group-living lizard, *Cordylus cataphractus* (Cordylidae). *African Zool.* 35:1–7.

Murphy, G. I. 1968. Pattern in life history and environment. *Amer. Natur.* 102:390–404.

Murphy, J. B., and L. A. Mitchell. 1974. Ritualized combat behavior of the pygmy Mulga monitor lizard, *Varanus gilleni* (Sauria: Varanidae). *Herpetologica* 3:90–97.

Nagy, K. A. 1982. Field studies of water relations. *In Biology of the Reptilia,* vol. 12: *Physiology C, Physiological ecology,* ed. C. Gans and F. H. Pough, 483–501. New York: Academic Press.

Nagy, K. A., and V. H. Shoemaker. 1984. Field energetics and food consumption of the Galápagos marine iguana, *Amblyrhynchus cristatus. Physiol. Zool.* 57:281–290.

Necas, P. 1999. *Chameleons: Nature's hidden jewels.* Frankfurt: Chimaira.

Niewiarowski, P. H. 1994. Understanding geographic life-history variation in lizards. In *Lizard ecology: Historical and experimental perspectives,* ed. L. J. Vitt and E. R. Pianka, 31–49. Princeton: Princeton University Press.

Niewiarowski, P. H., J. D. Congdon, A. E. Dunham, L. J. Vitt, and D. W. Tinkle. 1997. Tales of lizard tails: Effects of tail autotomy on subsequent survival and growth of free-ranging hatchling *Uta stansburiana. Can. J. Zool.* 75:542–548.

Noble, G. K., and H. T. Bradley. 1933. The mating behavior of lizards; its bearing on the theory of sexual selection. *Annals N.Y. Acad. Sci.* 35:25–100.

Noble, G. K., and E. R. Mason. 1933. Experiments on the brooding habits of the lizards *Eumeces* and *Ophisaurus. Amer. Mus. Novit.* 619:1–29.

Norman, D. R. 1987. Man and tegu lizards in eastern Paraguay. *Biol. Conserv.* 41:39–56.

Norrell, M. A., M. C. McKenna, and M. J. Novacek. 1992. *Estesia mongolensis,* a new fossil varanoid from the late Cretaceous Barun Goyot Formation of Mongolia. *Amer. Mus. Novitates* 3045:1–24.

Norris, K. S. 1953. The ecology of the desert iguana, *Dipsosaurus dorsalis. Ecology* 34: 265–287.

———. 1958. The evolution and systematics of the iguanid genus *Uma* and its relation to the evolution of other North American desert reptiles. *Bull. Amer. Mus. Nat. Hist.* 114:247–326.

Nouira, S., and Y.-P. Mou. 1982. Régime alimentaire d'un Lacertidae *Eremias olivieri* (Audouin) des îles Kerkennah en Tunisie. *Rev. Ecol. (Terre Vie)* 36:621–631.

Nydam, R. L. 2000. New records of early, medial, and late Cretaceous lizards and the evolution of the Cretaceous lizard fauna of North America. Ph.D. diss., University of Oklahoma, Norman.

Olsson, M., R. Shine, and E. Bak-Olsson. 2000. Locomotor impairment of gravid

lizards: Is the burden physical or physiological? *J. Evol. Biol.* 13:263–268.

Ott, M., and F. Schaeffel. 1995. A negatively powered lens in the chameleon. *Nature* 373:692–694.

Ott, M., F. Schaeffel, and W. Kirmse. 1998. Binocular vision and accommodation in prey-catching chameleons. *J. Comp. Physiol. A* 182:319–330.

Overall, K. L. 1994. Lizard egg environments. In *Lizard ecology: Historical and experimental perspectives,* ed. L. J. Vitt and E. R. Pianka, 51–72. Princeton: Princeton University Press.

Owen, H. 1976. Continental displacement and expansion of the earth during the Mesozoic and Cenozoic. *Phil. Trans. Roy. Soc. Lond. A* 281:223–291.

Packard, M. J., and G. C. Packard. 1986. Effect of water balance on growth and calcium mobilization of embryonic painted turtles (*Chrysemys picta*). *Physiol. Zool.* 59:398–405.

Packard, M. J., G. C. Packard, and T. J. Boardman. 1982. Structure of eggshells and water relations of reptilian eggs. *Herpetologica* 38:136–155.

Papenfuss, T. J. 1982. The ecology and systematrics of the amphisbaenian genus *Bipes. Occ. Pap. California Acad. Sci.* 136: 1–42.

Papenfuss, T. J., J. R. Macey, and J. A. Schulte II. 2001. A new lizard species in the genus *Xantusia* from Arizona. *Sci. Pap. Nat. Hist. Mus. Univ. Kansas* 23:1–9.

Parker, W. S., and E. R. Pianka. 1974. Further ecological observations on the western banded gecko, *Coleonyx variegatus. Copeia* 1974:528–531.

———. 1976. Ecological observations on the leopard lizard *Crotaphytus wislizeni* in different parts of its range. *Herpetologica* 32:95–114.

Patchell, F. C., and R. Shine. 1986. Feeding mechanism in pygopodid lizards: How can *Lialis* swallow such large prey? *J. Herpetol.* 20:59–64.

Pefaur, J. E., and W. E. Duellman. 1980. Community structure in high Andean herpetofaunas. *Trans. Kansas Acad. Sci.* 83: 45–65.

Perry, G. 1999. The evolution of search modes: Ecological versus phylogenetic perspectives. *Amer. Natur.* 153:98–109.

Perry, G., and E. R. Pianka. 1997. Animal foraging: Past, present, and future. *Trends Ecol. Evol.* 12:360–364.

Perry, G., I. Lampl, A. Lerner, D. Rothenstein, E. Shani, N. Sivan, and Y. L. Werner. 1990. Foraging mode in lacertid lizards: Variation and correlates. *Amph.-Rept.* 11:373–384.

Peterson, E. H. 1992. Retinal structure. In

Biology of the Reptilia, vol. 17: *Neurology C, Sensorimotor integration,* ed. C. Gans and P. S. Ulinski, 1–135. Chicago: University of Chicago Press.

Peterson, J. A., and E. E. Williams. 1981. A case history in retrograde evolution: The *onca* lineage in anoline lizards. II. Subdigital fine structure. *Bull. Mus. Comp. Zool.* 149:215–268.

Petren, K., and T. J. Case. 1996. An experimental demonstration of exploitative competition in an ongoing invasion. *Ecology* 77:118–132.

Pianka, E. R. 1966. Convexity, desert lizards, and spatial heterogeneity. *Ecology* 47:1055–1059.

———. 1967. On lizard species diversity: North American flatland deserts. *Ecology* 48:333–351.

———. 1968. Notes on the biology of *Varanus eremius. West. Austr. Natur.* 11:39–44.

———. 1969a. Notes on the biology of *Varanus caudolineatus* and *Varanus gilleni. West. Austr. Natur.* 11:76–82.

———. 1969b. Sympatry of desert lizards (*Ctenotus*) in Western Australia. *Ecology* 50:1012–1030.

———. 1970a. Comparative autecology of the lizard *Cnemidophorus tigris* in different parts of its geographic range. *Ecology* 51:703–720.

———. 1970b. Notes on *Varanus brevicauda. West. Austr. Natur.* 11:113–116.

———. 1970c. On *r*- and *K*-selection. *Amer. Natur.* 104:592–597.

———. 1970d. Notes on the biology of *Varanus gouldi flavirufus. West. Austr. Natur.* 11:141–144.

———. 1971a. Comparative ecology of two lizards. *Copeia* 1971:129–138.

———. 1971b. Ecology of the agamid lizard *Amphibolurus isolepis* in Western Australia. *Copeia* 1971:527–536.

———. 1971c. Lizard species density in the Kalahari Desert. *Ecology* 52:1024–1029.

———. 1971d. Notes on the biology of *Varanus tristi*s. *West. Austr. Natur.* 11:180–183.

———. 1976. Natural selection of optimal reproductive tactics. *Am. Zool.* 16:775–784.

———. 1982. Observations on the ecology of *Varanus* in the Great Victoria desert. *West. Austr. Natur.* 15:37–44.

———. 1985. Some intercontinental comparisons of desert lizards. *Natl. Geog. Soc. Res. Rep.* 1:490–504.

———. 1986. *Ecology and natural history of desert lizards: Analyses of the ecological niche and community structure.* Princeton: Princeton University Press.

———. 1989. Desert lizard diversity: additional comments and some data. *Amer. Natur.* 134:344–364.

———. 1992. Reproductive tactics. In *Sex origin and evolution: Proceedings of International Symposium on Origin and Evolution of Sex,* ed. R. Dallai, 189–209. Siena, Italy: Mucchi Editore.

———. 1993. The many dimensions of a lizard's ecological niche. Chapter 9 in *Lacertids of the Mediterranean Basin,* ed. E. D. Valakos, W. Böhme, V. Pérez-Mellado, and P. Maragou, 121–154. Athens, Greece: Hellenic Zoological Society. University of Athens.

———. 1994a. Comparative ecology of *Varanus* in the Great Victoria desert. *Austr. J. Ecol.* 19:395–408.

———. 1994b. Biodiversity of Australian desert lizards. In *Biodiversity and terrestrial ecosystems,* C.-I. Peng and C. H. Chou, 259–281. Taipei: Institute of Botany, Academia Sinica Monograph Series No. 14.

———. 1994c. *The lizard man speaks.* Austin: University of Texas Press.

———. 1995. Evolution of body size: Varanid lizards as a model system. *Amer. Natur.* 146:398–414.

———. 1996. Long-term changes in lizard assemblages in the Great Victoria Desert: Dynamic habitat mosaics in response to wildfires. In *Long-term studies of vertebrate communities,* ed. M. L. Cody and J. A. Smallwood, 191–215. San Diego: Academic Press.

———. 1997. Australia's thorny devil. *Reptiles* 5:14–23.

———. 2001. The role of phylogenetics in evolutionary ecology. In *Herpetologia Candiana,* ed. P. Lymberakis, E. Valakos, P. Pafilis, and M. Mylonas, 1–20. Irakleio, Crete, Greece: Natural History Museum of Crete, Societas Europea Herpetologica.

Pianka, E. R., and W. F. Giles. 1982. Notes on the biology of two species of nocturnal skinks, *Egernia inornata* and *Egernia striata,* in the Great Victoria Desert. *West. Austr. Natur.* 15:44–49.

Pianka, E. R., and W. L. Hodges. 1998. Horned lizards. *Reptiles* 6:48–63.

Pianka, E. R., and R. B. Huey. 1978. Comparative ecology, niche segregation, and resource utilization among gekkonid lizards in the southern Kalahari. *Copeia* 1978: 691–701.

Pianka, E. R., and W. S. Parker. 1972. Ecology of the iguanid lizard *Callisaurus draconoides. Copeia* 1972:493–508.

———. 1975a. Age-specific reproductive tactics. *Amer. Natur.* 109:453–464.

———. 1975b. Ecology of horned lizards: A review with special reference to *Phrynosoma platyrhinos. Copeia* 1975:141–162.

Pianka, E. R., and H. D. Pianka. 1970. The ecology of *Moloch horridus* (Lacertilia: Agamidae) in Western Australia. *Copeia* 1970:90–103.

———. 1976. Comparative ecology of twelve species of nocturnal lizards (Gekkonidae) in the Western Australian desert. *Copeia* 1976:125–142.

Pianka, E. R., R. B. Huey, and L. R. Lawlor. 1979. Niche segregation in desert lizards. Chapter 4 in *Analysis of ecological systems,* ed. D. J. Horn, R. Mitchell, and G. R. Stairs, 67–115. Columbus: Ohio State University Press.

Pianka, G. A., E. R. Pianka, and G. G. Thompson. 1996. Egg laying by thorny devils *(Moloch horridus)* under natural conditions in the Great Victoria Desert. *J. Roy. Soc. West. Austr.* 79: 195–197.

———. 1998. Natural history of thorny devils *Moloch horridus* (Lacertilia: Agamidae) in the Great Victoria Desert. *J. Royal Soc. West. Austr.* 81:183–190.

Pietruszka, R. D. 1986. Search tactics of desert lizards: How polarized are they? *Anim. Behav.* 34:1742–1758.

Pough, F. H., R. M. Andrews, J. E. Cadle, M. L. Crump, A. H. Savitzky, and K. D. Wells. 1998. *Herpetology.* Upper Sadler River, N.J.: Prentice-Hall.

Pregill, G. K. 1992. Systematics of the West Indian lizard genus *Leiocephalus* (Squamata: Iguania: Tropiduridae). *Occ. Pap. Mus. Nat. Hist. Univ. Kansas* 84:1–69.

Pregill, G. K., J. A. Gauthier, and H. W. Greene. 1986. The evolution of helodermatid squamates, with description of a new taxon and an overview of Varanoidea. *Trans. San Diego Soc. Nat. Hist.* 21:167–202.

Presch, W. 1974. Evolutionary relationships and biogeography of the macroteiid lizards (family Teiidae, subfamily Teiinae). *Bull. Southern California Acad. Sci.* 73: 23–32.

———. 1980. Evolutionary history of the South American microteiid lizards (Teiidae: Gymnophthalminae). *Copeia* 1980:36–56.

Proud, K. R. S. 1978. Some notes on a captive earless monitor lizard, *Lanthanotus borneensis. Sarawak Mus. J.* 24:235–242.

Qualls, C. P. 1996. Influence of the evolution of viviparity on eggshell morphology in the lizard *Lerista bougainvillii. J. Morph.* 228:119–125.

Qualls, C. P., and R. G. Jaeger. 1991. Dear enemy recognition in *Anolis carolinensis. Herpetologica* 25:361–363.

Radder, R. S., B. A. Shanbhag, and S. K. Saidapur. 1998. Prolonged oviductal egg retention arrests embryonic growth at stage 34 in captive *Calotes versicolor. Herp. Rev.* 29:217–218.

Radtkey, R. R., S. M. Fallon, and T. J. Case. 1997. Character displacement in some *Cnemidophorus* lizards revisited: A phylogenetic analysis. *Proc. Natl. Acad. Sci.* 94: 9740–9745.

Rand, A. S. 1967. Predator-prey interactions and the evolution of aspect diversity. *Atas do Simpósio sôbre a Biota Amazônica* 5: 73–83.

———. 1968. A nesting aggregation of iguanas. *Copeia* 1968:552–561.

Raufman, J. P. 1996. Bioactive peptides from lizard venoms. *Regulatory Peptides* 61:1–18.

Raxworthy, C. J. 1991. Field observations on some dwarf chameleons (*Brookesia* spp.) from rainforest areas of Madagascar, with description of a new species. *J. Zool., Lond.* 224:11–25.

Reboucas-Spieker, R., and P. E. Vanzolini. 1978. Parturition in *Mabuya macrorhyncha* Hoge, 1946 (Sauria, Scincidae), with a note on the distribution of maternal behavior in lizards. *Pap. Avul. Zool., S. Paulo* 32:95–99.

Reilly, S. M. 1995. Quantitative electromyography and muscle function of the hind limb during quadrapedal running in the lizard *Sceloporus clarkii. Zoology* 98:263–277.

Reilly, S. M., and M. J. DeLancey. 1997. Sprawling locomotion in the lizard *Sceloporus clarkii:* Quantitative kinematics of a walking trot. *J. Exp. Biol.* 200:753–765.

Rich, T. H. 1985. *Megalania prisca* Owen, 1859 The Giant Goanna. In *Kadimakara: Extinct vertebrates of Australia,* ed. P. V. Rich and G. F. van Tets, 152–155. Lilydale, Vic., Austr.: Pioneer Design Studio.

Richardson, K. C., and P. M. Hinchliffe. 1983. Caudal glands and their secretions in the western spiny-tailed gecko, *Diplodactylus spinigerus. Copeia* 1983:161–169.

Rieppel, O. 1988. A review of the origin of snakes. *Evol. Biol.* 22:37–130.

———. 1994. The Lepidosauromorpha: An overview with special emphasis on the Squamata. In *In the shadow of the dinosaurs: Early Mesozoic tetrapods,* ed. N. C. Fraser and H.-D. Sues, 23–37. New York: Cambridge University Press.

Riley, J., J. M. Winch, A. F. Stimson, and R. D. Pope. 1986. The association of *Amphisbaena alba* (Reptilia: Amphisbaenidae) with the leaf-cutting ant *Atta cephalotes* in Trinidad. *J. Nat. Hist.* 20:459–470.

Roberts, A., and C. P. Mountford. 1965. *The dreamtime: Australian aboriginal myths in paintings.* Adelaide, S.A., Austr.: Rigby.

———. 1969. *The dawn of time.* Adelaide, S.A., Austr.: Rigby.

———. 1971. *The first sunrise.* Adelaide, S.A., Austr.: Rigby.

Roberts, A., and D. Roberts. 1989. *Shadows in the mist.* N.p.

Robinson, M. D. 1979. Systematics of skinks of the *Eumeces brevirostris* species group in western Mexico. *Contr. Sci. Nat. Hist. Mus. Los Angeles County* 319:1–13.

Robinson, M. D., and A. B. Cunningham. 1978. Comparative diet of two Namib Desert sand lizards (Lacertidae). *Madoqua* 11:411–53.

Robinson, P L. 1976. How *Sphenodon* and *Uromastyx* grow their teeth and use them. In *Morphology and biology of reptiles,* ed. A. Bellairs and C. B. Cox. 43–64. London: Linnean Society of London.

Rocha, C. F. D. 1992. Reproductive and fat body cycles of the tropical sand lizard (*Liolaemus lutzae*) of southeastern Brazil. *J. Herpetology* 26:17–23.

Rocha, C. F. D., and H. G. Bergallo. 1992. Population decrease: The case of *Liolaemus lutzae,* an endemic lizard of southeastern Brazil. *Ciência e Cultura* 44:52–54.

Rodrigues, M. T. 1987. Sistemática, ecologia e zoogeografia dos *Tropidurus* do grupo *torquatus* ao Sul do Rio Amazonas (Sauria, Iguanidae). *Arq. Zool., São Paulo* 31: 105–230.

———. 1991. Herpetofauna das dunas interiores do Rio São Francisco, Bahia, Brasil. I. Introdução á área e descrição de um novo género de microteiideos (*Calyptommatus*) com notas sobre sua ecologia, distribuição e especiação (Sauria, Teiidae). *Pap. Avul. Zool., S. Paulo* 37:285–320.

Rodrigues, M. T., Y. Yonenaga-Yassuda, and S. Kasahara. 1989. Notes on the ecology and karyotypic description of *Strobilurus torquatus* (Sauria, Iguanidae). *Rev. Brasil. Genet.* 12:747–759.

Rolston, H. 1985. Duties to endangered species. *BioScience* 35:718–726.

Rose, D. 1992. Free trade and wildlife trade. *Conserv. Biol.* 6:148–150.

Rose, K. A., L. W. Brewer, L. W. Barnthouse, G. A. Fox, N. W. Gard, M. Mendonca, K. R. Munkittrick, and L. J. Vitt. 1999. Ecological responses of oviparous vertebrates to contaminant effects on reproduction and development. In *Reproduction and developmental effects of contaminants in oviparous vertebrates,* ed. R. T. Di Giulio and D. E. Tillet, 225–281. Pensacola, Fla.: SETAC Press.

Rose, W. 1962. *The reptiles and amphibians of southern Africa.* Cape Town, South Africa: Maskew Miller.

Rosenberg, H. I., and A. P. Russell. 1980. Structural and functional aspects of tail squirting: A unique defense mechanism of *Diplodactylus* (Reptilia: Gekkonidae). *Can. J. Zool.* 58:865–881.

Rosenberg, H. I., A. P. Russell, and M. Kapoor. 1984. Preliminary characterization of the defensive secretion of *Diplodactylus* (Reptilia: Gekkonidae). *Copeia* 1984: 1025–1028.

Ruby, D. E. 1977. The function of shudder displays in the lizard, *Sceloporus jarrovi.* *Copeia* 1977:110–114.

———. 1978. Seasonal changes in the territorial behavior of the iguanid lizard *Sceloporus jarrovi. Copeia* 1978:430–438.

———. 1981. Phenotypic correlates of male reproductive success in the lizard *Sceloporus jarrovi.* In *Natural selection and social behavior,* ed. R. D. Alexander and D. W. Tinkle, 96–107. New York: Chiron Press.

———. 1984. Male breeding success and differential access to females in *Anolis carolinensis. Herpetologica* 40:272–280.

Ruby, D. E., and D. I. Baird. 1993. Effects of sex and size on agonistic encounters between juvenile and adult lizards, *Sceloporus jarrovi. J. Herpetol.* 27:100–103.

Ruby, D. E., and A. E. Dunham. 1987. Variation in home range size along an elevational gradient in the iguanid lizard *Sceloporus merriami. Oecologia* 71:473–480.

Russell, A. P. 1975. A contribution to the functional morphology of the foot of the tokay, *Gekko gecko* (Reptilia, Gekkonidae). *J. Zool., Lond.* 176:437–476.

———. 1979. The origin of parachuting locomotion in the gekkonid lizards (Reptilia: Gekkonidae). *Zool. J. Linn. Soc.* 65: 233–249.

Russell, A. P., and A. M. Bauer. 1987. Caudal morphology of the knob-tailed geckos, genus *Nephrurus* (Reptilia: Gekkonidae), with special reference to the tail tip. *Austr. J. Zool.* 35:541–551.

Sampedro Marin, A., V. B. Alvarez, and O. T. Fundora. 1979. Habitat, alimentación y actividad de dos especies de *Leiocephalus* (Sauria: Iguanidae) en dos localidades de la región suroriental de Cuba. *Cien. Biol.* 3:129–139.

Sartorius, S., L. Vitt, and G. R. Colli. 1999. Use of naturally and anthropogenically disturbed habitats in Amazonian rainforest by the teiid lizard *Ameiva ameiva. Biol. Conserv.* 90:91–101.

Savage, J. M., and C. Guyer. 1989. Infrageneric classification and species composition of the anole genera, *Anolis, Ctenonotus, Dactyloa, Norops,* and *Semiurus* (Sauria: Iguanidae). *Amphibia-Reptilia* 10:105–116.

Saville-Kent, W. 1897. *The naturalist in Australia.* London: Chapman & Hall.

Savitsky, A. H. 1981. Hinged teeth in snakes: An adaptation for swallowing hard-bodied prey. *Science* 212:346–349.

Schall, J. J. 1990. Aversion of whiptail liz-

ards (*Cnemidophorus*) to a model alkaloid. *Herpetologica* 46:34–39.

Schall, J. J., and E. R. Pianka. 1980. Evolution of escape behavior diversity. *Amer. Natur.* 115:551–566.

Schall, J. J., and S. Ressel. 1991. Toxic plant compounds and the diet of the predominantly herbivorus whiptail lizard, *Cnemidophorus arubensis. Copeia* 1991:111–119.

Schmidt, K. P., and R. F. Inger. 1957. *Living reptiles of the world.* Garden City, N.Y.: Hanover House.

Schoener, T. W. 1979. Inferring the properties of predation and other injury-producing agents from injury frequencies. *Ecology* 60:1110–1115.

Schoener, T. W., and A. Schoener. 1980. Ecological and demographic correlates of injury rates in some Bahamian *Anolis* lizards. *Copeia* 1980:839–850.

Schoener, T. W., J. B. Slade, and C. H. Stinson. 1982. Diet and sexual dimorphism in the very catholic lizard genus *Leiocephalus* of the Bahamas. *Oecologia* 53:160–169.

Schreiber, M. C., R. Powell, J. J. S. Parmerlee, A. Lathrop, and D. D. Smith. 1993. Natural history of a small population of *Leiocephalus schreibersii* (Sauria: Tropiduridae) from altered habitat in the Dominican Republic. *Biol. Sci.* 56:82–90.

Schwartz, A., and R. W. Henderson. 1991. *Amphibians and reptiles of the West Indies: Descriptions, distributions, and natural history.* Gainesville: University of Florida Press.

Schwarzkopf, L. 1994. Measuring tradeoffs: A review of studies of costs of reproduction in lizards. In *Lizard ecology: Historical and experimental perspectives,* ed. L. J. Vitt and E. R. Pianka, 7–29. Princeton: Princeton Univ. Press.

Schwenk, K. 1993. The evolution of chemoreception in squamate reptiles: A phylogenetic approach. *Brain Behav. Evol.* 41: 124–137.

———. 1994a. Systematics and subjectivity: The phylogeny and classification of iguanian lizards revisited. *Herp. Rev.* 25:53–57.

———. 1994b. Why snakes have forked tongues. *Science* 263:1573–1577.

———. 1995. Of tongues and noses: Chemoreception in lizards and snakes. *Trends Ecol. Evol.* 10:7–12.

———. 2000. Feeding in lepidosaurs. In *Feeding: Form, function and evolution in tetrapod vertebrates,* ed. K. Schwenk, 175–291. San Diego: Academic Press.

Schwenk, K., and D. A. Bell. 1988. A cryptic intermediate in the evolution of chameleon tongue projection. *Experentia* 44: 697–700.

Schwenk, K., and G. S. Throckmorton.

1989. Functional and evolutionary morphology of lingual feeding in squamate reptiles: Phylogenetics and kinematics. *J. Zool., Lond.* 219:153–175.

Seeger, A. 1981. *Nature and society in central Brazil.* Cambridge, Mass.: Harvard University Press.

Seppa, N. 2001. Reptilian drug may help treat diabetes. *Science News* 160(3):47.

Sherbrooke, W. C. 1981. *Horned lizards, unique reptiles of western North America.* Globe, Ariz.: Southwest Parks & Monuments Association.

Sherbrooke, W. C., and R. R. Montanucci. 1988. Stone mimicry in the round-tailed horned lizard, *Phrynosoma modestum* (Sauria: Iguanidae). *J. Arid Environ.* 14:275–284.

Shine, R. 1980. Costs of reproduction in reptiles. *Oecologia* 46:92–100.

———. 1983. Reptilian viviparity in cold climates: Testing the assumptions of an evolutionary hypothesis. *Oecologia* 57: 397–405.

———. 1985. The evolution of viviparity in reptiles: An ecological analysis. In *Biology of the Reptilia,* vol. 15: *Developmental biology B,* ed. C. Gans and F. Billett, 677–680. New York: John Wiley.

———. 1986a. Food habits, habitats, and reproductive biology of four sympatric species of varanid lizards in tropical Australia. *Herpetologica* 42:346–360.

———. 1986b. Evolutionary advantages of limblessness: Evidence from the pygopodid lizards. *Copeia* 1986:525–529.

Shine, R., and J. J. Bull. 1979. The evolution of live bearing in lizards and snakes. *Amer. Natur.* 113:905–923.

Shine, R., and L. Schwarzkopf. 1992. The evolution of reproductive effort in lizards and snakes. *Evolution* 46:62–75.

Shine, R., S. Keogh, P. Doughty, and H. Giragossyan. 1998. Costs of reproduction and the evolution of sexual dimorphism in a "flying lizard," *Draco melanopogon* (Agamidae). *J. Zool., Lond.* 246:203–213.

Sick, H. 1951. Beobachtungen an dem Stachelschwanzleguan, *Hoplocercus spinosus. Natur und Volk* 81:30–35.

Sinervo, B., and R. B. Huey. 1990. Allometric engineering: An experimental test of the causes of interpopulational differences in performance. *Science* 248: 1106–1109.

Sites, J. W., S. K. Davis, T. Guerra, J. B. Iverson, and H. L. Snell. 1996. Character congruence and phylogenetic signal in molecular and morphological data sets: A case study in the living iguanas (Squamata, Iguanidae). *Mol. Biol. Evol.* 13:1087–1105.

Smith, A., and J. Bryden. 1977. *Mesozoic and Cenozoic palocontinental maps.* Cambridge Univ. Press, 70 pp.

Smith, G. R. 1992. Sexual dimorphism in the curly-tailed lizard *Leiocephalus psammodromus. Carib. J. Sci.* 28:99–101.

Smith, G. R., and J. B. Iverson. 1993. Reproduction in the curly-tailed lizard *Leiocephalus psammodromus* from the Caicos Islands. *Can. J. Zool.* 71:2147–2151.

Smith, G. R., J. A. Lemos-Espinal, and R. E. Ballinger. 1997. Sexual dimorphism in two species of knob-scaled lizards (genus *Xenosaurus*) from Mexico. *Herpetologica* 53:200–205.

Smith, K. K. 1980. Mechanical significance of streptostyly in lizards. *Nature* 283: 778–779.

———. 1982. An electromyographic study of the function of the jaw adducting muscles in *Varanus exanthematicus* (Varanidae). *J. Morph.* 173:137–158.

Smole, W. J. 1976. *The Yanoama Indians: a cultural geography.* Austin: University of Texas Press.

Smyth, M., and M. J. Smith 1974. Aspects of the natural history of three Australian skinks, *Morethia boulengeri, Menetia greyii,* and *Lerista bouganvillii. J. Herpetol.* 8: 329–335.

Snyder, R. C. 1952. Quadrupal and bipedal locomotion in lizards. *Copeia* 1952:64–70.

Somma, L. A., and J. D. Fawcett. 1989. Brooding behaviour of the prairie skink, *Eumeces septentrionalis,* and its relationship to the hydric environment of the nest. *Zool. J. Linn. Soc.* 95:245–256.

Sporn, C. C. 1955. The breeding of the mountain devil in captivity. *West. Austr. Natur.* 5:1–5.

———. 1958. Further observations on the mountain devil in captivity. *West. Austr. Natur.* 6:136–137.

———. 1965. Additional observations on the life history of the mountain devil, *Moloch horridus,* in captivity. *West. Austr. Natur.* 9:157–159.

Sprackland, R. G. 1970. Further notes on *Lanthanotus. Sarawak Mus. J.* 18:412–413.

———. 1972. A summary of observations of the earless monitor, *Lanthanotus borneenesis. Sarawak Mus. J.* 20:323–327.

Stamps, J. A. 1973. Displays and social organization in female *Anolis aeneus. Copeia* 1973:264–272.

———. 1977. Social behavior and spacing patterns in lizards. In *Biology of the Reptilia,* vol. 7: *Ecology and behavior,* ed. C. Gans and D. W. Tinkle, 265–334. New York: Academic Press.

———. 1978. A field study on the ontogeny of social behavior in the lizard *Anolis aeneus. Behaviour* 64:1–31.

———. 1983. Sexual selection, sexual dimorphism, and territoriality. In *Lizard ecology: Studies of a model organism,* ed. R. B. Huey, E. R. Pianka, and T. W. Schoener, 169–204. Cambridge, Mass: Harvard University Press.

Stearns, S. C. 1976. Life-history tactics: A review of the ideas. *Q. Rev. Biol.* 51:3–47.

———. 1977. The evolution of life history traits: A critique of the theory and a review of the data. *Ann. Rev. Ecol. Syst.* 8: 145–171.

Stebbins, R. C. 1944. Some aspects of the ecology of the iguanid genus *Uma. Ecol. Monogr.* 14:311–332.

Steward, J. W. 1965. Territorial behavior in the wall lizard, *Lacerta muralis. Brit. J. Herpetol.* 3:224–229.

Stewart, J. R., and D. G. Blackburn. 1988. Reptilian placentation: Structural diversity and terminology. *Copeia* 1988:839–852.

Storr, G. M. 1964. *Ctenotus,* a new generic name for a group of Australian skinks. *West. Austr. Natur.* 9:84–89.

———. 1968. The genus *Ctenotus* (Lacertilia, Scincidae) in the eastern division of Western Australia. *J. Royal Soc. West. Austr.* 51:97–109.

Storr, G. M., L. A. Smith, and R. E. Johnstone. 1999. *Lizards of Western Australia. I. Skinks.* Perth: Western Australian Museum.

Sulimski, A. 1975. Macrocephalosauridae and Polyglyphanodontinae (Sauria) from the Upper Cretaceous of Mongolia. *Paleo. Polonica* 38:43–56.

Szczerbak, N. N., and M. L. Golubev. 1996. *Gecko fauna of the USSR and contiguous regions.* Transl. M. L. Golubev and S. A. Malinsky. Ithaca, N.Y.: Society for the Study of Amphibians and Reptiles.

Taylor, E. H. 1935. A taxonomic study of the cosmopolitan scincoid lizards of the genus *Eumeces* with an account of the distribution and relationships of its species. *Bull. Univ. Kansas* 36:1–642.

Tchernov, E., O. Rieppel, H. Zaher, M. J. Polcn, and L. J. Jacobs. 2000. A new fossil snake with limbs. *Science* 287:2010–2012.

Thompson, G. 1993. Daily movement patterns and habitat preferences of *Varanus caudolineatus* (Reptilia: Varanidae). *Wildl. Res.* 20:227–231.

Thompson, G. G., and D. R. King. 1995. Diet of *Varanus caudolineatus* (Reptilia: Varanidae) *West. Austr. Natur.* 20:199–204.

Thompson, G. G., and E. R. Pianka. 1999. Reproductive ecology of the black-headed goanna *Varanus tristis* (Squamata: Varanidae). *J. Roy. Soc. West. Austr.* 82:27–31.

Thompson, G. G., M. de Boer, and E. R. Pianka. 1999. Activity areas and daily movements of an arboreal monitor lizard, *Varanus tristis* (Squamata: Varanidae) during the breeding season. *Austr. J. Ecol.* 24:117–122.

Thompson, G. G., E. R. Pianka, and M. de Boer. 1999. Thermoregulation of an arboreal monitor lizard, *Varanus tristis* (Squamata: Varanidae) during the breeding season. *Amph.-Rept.* 20:82–88.

Thompson, G. G., P. C. Withers, and S. A. Thompson. 1992. The combat ritual of two monitor lizards, *Varanus caudolineatus* and *Varanus gouldii*. *West. Austr. Natur.* 19:21–25.

Thompson, J. E. S. 1963. *Maya archeologist.* Norman: University of Oklahoma Press.

Tinkle, D. W. 1967. The life and demography of the side-blotched lizard, *Uta stansburiana*. *Misc. Publ. Mus. Zool. Univ. Mich.* No. 132:1–182.

———. 1969. The concept of reproductive effort and its relation to the evolution of life histories in lizards. *Amer. Natur.* 103: 501–516.

Tinkle, D. W., and J. W. Gibbons. 1977. The distribution and evolution of viviparity in reptiles. *Misc. Publ. Mus. Zool. Univ. Mich.* 154:1–55.

Tinkle, D. W., and N. F. Hadley. 1975. Lizard reproductive effort: Caloric estimates and comments on its evolution. *Ecology* 56:427–434.

Tinkle, D. W., H. M. Wilbur, and S. G. Tilley. 1970. Evolutionary strategies in lizard reproduction. *Evolution* 24:55–74.

Tokunaga, S. 1984. Morphological variation and sexual dimorphism in *Gecko japonicus* from Fukuoka, northern Kyushu, Japan. *Jap. J. Herpetol.* 10:80–88.

Tracy, C. R. 1999. Differences in body size among chuckwalla (*Sauromalus obesus*) populations. *Ecology* 80:259–271.

Trauth, S. E., W. E. Cooper, Jr., L. J. Vitt, and S. A. Perrill. 1987. Cloacal anatomy of the broad-headed skink, *Eumeces laticeps,* with a description of a female pheromonal gland. *Herpetologica* 43:458–466.

Trillmich, K. 1979. Feeding behaviour and social behaviour of the marine iguana. *Noticias Galápagos* 29:17–20.

Turner, F. B., P. A. Medica, and D. D. Smith. 1973. Reproduction and survivorship of the lizard *Uta stansburiana* and the effects of winter rainfall, density, and predation on these processes. *Reports of 1973, Progress 3: Process Studies:* 117–128.

Valido, A., and M. Nogales. 1994. Frugivory and seed dispersal by the lizard *Gallotia galloti* (Lacertidae) in a xeric habitat of the Canary Islands. *Oikos* 70:403–411.

Van Devender, R. W. 1982. Growth and ecology of spiny-tailed and green iguanas in Costa Rica, with comments on the evolution of herbivory and large body size. In *Iguanas of the world,* ed. G. M. Burghardt and A. S. Rand, 162–183. Park Ridge, N.J.: Noyes.

van Wyk, J. H. 2000. Seasonal variation in stomach contents and diet composition in the large girdled lizard, *Cordylus giganteus* (Reptilia: Cordylidae), in the Highveld grasslands of the northeastern Free State, South Africa. *African Zool.* 35:9–27.

Vanzolini, P. E. 1956. Notas sôbre a zoologia dos índios canela. *Rev. Mus. Paulista* 10: 155–171.

———. 1961. Notas bionómicas sôbre *Dracaena guianensis* no Pará (Sauria, Teiidae). *Pap. Avul. Zool., S. Paulo* 14: 237–241.

———. 1968a. Geography of the South American Gekkonidae (Sauria). *Arq. Zool., S. Paulo* 17:85–112.

———. 1968b. Lagartos brasileiros da família Gekkonidae (Sauria). *Arq. Zool., S. Paulo* 17:1–84.

———. 1999. On *Anops* (Reptilia: Amphisbaenia: Amphibaenidae). *Pap. Avul. Zool., S. Paulo* 41:1–37.

Vaughn, L. K., H. A. Bernheim, and M. J. Kluger. 1974. Fever in the lizard *Dipsosaurus dorsalis*. *Nature* 252:473–474.

Vences, M., F. Glaw, and W. Böhme. 1997/98. Evolutionary correlates of microphagy in alkaloid-containing frogs (Amphibia: Anura). *Zool. Anzeiger* 236:217–230.

Vinson, J., and J.-M. Vinson. 1969. The saurian fauna of the Mascarene Islands. *Mauritius Inst. Bull.* 6:203–320.

Vitt, L. J. 1973. Reproductive biology of the anguid lizard *Gerrhonotus coeruleus principis*. *Herpetologica* 29:176–184.

———. 1978. Caloric content of lizard and snake (Reptilia) eggs and bodies and the conversion of weight to caloric data. *J. Herpetol.* 13:65–72.

———. 1981. Lizard reproduction: Habitat specificity and constraints on relative clutch mass. *Amer. Natur.* 117:506–514.

———. 1983. Reproduction and sexual dimorphism in the tropical teiid lizard *Cnemidophorus ocellifer*. *Copeia* 1983: 359–366.

———. 1985. On the biology of the little-known anguid lizard *Diploglossus lessonae* in northeast Brazil. *Pap. Avul. Zool., S. Paulo* 36:69–76.

———. 1986. Reproductive tactics of sympatric gekkonid lizards with a comment on the evolutionary and ecological consequences of invariant clutch size. *Copeia* 1986:773–786.

———. 1991. Ecology and life history of the scansorial arboreal lizard *Plica plica* (Iguanidae) in Amazonian Brazil. *Can. J. Zool.* 69:504–511.

———. 1992a. Diversity of reproduction strategies among Brazilian lizards and snakes: The significance of lineage and adaptation. In *Reproductive biology of South American vertebrates,* ed. W. C. Hamlett, 135–149. New York: Springer-Verlag.

———. 1992b. Mimicry of millipedes and centipedes by elongate terrestrial vertebrates. *Natl. Geographic Res. Expl.* 8:76–95.

———. 1993. Ecology of isolated open-formation *Tropidurus* (Reptilia: Tropiduridae) in Amazonian lowland rain forest. *Can. J. Zool.* 71:2370–2390.

———. 1995. The ecology of tropical lizards in the caatinga of northeast Brazil. *Occ. Pap. Oklahoma Mus. Nat. Hist.* 1:1–29.

Vitt, L. J., and R. E. Ballinger. 1982. The adaptive significance of a complex caudal adaptation in the tropical gekkonid lizard *Lygodactylus klugei*. *Can. J. Zool.* 60:2582–2587.

Vitt, L. J., and D. G. Blackburn. 1983. Reproduction in the lizard *Mabuya heathi* (Scincidae): A commentary on viviparity in New World *Mabuya*. *Can. J. Zool.* 61: 2798–2806.

Vitt, L. J., and C. M. Carvalho. 1992. Life in the trees: The ecology and life history of *Kentropyx striatus* (Teiidae) in the lavrado area of Roraima, Brazil, with comments on the life histories of tropical teiid lizards. *Can. J. Zool.* 70:1995–2006.

———. 1995. Niche partitioning in a tropical wet season: Lizards in the lavrado area of northern Brazil. *Copeia* 1995:305–329.

Vitt, L. J., and G. R. Colli. 1994. Geographical ecology of a neotropical lizard: *Ameiva ameiva* (Teiidae) in Brazil. *Can. J. Zool.* 72:1986–2008.

Vitt, L. J., and J. D. Congdon. 1978. Body shape, reproductive effort, and relative clutch mass in lizards: Resolution of a paradox. *Amer. Natur.* 112:595–608.

Vitt, L. J., and W. E. Cooper, Jr. 1985a. The relationship between reproduction and lipid cycling in the skink *Eumeces laticeps* with comments on brooding ecology. *Herpetologica* 41:419–432.

———. 1985b. The evolution of sexual dimorphism in the skink *Eumeces laticeps:* An example of sexual selection. *Can. J. Zool.* 63:995–1002.

———. 1986. Tail loss, tail color, and predator escape in *Eumeces* (Lacertilia: Scincidae): Age-specific differences in costs and benefits. *Can. J. Zool.* 64:583–592.

———. 1988. Feeding responses of broad-headed skinks (*Eumeces laticeps*) to velvet ants (*Dasymutilla occidentalis*). *J. Herpetol.* 22:485–488.

————. 1989. Maternal care in skinks (*Eumeces*). *J. Herpetol.* 23:29–34.

Vitt, L. J., and J. T. E. Lacher. 1981. Behavior, habitat, diet, and reproduction of the iguanid lizard *Polychrus acutirostris* in the caatinga of northeastern Brazil. *Herpetologica* 37:53–63.

Vitt, L. J., and E. R. Pianka, eds. 1994. *Lizard ecology: Historical and experimental perspectives.* Princeton: Princeton University Press.

Vitt, L. J., and H. J. Price. 1982. Ecological and evolutionary determinants of relative clutch mass in lizards. *Herpetologica* 38: 237–255.

Vitt, L. J., and P. A. Zani. 1996a. Organization of a taxonomically diverse lizard assemblage in Amazonian Ecuador. *Can. J. Zool.* 74:1313–1335.

————. 1996b. Ecology of the elusive tropical lizard *Tropidurus* [= *Uracentron*] *flaviceps* (Tropiduridae) in lowland rain forest of Ecuador. *Herpetologica* 52:121–132.

————. 1996c. Ecology of the South American lizard *Norops chrysolepis* (Polychrotidae). *Copeia* 1996:56–68.

————. 1996d. Ecology of the lizard *Ameiva festiva* (Teiidae) in southeastern Nicaragua. *J. Herpetol.* 30:110–117.

————. 1997. Ecology of the nocturnal lizard *Thecadactylus rapicauda* (Sauria: Gekkonidae) in the Amazon region. *Herpetologica* 53:165–179.

————. 1998a. Prey use among sympatric lizard species in lowland rain forest of Nicaragua. *J. Trop. Ecol.* 14:537–559.

————. 1998b. Ecological relationships among sympatric lizards in a transitional forest in the northern Amazon of Brazil. *J. Trop. Ecol.* 14:63–86.

Vitt, L. J., T. C. S. Avila-Pires, and P. A. Zani. 1996. Observations on the ecology of the rare Amazonian lizard, *Enyalius leechii* (Polychrotidae). *Herp. Nat. Hist.* 4:77–82.

Vitt, L. J., J. P. Caldwell, P. A. Zani, and T. A. Titus. 1997. The role of habitat shift in the evolution of lizard morphology: Evidence from tropical *Tropidurus*. *Proc. Natl. Acad. Sci.* 94:3828–3832.

Vitt, L. J., J. D. Congdon, and N. A. Dickson. 1977. Adaptive strategies and energetics of tail autotomy in lizards. *Ecology* 58:326–337.

Vitt, L. J., S. S. Sartorius, T. C. S. Avila-Pires, and M. C. Espósito. 1998. Use of time, space, and food by the gymnophthalmid lizard *Prionodactylus eigenmanni* from the western Amazon of Brazil. *Can. J. Zool.* 76:1681–1688.

————. 2001. Life on the leaf litter: The ecology of *Anolis nitens tandai* in the Brazilian Amazon. *Copeia* 2001:401–412.

Vitt, L. J., S. S. Sartorius, T. C. S. Avila-Pires, M. C. Espósito, and D. B. Miles. 2000. Niche segregation among sympatric Amazonian teiid lizards. *Oecologia* 122:410–420.

Vitt, L. J., R. A. Souza, S. S. Sartorius, T. C. S. Avila-Pires, and M. C. Espósito. 2000. Comparative ecology of sympatric *Gonatodes* (Squamata: Gekkonidae) in the western Amazon of Brazil. *Copeia* 2000: 83–95.

Vitt, L. J., R. C. van Loben Sels, and R. D. Ohmart. 1981. Ecological relationships among arboreal desert lizards. *Ecology* 62:398–410.

Vitt, L. J., P. A. Zani, and T. C. S. Avila-Pires. 1997. Ecology of the arboreal tropidurid lizard *Tropidurus* (= *Plica*) *umbra* in the Amazon region. *Can. J. Zool.* 75: 1876–1882.

Vitt, L. J., P. A. Zani, and A. A. M. Barros. 1997. Ecological variation among populations of the gekkonid lizard *Gonatodes humeralis* in the Amazon Basin. *Copeia* 1997:32–43.

Vitt, L. J., P. A. Zani, and J. P. Caldwell. 1996. Behavioural ecology of *Tropidurus hispidus* on isolated rock outcrops in Amazonia. *J. Trop. Ecol.* 12:81–101.

Vitt, L., P. Zani, and M. C. Espósito. 1999. Historical ecological of Amazonian lizards: Implications for community ecology. *Oikos* 87:286–294.

Vitt, L. J., P. A. Zani, and A. C. M. Lima. 1997. Heliotherms in tropical rainforest: The ecology of *Kentropyx calcarata* (Teiidae) and *Mabuya nigropunctata* (Scincidae) in the Curuá Una of Brazil. *J. Trop. Ecol.* 13:199–220.

Vitt, L. J., P. A. Zani, T. C. S. Avila-Pires, and M. C. Espósito. 1998. Geographical ecology of the gymnophthalmid lizard *Neusticurus ecpleopus* in the Amazon rain forest. *Can. J. Zool.* 76:1671–1680.

Vitt, L. J., P. A. Zani, J. P. Caldwell, and R. D. Durtsche. 1993. Ecology of the whiptail lizard *Cnemidophorus deppii* on a tropical beach. *Can. J. Zool.* 71:2391–2400.

Vitt, L. J., P. A. Zani, J. P. Caldwell, M. C. Araújo, and W. E. Magnusson. 1997. Ecology of whiptail lizards (*Cnemidophorus*) in the Amazon region of Brazil. *Copeia* 1997:745–757.

Vorhies, C. T. 1948. Food items of rattlesnakes. *Copeia* 1948:302–303.

Wall, G. L. 1942. *The vertebrate eye and its adaptive radiation.* New York: Hafner.

Weekes, H. C. 1935. A review of placentation among reptiles with particular regard to the function and evolution of the placenta. *Proc. Zool. Soc. London* Part 3:625–645.

Welch, J. R. 1993. *O ye legendary Texas horned frog!* Irving, Tex.: Yellow Rose Press.

Werner, D. I. 1982. Social organization and ecology of land iguanas, *Conolophus subscritatus,* on Isla Fernandina, Galápagos. In *Iguanas of the world,* ed. G. M. Burghardt and A. S. Rand, 342–365. Park Ridge, N.J.: Noyes.

————. 1983. Reproduction in the iguana *Conolophus subscritatus* on Isla Fernandina, Galápagos: Clutch size and migration costs. *Amer. Natur.* 121:757–775.

Werner, Y. 1969. Eye size in geckos of various ecological types (Reptilia: Gekkonidae and Sphaerodactylidae). *Isr, J. Zool.* 18:291–316.

Whiting, M. J. 1999. When to be neighbourly: Differential agonistic responses in the lizard *Platysaurus broadleyi.* *Behav. Ecol. Sociobiol.* 46:210–214.

Whiting, M. J., and P. W. Bateman. 1999. Male preference for large females in the lizard *Platysaurus broadleyi.* *J. Herpetol.* 33:309–312.

Whiting, M. J., and J. M. Greeff. 1997. Facultative frugivory in the Cape flat lizard, *Platysaurus capensis* (Sauria: Cordylidae). *Copeia* 1997:811–818.

————. 1999. Use of heterospecific cues by the lizard *Platysaurus broadleyi* for food location. *Behav. Ecol. Sociobiol.* 45:420–423.

Wiens, J. J. 2000. Decoupled evolution of display morphology and display behaviour in phrynosomatid lizards. *Biol. J. Linn. Soc.* 70:597–612.

Wiens, J. J., and B. D. Hollingsworth. 2000. War of the iguanas: Conflicting molecular and morphological phylogenies and long-branch attraction in iguanid lizards. *Syst. Biol.* 49:143–200.

Wiens, J. J., and T. W. Reeder. 1997. Phylogeny of the spiny lizards (*Sceloporus*) based on molecular and morphological evidence. *Herp. Monogr.* 11:1–101.

Williams, E. E. 1969. The ecology of colonization as seen in the zoogeography of anoline lizards on small islands. *Q. Rev. Biol.* 44:345–389.

————. 1983. Ecomorphs, faunas, island size, and diverse end points in island radiations of *Anolis.* In *Lizard ecology: Studies of a model organism,* ed. R. B. Huey, E. R. Pianka, and T. W. Schoener, 326–370. Cambridge, Mass.: Harvard University Press.

Williams, E. E., and J. A. Peterson. 1982. Convergent and alternative designs in the digital adhesive pads of scincid lizards. *Science* 215:1509–1511.

Wilson, S. K., and D. G. Knowles. 1988. *Australia's reptiles.* Sydney: Collins.

Winemiller, K. O., and E. R. Pianka. 1990. Organization in natural assemblages of desert lizards and tropical fishes. *Ecol. Monogr.* 60:27–55.

Winkler, D. A., P. A. Murray, and L. L. Jacobs. 1990. Early Cretaceous (Comanchean) vertebrates of central Texas. *J. Vert. Paleo.* 10:95–116.

Witten, G. J. 1993. Family Agamidae. In *Fauna of Australia,* vol. 2: *Amphibia Reptilia Aves,* ed. G. J. B. Ross, 240–252. Canberra: Australian Biological and Environmental Survey.

Wright, J. W. 1993. Evolution of the lizards of the genus *Cnemidophorus.* In *Biology of whiptail lizards (genus* Cnemidophorus*),* ed. J. W. Wright and L. J. Vitt, 27–81. Norman: Oklahoma Museum of Natural History.

Wright, J. W., and C. H. Lowe. 1968. Weeds, polyploids, parthenogenesis, and the geographical and ecological distribution of all-female species of *Cnemidophorus. Copeia* 1968:128–138.

Wright, J. W., and L. J. Vitt, eds. 1993. *Biology of whiptail lizards (genus* Cnemidophorus*).* Norman: Oklahoma Museum of Natural History.

Wu, X.-C., D. B. Brinkman, and A. R. Russell. 1997. *Sineoamphisbaena hexatabularis,* an amphisbaenian (Diaspsida: Squamata) from the Upper Cretaceous redbeds at Bayan Mandahu (Inner Mongolia, People's Republic of China), and comments on the phylogenetic relationships of the Amphisbaenia. *Can. J. Earth Sci.* 33: 541–577.

Wuethrich, B. 1997. Paleontology: How reptiles took wing. *Science* 275:1419–1420.

Yonenaga-Yassuda, Y., P. E. Vanzolini, M. T. Rodrigues, and C. M. Carvalho. 1995. Chromosome banding patterns in the unisexual microteiid *Gymnophthalmus underwoodi* and in two related species (Gymnophthalmidae, Sauria). *Cytogenet. Cell Genet.* 70:29–34.

Zani, P. A. 2000. The comparative evolution of lizard claw and toe morphology and clinging performance. *J. Evol. Biol.* 13: 316–325.

Zhang, Y. 1985. The Chinese crocodile lizard—a species threatened with extinction. *Chinese J. Zool.* 21:38–39.

Zug, G. R., L. J. Vitt, and J. P. Caldwell. 2001. *Herpetology: An introductory biology of amphibians and reptiles.* San Diego: McGraw-Hill.

Zweifel, R. G., and C. H. Lowe. 1966. The ecology of a population of *Xantusia vigilis,* the desert night lizard. *Amer. Mus. Novit.* 2247:1–57.

INDEX

Italic page numbers denote illustrations.

Designer: Nola Burger
Compositor: Integrated Composition Systems
Text: 10.5/13.5 Granjon
Display: Akzidenz Grotesk
Printer and binder: EuroGrafica SpA